Believing in Dawkins

Eric Steinhart

Believing in Dawkins

The New Spiritual Atheism

Eric Steinhart
Department of Philosophy
William Paterson University
Wayne, NJ, USA

ISBN 978-3-030-43051-1   ISBN 978-3-030-43052-8   (eBook)
https://doi.org/10.1007/978-3-030-43052-8

Cover credit: Mopic/shutterstock.com

This Palgrave Macmillan imprint is published by the registered company Springer Nature Switzerland
AG
The registered company address is: Gewerbestrasse 11, 6330 Cham, Switzerland

Acknowledgments

I thank Yujin Nagasawa for his persistent support, and for his vision of a better future for philosophical thought about religion. I thank Donald Crosby for encouraging me in my studies of religious naturalism. Thanks also to John Schellenberg and Paul Draper for their efforts to expand the philosophy of religion. Thanks go to Nancy Frankenberry for her early encouragement of this project. Thanks are due to Helen de Cruz and Stefani Ruper for their interviews which helped me to sharpen my ideas. I am very grateful to Stephanie Rivera Berruz for constantly challenging me to think in new ways. Brendan George, at Palgrave, continues to be a wonderful editor. My wife Kathleen Wallace deserves extraordinary credit for her support through all my projects. I gratefully appreciate the support of Dartmouth College and William Paterson University.

Praise for *Believing in Dawkins*

"Eric Steinhart is one of the most original and exciting philosophers today. In *Believing in Dawkins*, he rigorously elaborates and defends Richard Dawkins' work as a unique rational enterprise that explores questions not only about the material universe but also about value and meaning. The form of 'spiritual atheism' which Steinhart develops here should be welcomed as an attractive new option for a rapidly growing population of people who identify themselves as 'spiritual but not religious'. I wholeheartedly recommend this extraordinary book for its extraordinary contributions to philosophy, science, and spirituality."

—Yujin Nagasawa, *H.G. Wood Professor of the Philosophy of Religion, University of Birmingham, UK*

"Eric Steinhart provides an intriguing and imaginative systematic development of philosophical and spiritual themes that he finds in the writings and sayings of Richard Dawkins. Steinhart is careful not to attribute the elaborated system to Dawkins: the work is Steinhart's attempt to show one way in which Dawkins' work can be reconceived as a judiciously constructed philosophical theory of everything. Apart from anything else, the work serves as a provocation to those who do not care for 'spiritual' naturalism to wonder whether there are other ways in which they might similarly believe in Dawkins."

—Graham Oppy, *Professor of Philosophy at Monash University, Australia, and Chief Editor of the* Australasian Philosophical Review

Contents

List of Figures

Chapter 6

Abbreviations for Dawkins' Writings

Clinton Richard Dawkins has written many books, and I will frequently refer to them. Besides his books, I will use many of his articles. And, besides these writings, I have transcripts of some of his lectures and debates. I have watched many of his videos. But my focus will be on his published writings. I will refer to a few of his many newspaper articles. Since many of his short articles have been incorporated into his books, I will mainly refer to the books. To make reference to his works easier, I will use abbreviations followed by page numbers. For example, (GD 97) refers to page 97 in *The God Delusion*. Information about each book or article is in the References. To make my own writing flow more easily, I'll put these references into footnotes. If some paragraph has many references, *I'll list them all in the note for its first sentence.* Here is a list of some of his writings, along with the abbreviations I will use for them:

- SG (1976) *The Selfish Gene;*
- EP (1982) *The Extended Phenotype;*
- BW (1986) *The Blind Watchmaker;*
- EE (1988) The evolution of evolvability;

- AVL (1994) An atheist's view of life;
- ROE (1995a) *River out of Eden;*
- CMI (1996a) *Climbing Mount Improbable;*
- OTTR (1997) Obscurantism to the rescue;
- UR (1998a) *Unweaving the Rainbow;*
- ADC (2003a) *A Devil's Chaplain;*
- TL (2003b) *The Tanner Lectures;*
- SS (2004a) The sacred and the scientist;
- WHO (2004b) Who owns the argument from improbability?;
- IA (2006a) Intelligent aliens;
- SSSF (2007) Should science speak to faith? (with Krauss);
- FH (2007) *The Four Horsemen* (with Dennett, Harris, Hitchens);
- GD (2008) *The God Delusion;*
- GSE (2009) *The Greatest Show on Earth;*
- AK (2012a) Afterword to Krauss;
- MR (2012b) *The Magic of Reality;*
- BCD (2015a) *Brief Candle in the Dark;*
- AT (2016) *The Ancestor's Tale;*
- SITS (2017) *Science in the Soul.*

1

Introduction

1   Beyond Biology

At the start of his career, Richard Dawkins was famous for being a biologist; by the end, he was famous for being an atheist. As a biologist, he attracted a lot of academic attention; but as an atheist, he attracted attention both academic and public. He was reviled by many popular religious writers. He was attacked by academic theologians and theistic philosophers. However, as far as I know, no atheistic thinkers have risen to his defense. I am an atheist and I will defend Dawkins here. Consequently, I will be dealing here with his views on religion and spirituality. I will *not* be writing about his biology. I will *not* be doing philosophy of biology or philosophy of science. From an entirely atheistic perspective, I will be doing philosophy of religion and spirituality.

To justify the thesis that there is more to Dawkins than just biology, it will be helpful to briefly list his books. His most biological books are *The Ancestor's Tale* and *The Greatest Show on Earth*. They focus almost entirely on the details of evolutionary biology. Yet they also mention theological and philosophical issues. *The Selfish Gene* and *The Extended Phenotype* are still biological, but they are also more abstract. They both contain extensive discussions of ideas from computer science and information theory.

© The Author(s) 2020   **1**
E. Steinhart, *Believing in Dawkins*,
https://doi.org/10.1007/978-3-030-43052-8_1

They talk about philosophical topics closely associated with values and ethics. And they contain many thematic references to theological issues.

Then there are three books in which the abstract sciences play central roles: *The Blind Watchmaker*, *Climbing Mount Improbable*, and *River Out of Eden*. In these books, more purely mathematical structures are frequently in play. They deal with multi-dimensional spaces of possible organisms, adaptive landscapes, dynamical systems, optimization strategies, and so on. They talk about computers and programs, making great use of digital simulations. They discuss philosophical and theological topics. And, in these three books, these discussions are far more detailed and serious. They contain serious reflections on the nature of value, meaning, purpose, and larger cosmic themes. They are filled with ideas taken from Stoicism and Platonism.

Next comes *The God Delusion*, which explicitly gets into philosophical and theological issues. Most of *The God Delusion* is devoted to topics that lie entirely outside of biology or any of the empirical sciences. It discusses issues in ethics, in cosmology, in metaphysics. His two collections of essays, *A Devil's Chaplain* and *Science in the Soul*, also frequently address philosophical issues. Perhaps surprisingly, even his book *The Magic of Reality*, a book for children, gets into some interesting epistemology. All these books contain many Stoic and Platonic themes. After that comes I think his most philosophically interesting book: *Unweaving the Rainbow*. *Unweaving the Rainbow* is a fascinating meditation on the value and meaning of science. But *Unweaving the Rainbow* does far more than discuss these aspects of science. *Unweaving the Rainbow* is an intensely *spiritual* book. It develops a theory of the meaning and purpose of life. It discusses the metaphysical significance of beauty and truth. It pays deep homage to mathematics. Old Platonic themes are here on full display.

2   One Rational Magisterium

One great problem with Dawkins concerns his status as a celebrity atheist. This fame means that people often talk about his ideas without reading his writings. I have read all his books and many of his articles. And I will often refer to their pages. Very often. But since ending every sentence with a note makes for rough reading, I'll often use just one note for all the references in each paragraph. I'll put it after the first sentence. On the basis of his texts, I will argue that Dawkins uses science to do something beyond science. Consider his take on the relation between science and religion. Dawkins develops his own position on the science-religion relation by attacking the position of Stephen Jay Gould. Gould said science and religion are two *non-overlapping magisteria*. Dawkins refers to this as NOMA. NOMA means that science and religion deal with separate issues. Hence there are two sides to NOMA. On the one side, *science but not religion* deals with issues of fact; on the other side, *religion but not science* deals with issues of value and meaning. And, according to NOMA, neither side knows what the other is doing. The domain of fact does not intersect with that of value.

NOMA ties important concepts and values to God. It binds them to God by negative implications like these: if there is no God, then there is no objective morality; if there is no God, then there is no life after death; if there is no God, then there is no meaning. NOMA gives you this stark dichotomy: *either God or else nihilism*. Here's NOMA at work: "On the one hand, we can delude ourselves, clinging to the infantile illusion that some One, some Thing, is looking over us, somehow orchestrating the universe with each of us personally in mind. Or we can face, squarely, the reality that life is meaningless."[1] Of course, a *theist* would believe that—but why would an *atheist* believe it? The NOMA dichotomy is false. Atheism does not imply nihilism.

Dawkins attacks NOMA on both sides.[2] On the first side, he argues that religion and science overlap on many questions of fact. Religions make empirical claims: a great flood covered the entire earth; Jesus came back to life after he died; the Virgin Mary was raised bodily into the sky. Dawkins correctly says these are entirely scientific claims. On the second side, he agrees that science by itself has little to say about morality and meaning. But now Dawkins makes two important points. The first is that

---

[1]Barash (2006: 257).
[2]NOMA on both sides (OTTR; GD 77–85). Rational moral philosophy (SITS 271; OTTR 397–8; ADC ch. 1.4; AVL). Science cannot answer (GD 80, 185; SSSF). Rational universe (TL 73–4; see UR *xi*, 151; SSSF).

moral debates depend on facts. For example, facts about embryology are relevant to moral debates about abortion. The second is that religion has no moral authority. Moral authority comes from *rational moral philosophy*. Moreover, Dawkins recognizes that there may be deep questions which science cannot answer. Consider the rationality of the universe. Science reveals that the universe is *orderly*; but does it reveal that the universe is *rational*? You could argue either way. Yet Dawkins says he has a profound faith in the rationality of the universe. His profound faith is not scientific—to decide that the universe is rational is to make a *philosophical* decision.

As a scientist, Dawkins obviously values evidence.[3] But Dawkins is no positivist. A positivist thinks that every meaningful statement is empirically decidable—it can be verified or falsified using evidence alone. Dawkins affirms that theistic design might be indistinguishable from evolution by natural selection. No empirical evidence can decide against a God who generates all the empirical evidence. Likewise empirical evidence cannot tell us whether or not we are living in some digital simulation. The arguments against a God who emulates evolution (like the arguments for or against simulation) will be rational philosophical arguments. On the one hand, there's no evidence that some things exist outside of our observable universe. On the other hand, there's no evidence against things existing outside of it. Evidence alone is not enough.

Dawkins uses both science and philosophy to build a single *rational magisterium*. His rational enterprise uses empirical science to answer factual questions about our universe. It uses philosophically extended science to answer questions about value and meaning. For the sake of completeness, it also uses philosophically extended science to answer questions about ultimacy—it covers metaphysics and ontology. Of course, to rationally extend science, it is not necessary to add any new types of entities to science. Scientific documents contain symbols that refer to many kinds of objects. They refer, of course, to material things. But they also refer to properties and relations, to abstract laws and

---

[3]Theistic design (BW 316). Simulation (GD 98).

patterns, and to purely mathematical objects. Theists (and atheists too) are fond of saying that scientific naturalism is just materialism. Anybody who says that has not studied much science. Science includes logic and pure mathematics.

This rational enterprise competes with the old theistic religions.[4] It pushes their answers to every question out into the irrational abyss. Old religious concepts are given new rational yet irreligious meanings. Consequently, as the rational enterprise starts doing the old jobs of religion, it begins to *resemble* religion even though it is *distinct from* religion. It has become popular to refer to this distinctive likeness as *spiritual but not religious*. Dawkins says he is a spiritual person. He says science is a spiritual enterprise. He insists that "religion is not the only game in town when it comes to being spiritual." Dawkins is building an *irreligious spirituality*.

Atheism can be as spiritual as any theism.[5] The atheist Iris Murdoch wrote *The Sovereignty of the Good* and *Metaphysics as a Guide to Morals*. The philosopher Andre Comte-Sponville has written *The Little Book of Atheist Spirituality*. The atheist Sam Harris has written *Waking Up: A Guide to Spirituality without Religion*.[6] These excellent books show that there are at least three ways to build existentially rich spiritualities outside of theistic religions. Atheists have written about the sacred, the holy, and the numinous. The philosopher Dan Dennett talks about sacred values. He declares that "This world is sacred." The atheist philosopher Quentin Smith has developed an atheistic conception of holiness. The writer Christopher Hitchens argues for developing irreligious conceptions of the numinous, ecstatic, and transcendent. I will argue that the Dawkinsian texts support a *spiritual atheism* of great existential power.

---

[4]Spiritual person (SITS 5). Science is spiritual (ADC 27; SS). Game in town (FH 49).

[5]Murdoch (1970, 1992). Comte-Sponville (2006). Harris (2014). Dennett (2006: 22–24; 1995: 520). Smith (1988). Hitchens and Blair (2011: 45–47).

[6]Harris practices Buddhist meditation. Haught (2008, ch. 1) argues that the New Atheism begins with its own versions of the Buddhist Four Noble Truths.

3   From the New Atheism to Spiritual Naturalism

Dawkins presents his atheism in many books and articles.[7] He focuses on it in *The God Delusion*, where he argues that God does not exist. When he talks about God, Dawkins means the God worshipped in the Abrahamic religions. These are mainly Judaism, Christianity, and Islam. More generally, his atheism excludes all supernatural gods and goddesses. For Dawkins, to say some deity is supernatural means that it did not evolve. Many theists attacked *The God Delusion*. Much public debate has focused on Dawkins' atheism. Theists have spent far more energy attacking Dawkins than they have attacking other recent atheists. Why? Perhaps because he speaks and writes with a deeply religious voice. His books are saturated with religious allusions, images, and concepts. But I will argue for a deeper explanation: theists find Dawkins deeply threatening, because his work supports an alternative spirituality.

When it comes to the spirituality of science, Dawkins resembles Carl Sagan.[8] Sagan often waxed poetic about the spiritual aspects of science. Sagan wrote that a religion that "stressed the magnificence of the Universe as revealed by modern science might be able to draw forth reserves of reverence and awe hardly tapped by the conventional faiths." Sagan thinks this religion will emerge. After telling us that Sagan's books "touch the nerve-endings of transcendent wonder" once touched by religion, Dawkins tells us that his books have the same aspiration. He argues that science can arouse wonder, awe, ecstasy, and other deep emotions. These emotions were once thought to belong to religion, rather than to science. But Dawkins argues that science does a better job of arousing and satisfying those emotions than religion.

Dawkins reports that people often used to tell him that science is just another religion.[9] He used to emphatically deny the charge. But then he started to wonder what might happen if he accepted it. If science

---

[7]God is Abrahamic (TL 64, GD 15, 33–41, 56–7, 84, 184, chs. 4, 8, 9). Supernatural means not evolved (GD 96–9). Attacks (BCD 174).

[8]Conventional faiths (Sagan 1995: 52). Dawkins quotes Sagan (TL 58–9; GD 32–3; SITS 80). Transcendent wonder (GD 33). Science does a better job (SITS 269).

[9]SITS 269.

is a religion, then it should be taught in religious studies classes. He proposed a religious studies curriculum, which includes the empirical sciences. It teaches rational moral philosophy and rational metaphysics. It will instill in its students a deep faith in the rationality of existence. It will celebrate the mathematical beauty of nature. It will cultivate in its students a profoundly spiritual way of life. Dawkins argues that his religious curriculum would stand out as a superior alternative to the old theisms on every point. Should this new curriculum be called a rational religion? A scientific religion? Perhaps this is the new religion predicted by Carl Sagan. Dawkins *almost* endorses a kind of scientific religion.

Dawkins often talks about *Einsteinian religion*, which he contrasts with supernatural religion.[10] And while he rejects supernatural religion, his view of Einsteinian religion is more positive. Thus Sideris argues that Dawkins is *consecrating* science—he is trying to turn it into a new religion. But Dawkins quickly points out that Einsteinian religion isn't really *religion*. He says that science can be "religious in a non-supernatural sense of the word." But he does not prefer that sense. Religion conventionally involves humans trying to relate to alleged supernatural deities. Dawkins rejects unconventional uses of the term "religious". The deep aesthetic and emotional appreciation of nature is not religious. He thinks "religious naturalism" is confusing. He likewise rejects unconventional uses of the term "God". He rejects the pantheistic use of the term "God" to refer either to the whole universe or to some special aspect of it.

Science is *not* a religion.[11] On this point Dawkins is right. And religious naturalism just isn't religious. Thus Dawkins correctly rejects both religious naturalism and Einsteinian religion. But his rejection points to the open space between science and religion. Here again we can refer to being *spiritual but not religious*. Dawkins affirms that science can be spiritual in an irreligious sense. So perhaps his Einsteinian religion should be called *Einsteinian spirituality*. However, calling it Einsteinian

---

[10]Einsteinian religion (TL 58–64; GD 33–40). Einsteinian religion resembles *dark green religion* (Taylor 2010). Sideris (2017). Non-supernatural (ADC 27; ROE 33). Uses of "religious" (GD 33–5; ADC 147). Against religious naturalism (TL 60–1; GD 34; ADC 146). Against pantheism (GD 40–1).

[11]Science not a religion (TL 59; SITS 273). Religious naturalism (Oppy 2018: ch. 4). Science is spiritual (ADC 27; SITS 5). de Botton (2012). Rosenberg (2011).

ties it too closely to the views of one historical figure. Many others have contributed to this naturalistic spirituality. To refer to this Dawkinsian alternative to religion, I will use the name *spiritual naturalism*. Here we are *spiritual naturalists*. Our spiritual naturalism contains empirical science extended by rational philosophy. It affirms both moral and mathematical objectivity. Against Alain de Botton, we reject nostalgia for religion. Against Alex Rosenberg, we reject both scientism and nihilism.

As one of its many projects, spiritual naturalism shows that the jobs once done by God can be done by natural entities. What jobs did God do? To map out the divine job description, we will use the classical *Five Ways*. These are the five proofs which Thomas Aquinas thought revealed the existence of God. According to him, each proof shows that some special object plays some ultimate role in reality. He identifies each of these five objects with God. Although atheists love to find faults with the Five Ways, that will not be our strategy. We will revise those Ways by adding a twist at the end—the ultimate objects revealed by the Five Ways are not God. They are natural things. Rather than rejecting the Five Ways as invalid or unsound, we *naturalize* them.

The first and second ways in Aquinas are usually lumped together, and referred to as cosmological arguments.[12] They argue for a *first cause*. Since his entire theory of complexity requires that existence begins in simplicity, Dawkins explicitly affirms that there must be some simple first cause of all things. When he discusses cosmological arguments, Dawkins objects only to the step that finally identifies the first cause with God. The first cause is not God. On the contrary, Dawkins suggests naturalizing it by identifying it with the big bang or some other physical entity. To naturalize these first two ways, we will find the naturalistic first cause. The third way argues for some *ultimate necessary being*, and it is also said to be a cosmological argument. We will naturalize this third way by arguing for a naturalistic necessary being.

The fourth way is known as the degrees of perfection argument, which often gets grouped with the ontological argument by Anselm. It stands out because it is an argument from pure reason. But the *most perfect being* is not God. The fifth way is the argument from design. It has

---

[12]Dawkinsian first cause (GD 184). Is not God (GD 101–2, 184; BCD 420).

organic versions that deal with the structure of organisms and ecosystems. Dawkins shows that the *organic designer* is not God; it is evolution by natural selection. But the design argument also has cosmic versions that deal with the structure of the whole universe (for instance, the so-called fine tuning argument). Dawkins suggests that the *cosmic designer* should also be replaced with some kind of cosmic evolution (though probably not Darwinian). Here then are five natural things: the first cause; the necessary being; the most perfect being; the organic designer; and the cosmic designer. They are neither identical with God nor with each other.

Each specific job that God allegedly did generates a *local problem* for spiritual naturalism: find the natural entity that really does that job. Each natural entity is a local solution to its own local problem. On the metaphysical assumption that nature is rational, and thus self-consistent, these local solutions need to be fitted together into a harmonious whole. The local problems generate a *global problem:* find that system of local solutions which best satisfies constraints like self-consistency. Of course, there may be other constraints. Dawkins often says that simplicity is a constraint. On that metaphysical assumption, we need to find the simplest system of local solutions.

4 A Sanctuary for Spiritual Naturalists

As I read Dawkins, I see a collection of fragmentary sketches for a large system of concepts and practices. This large system is our spiritual naturalism. But these sketches are incomplete and not entirely self-consistent. These fragmentary outlines raise many problems for an atheistic spirituality. If they can be solved, the result will be an atheistic spirituality which can compete more successfully against theism. It should be able to provide the *theoretical benefits* of theism (but without God). It should even be able to provide the *practical benefits* of theism (but without God). As a Dawkinsian atheist, I accept his arguments that theism is both false and harmful. And as a philosopher, I am obligated to try to develop an alternative that is true and beneficial.

It is helpful to think of the Dawkinsian texts in architectural terms. His fragmentary sketches for spiritual naturalism are like architectural drawings. Sometimes they depict little windows, while other times they portray enormous spires reaching towards the stars. The edifice is vast. But these architectural diagrams are often unclear, incomplete, and inconsistent. I want to clarify them, fill in their missing parts, and resolve their conflicts. So I'm using his writings to construct a novel building. Sometimes I'm just completing a window or a door in ways they were clearly intended to be finished. But what about the immense towers and grand interiors? Dawkins outlines them, but he rarely says how they should be supported. Massive engineering is needed. To fit his fragmentary plans into a coherent and self-supporting structure, I will add some large-scale frameworks. I aim to complete our spiritual naturalism in a systematic way.

Given the Dawkinsian fragments and foundations, I prefer to think of myself as building a *sanctuary*. It is a sacred place, filled with joy and light. It is a spiritual refuge, a gleaming city. This sanctuary is made of thoughts, and we are constructing it together inside of our minds. We build it by systematically completing the Dawkinsian rational magisterium. Although religion is excluded from this Dawkinsian sanctuary, religious *people* are more than welcome. Their faiths they must leave at the door. As we build this sanctuary, we will not "debate God." We agree with Dawkins that God does not exist, and we have no interest in talking about non-existent things. Apart from an obligatory section dismissing the God hypothesis (Sect. 2.1 in Chapter 4), we will ignore God. Spiritual naturalism assumes that the meanings of spiritual terms are constituted by associated practices. Hence our Sanctuary permits no theistic practices—it permits neither worship nor prayer. As spiritual naturalists, we have our own practices. I will refer to this shining structure as the *Sanctuary for Spiritual Naturalists*.

The Dawkinsian architectural outlines for this Sanctuary suggest a building every bit as wild and glorious as the *Basilica i Temple Expiatori de la Sagrada Familia*. The Sagrada Familia was outlined by an architectural genius, namely, Antoni Gaudi. But Gaudi often didn't go into the specifics. Some of these specifics were profound—like how to support the enormous spires. And so later architects had to use their own imaginations and skills to complete the Sagrada Familia. The same holds true here. While Dawkins is an architectural genius, with a vision of grand sanctuary, his critics are right that he doesn't have the skills to make it all work. His own edifice would collapse under its own weight due to internal structural problems.

The title of this book is *Believing in Dawkins*. But believing in Dawkins does not mean treating his works like inerrant scriptures. Dawkins makes many mistakes, and he celebrates the scientific duty for self-correction.[13] Believing in Dawkins resembles believing in Antoni Gaudi. Just as *believing in Gaudi* means continuing his architectural enterprise, so *believing in Dawkins* means continuing his rational enterprise. It means correcting his mistakes, making his larger ideas mutually consistent, completing his arguments, filling in the details of his sketches. Believing in Dawkins means using his texts to build the Sanctuary for Spiritual Naturalists. On the basis of the Dawkinsian texts, I'm building this Sanctuary. I'm *building on Dawkins*.

Nevertheless, to ensure that the Sanctuary for Spiritual Naturalists stands up, I will often need to go beyond those Dawkinsian texts. I will often take ideas from the Stoic and Platonic philosophers. And I will often have to use my own philosophical training and imagination. Hence the faults in this Sanctuary are mine. When I add new material to this sanctuary, I will try to indicate that it comes from other philosophers, or from myself. And since the Dawkinsian texts are often ambiguous, they can be developed in many ways. This Sanctuary is only of many possible sanctuaries.

5   The Stoic Framework

When Dawkins wrote *Science in the Soul*, he gave it the subtitle *Selected Writings of a Passionate Rationalist*.[14] Consequently, spiritual naturalism is a passionate rationalism. It contains all the psychology needed to analyze our responses to the universe. Dawkins often says that many

---

[13]UR 31.
[14]Aesthetic and emotional (UR; SS). Sacred and holy (SS). Soul (SITS 212–15). Gratitude (2010). Exuberance (2013).

apparently religious responses to nature are really just aesthetic and emotional. The *sacred* and the *holy* are merely emotional responses to the overwhelming vastness of the universe. The *soul* is just the mind. *Gratitude* for your very being is merely a misfiring social emotion. *Spirituality* is just exuberance.

This reduction of religious responsiveness to personal psychology presents a problem.[15] While awe is one way of responding to the vastness of the universe, horror and terror are others. While the scientific picture of the universe brings spiritual uplift and ecstatic transport to some, it brings insomnia and nihilistic pessimism to others. Among the many conflicting ways of responding to the universe, which ways are right? If all the responses are equally valid, then Dawkinsian spirituality degenerates into relativism. Dawkins persistently objects to relativism. Likewise spiritual naturalism rejects the idea that all aesthetic-emotional responses are equally valid.

Against relativism, Dawkins argues that some aesthetic-emotional responses to the universe are *appropriate* while others are *inappropriate*.[16] It is inappropriate to be depressed or to complain about life. It is wrong to seek refuge in the comforting delusions of religion. It is ethically immature to fail to take responsibility for your own life. It is appropriate to rejoice in the scientific revelation of the structure of nature. It is right to courageously embrace your fate. It is noble and mature to make your own meaning in life. He says "it's such a wonderful experience to live in the world." And he declares that our universe is "a grand, beautiful, wonderful place." His distinction between appropriate and inappropriate responses to the universe parallels his distinction between good and bad poetic science. When scientifically-inspired poetry responds appropriately to nature, it is good; otherwise, it is bad.

However, if some responses are appropriate while others are not, then there must be some *standards* of appropriateness. They do not come from science. They seem to come from something like morality: you

---

[15]Opposite responses (SITS 269; UR *ix*). Objects to relativism (ROE 31–2; UR 21, ch. 6; ADC chs. 1.2, 1.7; GD 18–9, 319–20; GSE ch. 1).

[16]Depressed (UR *ix-xi*, 1–3; GD 404–5). Refuge (GD 20–22). Responsibility (GD 403–4). Rejoice (UR 6, 36). Make meaning (GD 404; FH 24–5). Wonderful life (FH 99). Good and bad poetic science (SITS 33, 150).

ought to respond this way rather than that way. It therefore looks like the aesthetic-emotional responses depend for their correctness on moral standards. After all, concepts like sacredness, holiness, and gratitude all revolve around values. When Dawkins naturalizes religious concepts by reducing them to merely psychological concepts, he isn't being fully clear. His reduction depends on hidden assumptions about value. The aesthetic-emotional responses to reality are more like moral responses regulated by deep standards of value.

For example, to say that something is *sacred* means that it is precious to somebody.[17] Dawkins points out that the Grand Canyon is sacred to many Amerindian tribes. He says the Grand Canyon confers stature on the Amerindian religions. And he says that "if I were forced to chose a religion, that's the kind of religion I could go for." He lists many things that are sacred to him. Thus sacredness is *extrinsic* preciousness; it is preciousness *to* somebody. But extrinsic preciousness depends on *intrinsic* preciousness. And intrinsic preciousness is *holiness*. Things are sacred because they are holy. Hence the holy and the sacred involve value. Since they involve value, they do not belong to the purely empirical part of Einsteinian religion—they belong to rational moral philosophy. Dawkins affirms that many things are sacred and holy. If we are going to take him seriously, then we cannot avoid metaphysics: holy things are like mirrors which reflect some ultimate source of value. This ultimate source of value is entirely natural. And, because holy things reflect it, we ought to treat them with reverence and respect. We ought to revere the fossils in the National Museum in Kenya. We ought to revere the giant redwoods in the Muir Woods. Because they shine with holy light, we ought to regard them as sacred. We should *consecrate* them.

Of course, since we are talking about how humans ought to respond to nature, we aren't really talking about morality. Morality specifies how intelligent social agents ought to behave towards each other. So the Dawkinsian standards point to duties deeper than morality. If we ought to respond to nature in some ways, then we have *duties to nature*. Since existence itself is natural, we have *duties to existence*, or *obligations to being*. To behave virtuously towards being is to respond to it

---

[17]Amerindian religions (SITS 1). Things sacred and holy (SS).

appropriately; to behave viciously towards being is to respond to it inappropriately. But if these duties and virtues are not moral, then what are they? Philosophers use the term *axiology* to refer to the most general study of value. Axiology includes but exceeds morality. We have duties to nature because of its non-moral axiological features. Dawkinsian principles require that those axiological features are objective. And if nature has objective axiological features, then it is even possible for nature to have duties to itself. This is consistent with the Dawkinsian thesis that Einsteinian religion includes rational moral philosophy. Rational moral philosophy is just part of rational axiology. But rational axiology is not part of empirical science; it is a part of rational metaphysics.

Axiological standards can be expressed as if-then rules. Here are two axiological rules: on the one hand, if something generates happiness, then the appropriate response to it is existential positivity (joy, love, gratitude, optimism, etc.); on the other hand, if something generates suffering, then the appropriate response to it is existential negativity (despair, hatred, revulsion, pessimism, etc.). These two rules express the *utilitarian* approach to value. Utilitarians say that suffering is evil, while happiness is good. Nagasawa uses utilitarianism to motivate the *atheistic problem of evil*.[18] It can be expressed like this: (1) Dawkins says that nature is saturated with suffering; (2) if nature is saturated with suffering, then the appropriate response to it is existential negativity; (3) but Dawkins says the appropriate response to it is existential positivity; (4) hence Dawkins contradicts himself. Here believing in Dawkins means rescuing him from this apparent contradiction. To carry out this rescue operation, I must employ some deep metaphysical principles. Dawkins never discusses them. They need to be added to his rational magisterium. By adding them I am building on Dawkins.

When Dawkins discusses appropriate and inappropriate ways of responding to nature, he is relying on some deeply buried metaphysical assumptions.[19] He inherits many of them from Nietzsche. When Dawkins says that atheism is life-affirming and life-enhancing, he is

---

[18] Nagasawa (2018).
[19] Channeling Nietzsche (GD 405). Live dangerously (Nietzsche, *Gay Science*, sec. 238). Old teacher (ADC 13). Holy enough (Nietzsche, *The Will to Power*, sec. 1052). Suffering in evolutionary world-view (ROE 131–3; GSE 390–6).

channeling Nietzsche. Nietzsche distinguished between the Christian and the Dionysian ways of life. The Dionysian way of life includes *amor fati*, the courageous love of fate. Dawkins often endorses *amor fati* (see Sect. 4 in Chapter 3). The Dionysian way of life entails that we ought to live dangerously (but not stupidly). Here Dawkins enthusiastically quotes an old teacher, who explicitly endorsed this Dionysian way of life. Nietzsche said that the Dionysian person regards existence as being "*holy enough* to justify even a monstrous amount of suffering." Dawkins does discuss the problem of suffering in an evolutionary world-view. And what he says about suffering points towards a Dionysian solution: it points towards the holiness of being, it points towards Nietzsche. Dawkins makes much more sense if what Nietzsche said about being is right. The fact that Nietzsche refers to the *holiness of existence* doesn't make him less of an atheist; on the contrary, it makes him even more deeply atheistic. The holiness of existence is the holiness of nature; it is an objective axiological feature of nature. *Amor fati* is the objectively appropriate emotional response to this holiness. To avoid self-contradiction, that is, to solve the atheistic problem of evil, Dawkins needs this Nietzschean holiness of nature. His value-theory needs to be Dionysian rather than utilitarian. But Nietzsche got his Dionysian ideas from the Stoics.

The Stoics had faith in the rationality of nature. Dawkins affirms a similar faith. Both the Stoics and Dawkins argue for the necessity of rational moral philosophy. They both affirm *amor fati*. The Stoics affirmed it because they thought the universe was regulated by the *Logos*. The Logos is the rational and providential ordering of nature. The Stoics reasoned like this: since the Logos is providential, it ensures that everything is *for the cosmic best;* since you are a part of the universe, everything that happens to you is for the cosmic best; since it is appropriate to love the cosmic best, you can and should love everything that happens to you. On the one hand, if nature is regulated by the Logos, then love of fate is the right response to it. On the other hand, if it is not, then *amor fati* is not the right response. So either Dawkinsian principles entail something like the Logos, or else Dawkins is wrong to recommend *amor fati*. Either Dawkins embraces something like the Logos, or else he contradicts himself.

I will argue that Dawkinsian principles *do* entail something like a modernized and naturalized Stoic Logos.[20] The old Logos doctrine, when thought of in modern physical terms, says that *nature is purely structural*. The updated Logos doctrine closely resembles the recent philosophical doctrine called *structural realism*. To say that nature is providential now just means that nature is a structure which is *intrinsically* for the best. It is a structure which is for the best *in itself*. Nature is holy. But if nature is for the best in itself, it does *not* follow that it optimizes your happiness, nor that it optimizes human happiness. After all, the Stoics were not utilitarians. They did not reduce goodness to happiness nor evil to suffering. The Stoic Logos works to maximize the virtue that emerges in competitive struggle. It works to maximize the *arete* that emerges in the strife-torn *agon*. Any suffering or happiness that emerges from this drive to maximize *arete* is an axiologically irrelevant by-product. The solution to the atheistic problem of evil is to reject utilitarianism. The correct axiological rule is this: if nature maximizes *arete*, then the appropriate response to it is existential positivity.

Dawkins frequently relies on Stoic values and principles. He inherits many Stoic doctrines. Consequently, as I build the Sanctuary for Spiritual Naturalists, I will use many Stoic ideas. As I incorporate them into the Sanctuary, I will ensure that they are modernized and naturalized. This naturalized Stoicism will provide the *outer framework* for the Sanctuary. It will allow me to make greater sense of many Dawkinsian ideas. It will allow me to fit many Dawkinsian fragments into a more coherent ethical and metaphysical whole. It will allow me to show how Dawkins solves the atheistic problem of evil. Of course, I do *not* claim that Dawkins is a Stoic; but I do claim that, without certain Stoic ideas, Dawkins doesn't make much sense. Believing in Dawkins means trying to make sense out of Dawkins, and Stoic ideas help me to make that sense. But is this Stoic framework architecturally adequate?

---

[20] Structural realism (French and Ladyman 2010). Not best for us (Mulgan 2015).

6   The Platonic Framework

Since Stoicism raises many questions which it cannot answer, the Stoic framework is not deep enough to support the Sanctuary. Since Dawkinsian spirituality includes many ideas that do not fit into any Stoic architecture, it is not strong enough either. So the Stoic framework is not adequate. Fortunately, Dawkins himself points to a stronger and deeper framework. He points to a kind of naturalized Platonism. I will argue that, apart from this Platonism, much of what Dawkins says about the meaning and value of science makes very little sense. But I aim to make sense of Dawkins.

When Dawkins talks about physical things, he frequently turns to information theory and computer science.[21] He built digital worlds for his biomorphs. He used artificial life programs to explain the evolution of spider webs and eyes. He is intrigued by the idea that our universe is a digital computation. His many treatments of things as programs makes his ontology look less like materialism and more like the *patternism* of Ray Kurzweil. According to this patternism, reality is fundamentally composed of structures of pure information. Dawkins turns to mathematics. He talks about abstract spaces of possible organisms, which he explicitly says are mathematical structures. He endorses the thesis, from Peter Atkins, that the entire universe is constructed out of the self-elaborations of the empty set. And, when it comes down to the real distinction between science and theology, Dawkins does not appeal to empiricism or to materialism. He appeals to mathematics. Not surprisingly, Dawkins argues that the mathematical structure of reality is objectively beautiful (Sects. 5.1 in Chapter 3 and 2.3 in Chapter 4).

Plato argued that the purpose of life is to get out of the cave.[22] Dawkins replaces the cave with the burka (a garment that almost blinds us), and says that the purpose of life is to take off the burka. He argues

---

[21]Biomorphs (BW ch. 3; EE; BCD 363–94). Webs and eyes (CMI chs. 2 & 5). Our digital universe (GD 98, 186; SITS 85). Kurzweil (2005: 5, 371, 386–388). Abstract spaces (BW ch. 3; CMI ch. 6; ADC ch. 2.2; AT 676). Are mathematical (CMI 200). Atkins empty set (BW 14; GD 143–4). Appeals to mathematics (AK 190; UR 63–4).

[22]Burka (GD 405–20). Sacred truth (SITS 22–8). Intensely Platonic (UR 1–5, 312–3). Blessed (UR 5; see ADC 12).

for the sacredness of truth and scientific objectivity. His book *Unweaving the Rainbow* celebrates the Platonic values of light and vision. Its beginning and ending are intensely Platonic. Evolution escapes from the cave as it runs from blindness to vision. Dawkins says that we are "hugely blessed" to be able to *open our eyes* on the cosmic spectacle. He says that *vision makes life worth living*. It makes life worth living because it enables us to get outside of the universe by building scientific models of it in our heads. Getting out of the universe resembles getting out of Plato's cave. Vision provides life with its ultimate sufficient reason, because it enables us to *reflect* the structure of the universe back to itself through scientific modeling. Of course, this vision is not merely empirical. Here again Dawkins turns to mathematics: to fully appreciate the grandeur and beauty of nature, we must look into it with mathematical eyes. This too is Platonic. But what makes vision a source of value?

Plato answers that as we step out of the cave, we become illuminated by the light of the sun.[23] Light is a powerful symbol. Light is closely associated with enlightenment, while darkness is linked with ignorance, superstition, irrationality, and evil. Dawkins often talks about light. He says we are illuminated by a spotlight. When some New Atheists tried to brand themselves as the *Brights*, Dawkins endorsed it. He says that as we mentally step out of the universe, we are illuminated by a light that makes life worth living. If we are blessed by vision, then we are illuminated by a holy light. For Plato, of course, this holy light shines from the Good. The Good is not a deity. On the contrary, the Good is a moral standard which can be used to judge the actions and characters of any deity. Much as Plato used it in his dialogs to condemn the Olympian deities, so Dawkins uses it in *The God Delusion* to condemn the Abrahamic deity. Of course, the ancient Platonists were not Abrahamic monotheists. They rejected Abrahamic theology: the Good is not God. As Christianity conquered Rome, they *opposed* it. For their paganism, they were often violently persecuted. Dawkins explicitly condemns the cultural violence done by the Abrahamic monotheisms.

---

[23]Spotlight (UR 5). Brights (2003c; GD 380). Worth living (UR 312). Plato on the Good (*Republic*, 508b–9b). The Good is not a deity (Murdoch 1992: 37–38, 475–477). Platonists used the term "God" to refer to their divine mind; they rarely identified it with the Good. Violently persecuted (Nixey 2018). Cultural violence (GD 282–3).

To reduce the Good to God is both conceptually and morally wrong. To reduce the Good to God is to commit an act of cultural violence. It is to participate in the mono-theo-normative *hijacking* of the Good.

Dawkins often talks about *hijacking*.[24] He says religion has hijacked our most precious concepts. For example, he says "The word 'spiritual' has been hijacked by religion." He says "don't let religious people hijack you ... because you call yourself a spiritual person." The list of hijacked concepts is long (spirituality, holiness, sacredness, transcendence, the abyss, ecstasy, being-itself, mystical experience, and so on). Theists hijacked these concepts by tying them to God, where have been held hostage for nearly two thousand years. NOMA facilitates this hijacking by insisting that these valuable concepts belong exclusively to the religious magisterium: if there is no God, there is no spirituality. Nevertheless, since words (and their concepts) are our servants and not our masters, we are free to separate them from religion. We are free to *reclaim* those concepts for spiritual naturalism. We can naturalize them by separating them from God. Following Dawkins, I aim to reclaim these hijacked concepts. To aid in this reclaiming, I will employ the old pagan philosophies of Stoicism and Platonism.

For Platonists, nature strives to maximize self-vision; it strives to see itself. Among the Platonists, I'm also including the so-called Neoplatonists. Thus Plotinus portrays nature as rapt in self-vision; and, as nature sees itself, it comes to see the Good.[25] For Platonists, nature strives to see itself because it has a *duty* to see itself. Nature does its duty. The logic of duty is known as *deontic logic*.[26] It is the logic of obligation, of what *ought* to be. Two qualifications follow. The first states this duty of nature is not a moral (or ethical) duty. Moral duties involve intelligent social agents; but nature is not intelligent or social; hence its duty towards itself is not moral. The duty of nature towards itself is *axiological*. Thus nature does its axiological duty. It satisfies its axiological

---

[24]Against hijacking (BW 131; FH 51; GSE 340, 376; AT 158). Precious concepts hijacked (UR 18, 114, 210; GD 354; SITS 77, 226). Word "spiritual" hijacked (2013: 0:05–12, 1:17–25, 1:41–53). Words our servants (BW 1).

[25]Plotinus (*Enneads*, 3.8).

[26]The deontic logic used here includes the axioms of standard deontic logic, and is close to Andersonian-Kangerian deontic logic.

obligations, it satisfies the demands of deontic logic. The second qualification recognizes that nature does not literally see itself. More accurately, nature is rapt in self-reflection. The holiness of nature is its *reflexivity*, its capacity to reflect itself back to itself through parts that mirror the whole. These parts include organisms (or computers) that reflect nature back to itself by doing science and mathematics. Dawkins supports these Platonic themes. What makes life worth living? Reflexivity.

Nature serves the Good by maximizing reflexivity. It serves the Good by bringing forth sharper eyes and smarter brains. What are the elevations on Mount Improbable? They are degrees of reflexivity. But maximizing vision also requires maximizing visible richness. It requires maximizing beauty. To maximize vision is to maximize both that which sees and that which is seen; hence to maximize reflexivity is to maximize both that which reflects and that which is reflected. The Darwinian lesson is that the best way to maximize reflexivity is to maximize the *arete* that emerges in the *agon*. Holiness shines through *arete*. All parts of nature serve the Good by striving for greater *arete* in their *agons*. Since we are parts of nature, we too are obligated to serve the Good. We have holy duties to maximize reflexivity. We have holy duties to build models of the cosmos inside the cosmos. You ought to love your fate because your fate serves the Good. As you strive for *arete* in your *agon*, you will suffer. And thus your suffering has meaning. But now your suffering is not a Christian condemnation of life; it is Dionysian affirmation of life. Still, what is the Good? It is a purely transcendental object. Doesn't Dawkins reject transcendence? The physicist Alan Lightman had a mystical experience. Discussing it with Lightman, Dawkins says "You can't out-transcendence me."[27] Dawkins doesn't deny transcendence. He reclaims it from theistic bondage. The modern atheist Iris Murdoch illustrates this reclamation. She replaces God with the Good. She replaces the supernatural Abrahamic deity with a natural pagan ideal.

As I build the Sanctuary for Spiritual Naturalists, I will incorporate many Platonic ideas. I will ensure that this Platonism gets naturalized, and I will give many arguments to support it. This naturalized Platonism will provide the *inner framework* for the Sanctuary. It will enable me to

---

[27]Lightman (2018). Dawkins on transcendence (Dawkins & Lightman 2018: 12: 28–40).

fit the Dawkinsian fragments into a highly coherent philosophical whole. Of course, I do *not* claim that Dawkins is a Platonist. Sometimes he appears to support Platonism, sometimes he seems to reject it.[28] But believing in Dawkins means building on Dawkins. It means trying to make greater sense out of Dawkins, and Platonism helps me make that sense.

References

Barash, D. (2006). What the whale wondered: Evolution, existentialism and the search for 'meaning'. In A. Grafen & M. Ridley (Eds.), *Richard Dawkins: How a Scientist Changed the Way We Think* (pp. 255–262). New York: Oxford.

Comte-Sponville, A. (2006). *The Little Book of Atheist Spirituality* (N. Huston, Trans.). New York: Viking.

de Botton, A. (2012). *Religion for Atheists: A Non-Believer's Guide to the Uses of Religion*. New York: Random House.

Dennett, D. (1995). *Darwin's Dangerous Idea: Evolution and the Meanings of Life*. New York: Simon & Schuster.

Dennett, D. (2006). *Breaking the Spell*. New York: Viking Penguin.

French, S., & Ladyman, J. (2010). In defence of ontic structural realism. In A. Bokulich & P. Bokulich (Eds.), *Scientific Structuralism* (pp. 25–42). New York: Springer.

Harris, S. (2014). *Waking Up*. New York: Simon & Schuster.

Haught, J. (2008). *God and the New Atheism: A Critical Response to Dawkins, Harris, and Hitchens*. Louisville, KY: Westminster John Knox Press.

Hitchens, C., & Blair, T. (2011). *Hitchens vs. Blair: Be It Resolved Religion Is a Force for Good in the World*. Berkeley, CA: Publishers Group West.

Kurzweil, R. (2005). *The Singularity Is Near: When Humans Transcend Biology*. New York: Viking.

Lightman, A. (2018). *Searching for Stars on an Island in Maine*. New York: Random House.

---

[28] For Platonism (BW 65–7; CMI 200, 218; ADC 82–3; AT 504; SITS 289). Against Platonism (GSE 21–7; SITS 287–97).

Mulgan, T. (2015). *Purpose in the Universe: The Moral and Metaphysical Case for Ananthropocentric Purposivism*. New York: Oxford University Press.

Murdoch, I. (1970). *The Sovereignty of the Good*. New York: Schocken.

Murdoch, I. (1992). *Metaphysics as a Guide to Morals*. London: Chatto & Windus.

Nagasawa, Y. (2018). The problem of evil for atheists. In N. Trakakis (Ed.), *The Problem of Evil: Eight Views in Dialogue* (pp. 151–175). New York: Oxford.

Nixey, C. (2018). *The Darkening Age: The Christian Destruction of the Classical World*. New York: Houghton Mifflin Harcourt.

Oppy, G. (2018). *Naturalism and Religion*. New York: Routledge.

Rosenberg, A. (2011). *The Atheist's Guide to Reality*. New York: W. W. Norton.

Sagan, C. (1995). *Pale Blue Dot*. London: Headline.

Sideris, L. (2017). *Consecrating Science: Wonder, Knowledge, and the Natural World*. Oakland: University of California Press.

Smith, Q. (1988). An analysis of holiness. *Religious Studies, 24*(4), 511–527.

Taylor, B. (2010). *Dark Green Religion: Nature Spirituality and the Planetary Future*. Berkeley, CA: University of California Press.

2

Complexity

1 The Complexity Liturgy

1.1 Scientific Liturgies

Our Platonism entails that we have a holy duty to study nature scientifically.[1] Much of *Unweaving the Rainbow* is devoted to showing the spiritual value of doing science. If doing science is spiritual, then learning it is also spiritual. One reason for the spirituality of science is that it reveals the sacred truth of nature. I will eventually discuss other reasons to affirm the spirituality of science (Sects. 4 and 5 in Chapter 3). Spiritual naturalists affirm that both doing and learning science are spiritual activities.

Spiritual naturalism assigns new meanings to old religious terms. To work through a course of scientific topics for the sake of spiritual development is to perform a *liturgy*. For example, to study complexity for its spiritual value is the *complexity liturgy*. Some aspects of this liturgy may be tedious or difficult—it involves setting some heavy stones into place. Since complexity plays a foundational role in Dawkinsian thinking, this

---

[1]Science is spiritual (ADC 27; SITS 5). Sacred truth (SITS 26, 326).

© The Author(s) 2020
E. Steinhart, *Believing in Dawkins*,
https://doi.org/10.1007/978-3-030-43052-8_2

liturgy is worthwhile. It will be followed by the four *physical liturgies*. These are the atomic, molecular, biological, and thermodynamic liturgies. After these liturgies, I turn to the cosmological liturgy in Chapters 4 and 5, then the ontological liturgy in Chapters 6 and 7. These liturgies are spiritual exercises. By doing these liturgies, you build the Sanctuary for Spiritual Naturalists inside of your own mind.

1.2   Combinatorial Complexity

When Dawkins defines complexity, he talks about the complexities of *types* of things, rather than the complexities of particular individuals.[2] He does this because one particular individual can be an instance of many types. The astronaut Neil Armstrong is an instance of the type *human*, the type *material thing*, and many other types. All these different types have different complexities. Dawkins starts his discussion of complexity with an observation about mountains and airliners. On the one hand, if you scramble the parts of a mountain, you almost always end up with a mountain. The type *mountain* doesn't care much about its internal structure. So the type *mountain* is simple. On the other hand, if you scramble the parts of an airliner, you almost never end up with an airliner. The type *airliner* cares greatly about its internal structure. So the type *airliner* is complex. How can we make these ideas more precise?

To find the raw complexity of some type of thing, start by decomposing some example of that type into some set of parts. All the possible arrangements of those parts makes the set of *arbitrary arrangements* of the type. Every arbitrary arrangement is a way of scrambling the parts of the type. Some of those arbitrary arrangements preserve the type while others destroy it. All the ways that preserve the type go into the set of *stable arrangements*. Each set has some size, which is its *multiplicity*.[3] The size of the arbitrary set is the *arbitrary multiplicity*, while the size of the stable set is the *stable multiplicity*. Now we need to compare these two multiplicities.

---

[2]BW 6–9; SITS 122.
[3]BW 7; CMI 77; GD 137–9; AT 688.

The arbitrary arrangements of a simple type tend to preserve it. So the stable multiplicity of a simple type is very similar to its arbitrary multiplicity. If we divide some number by a similar number, we get a small number. This small number indicates the low complexity of the simple type. Suppose there are one trillion ways of scrambling the mountain, and one billion of them preserve its type. The arbitrary multiplicity is one trillion while the stable multiplicity is one billion. One trillion divided by one billion is one thousand. This is the low complexity of the type *mountain*.

The arbitrary arrangements of a complex type tend to destroy the type. So the stable multiplicity of a complex type tends to be much smaller than its arbitrary multiplicity. If we divide some greater number by some much smaller number, we get a large number. The large number indicates high complexity. Suppose there are also one trillion ways of scrambling the airliner, but only one thousand ways preserves its type. One trillion divided by one thousand is one million. The type *airliner* has high complexity.

The *raw complexity* of any type is its arbitrary multiplicity divided by its stable multiplicity. One problem with this definition is that the complexities get very big very fast. Another problem is that complexity is often thought of as an informational quantity, to be measured in binary digits (bits). Both problems can be solved by taking the *logarithm* of the raw complexity.[4] More precisely, the complexity of an instance of some type is the logarithm of its arbitrary multiplicity divided by its stable multiplicity. The complexity of the type is the average of the complexities of its instances.

1.3 Spiritual Lessons from the Liturgy

The complexity liturgy teaches at least three immediate spiritual lessons. The first lesson is that *complexity is improbability*. Consider a field strewn with Boeing 747 parts.[5] Focus on the type *airliner*. So the stable multiplicity is the *airliner multiplicity*, which is the number of ways

---

[4]The number of bits needed to encode some number is its base-2 logarithm.
[5]BW 234; GD 137–9.

of combining those parts into an airliner. The *arbitrary multiplicity* is the number of arbitrary ways of combining them. If a whirlwind blows through that field, randomly arranging the parts, then what is the probability that an airliner emerges? What is the probability of an airliner coming from chance? It is the airliner multiplicity divided by the arbitrary multiplicity. The airliner probability is a tiny number divided by a huge number. But tiny divided by huge gets pretty close to zero.

More generally, for any type, its probability is the inverse of its complexity. *More complex things are less likely to arise by chance.* As the complexity of some type goes up, so too does its rareness in the abstract space of possible combinations. But rare things are not likely to be produced by random processes (such as a lottery). If chance is the only factor at work, then things will almost always stay simple. So if you see complexity rising, then *non-random factors* are at work.[6] Dawkins applies this lesson from atoms to universes. Why is this spiritually significant? Because atheism is often portrayed as reducing reality to randomness. On the contrary, atheism insists that reality is animated by extremely powerful and entirely natural non-random factors.

The second lesson says the Dawkinsian theory of complexity makes very little sense unless the *types* of things are objectively real. If types are just concepts or words, then the complexities of types are merely subjective. And since the theory of evolution depends on these complexities, that theory is also merely subjective. Dawkins correctly despises subjectivism.[7] Thus what Dawkins says seems to imply that types objectively exist. The thesis that types objectively exist goes back to Plato. Is Dawkins a Platonist? I don't know. But believing in Dawkins means trying to make sense out of his writings, and his writings make far more sense with Platonism than without it.

The third lesson concerns the distinction between science and theology. They both appeal to empirical evidence. For instance, the design arguments and cosmological arguments for God take empirical facts as their premises. Theists often say that people directly perceive God in mystical experience. So what is the difference between science

---

[6]GD 138.
[7]ROE 31–2; UR 21, ch. 6; ADC chs. 1.2, 1.7; GD 18–9, 319–20; GSE ch. 1.

and theology? Here Dawkins has a brilliant answer: *the difference is mathematics*.[8] Science uses mathematics, but theology does not. If mathematical truth is not objective, then the difference between science and theology is not objective; theology is just as valid as science. Of course, Dawkins emphatically rejects that equality. And if mathematical truth is objective, then mathematical existence is objective too. Why is science better than theology? Because numbers exist, and angels do not. By using numbers to measure complexity, the complexity liturgy reveals an important spiritual truth: the study of mathematics is a spiritual discipline. Like Plato, Dawkins says that mathematics helps to expand, discipline, and control the imagination.[9]

2   The Atomic Liturgy

2.1   The Library of All Possible Atoms

Our universe begins with the big bang, an eruption of radiant energy.[10] And while no God exists behind or within this eruption, I will later argue that this fiery radiation is holy (Sect. 5.1 in Chapter 6). As this fiery energy condenses, simple particles like quarks and electrons appear. Quarks conjoin to make protons and neutrons. They combine with electrons to make the first and simplest atoms, mostly hydrogen, helium, and lithium. The scientific study of atoms is a spiritual exercise. It is the *atomic liturgy*.

Dawkins often talks about libraries of possible objects.[11] The laws of physics in our universe define a *library of possible atoms*. Each book in this library defines a *kind* or *type* of atom. These books include all the atomic types in the familiar periodic table of the elements. But other atomic types are possible. How many atomic books exist? Suppose the upper bound on the number of protons in a possible atom is two hundred, and

[8]AK 190; UR 63–4.
[9]BW 74.
[10]Big bang (SG 13; UR 60; GD 101, 174). First atoms (SG 13; AT 2, 592–4).
[11]BW ch. 3; CMI ch. 6; ADC ch. 2.2; AT 676.

the upper bound on neutrons is the same. So the number of books in the atomic library is forty thousand. There are forty thousand possible kinds of atoms.

The atomic library has two dimensions or axes. Its vertical axis counts protons, while its horizontal axis counts neutrons. This two-dimensional space is sometimes called *Segré space*.[12] Of course, atoms also include electrons. But we can ignore them now. You can picture Segré space as a building with two hundred floors. Each floor contains a single shelf, with two hundred books. The books in each shelf are arranged in order of increasing neutron numbers. If we apply the Dawkinsian theory of complexity to atoms, it turns out that the complexity of an atom corresponds to its proton number.[13] So the books on higher floors are more complex atomic types.

Figure 1 illustrates part of the atomic library. Wherever two lines intersect in this Figure, there exists some possible atomic type. More technically, these types are called *isotopes*. Black dots mark the stable isotopes. Most of the atomic types are not stable, so the majority of intersecting lines are not marked by dots. The laws of physics define a narrow line of stability surrounded by a vast region of instability. Of the forty thousand books in the atomic library, only a few hundred actually contain atoms. Most of these atomic types are actually empty. Yet they are *possibly* occupied.

These possible kinds of atoms are not particular physical atoms. Although all uranium atoms belong to the same kind or type, that type is not itself an atom of uranium. What are these types? One traditional answer to this question is known as *Platonism* (Sect. 2 in Chapter 6). Platonism says that the possible types of atoms are abstract forms of atoms. Forms are also known as universals, essences, or natures. The form of an atom defines its structure in terms of the properties and relations of its parts. Every physical atom *is an instance of* or *is an example of* an abstract atomic form. Thus every physical oxygen atom *instantiates* or *exemplifies* the abstract oxygen-form.

---

[12]Magill and Galy (2005: 31–32).

[13]The n-th type of atom has n protons. A little mathematics shows that the complexity of the n-th type of atom is the logarithm of the n-th Bell number.

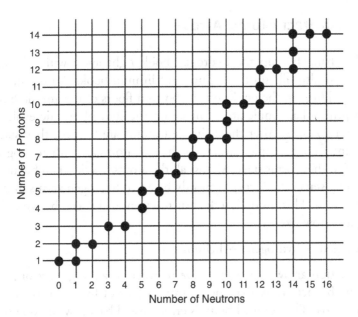

Fig. 1 Some floors in the library of possible atoms

Platonism says these atomic forms exist objectively. They are not concepts in our heads or words in our languages. Thus Segré space is an objectively existing abstract mathematical space. Should we affirm mathematical objects? As naturalists, we accept those kinds of things that appear in scientific theories. Since mathematical objects appear in those theories, we accept them. But our naturalistic Platonism does not support any dualism. Segré space is both physical and mathematical. Dawkins says Platonism works for simple things like geometric shapes; if it works for them, it also works for simple things like atoms.[14] Much of what Dawkins says makes very little sense outside of a Platonic metaphysics. As I go over the Dawkinsian texts, I will often use a Platonic framework to organize his philosophical fragments. I will use it to help raise the Sanctuary for Spiritual Naturalists. However, this Sanctuary is my own construction. Believing in Dawkins means building on Dawkins.

---

[14]SITS 289.

2.2   Abstract Atomic Arrows

The big bang fills up the hydrogen book right away with zillions of atoms.[15] It also fills up the helium and lithium books. So the universe starts by filling up the books on the bottom floors of the atomic library. These primordial simple atoms make up the start of atomic evolution: the atomic *alphas*. After these simplest atoms are formed, they start to evolve into more complex atoms. As the first atoms are gathered together by gravity, they forms stars. Almost all the atoms heavier than helium are formed by stars. The first stars mostly fuse two hydrogen atoms into one helium atom. This fusion is a nuclear reaction, which can be symbolized using an arrow like this:

hydrogen + hydrogen → helium.

This reaction arrow expresses a *type* of reaction among *types* of atoms. It is an abstract arrow that moves from a pair of hydrogen books to a helium book. The atomic library is criss-crossed by arrows linking books to books. These abstract arrows have concrete models. If two particular hydrogen atoms fuse into one helium atom, then those changing atoms physically model the abstract arrow. When we present reaction arrows, we'll omit some irrelevant details. Here are some reaction arrows:

helium + helium → beryllium;
beryllium + helium → carbon;
carbon + helium → oxygen;
carbon + carbon → neon + helium;
oxygen + oxygen → silicon + helium;
oxygen + oxygen → magnesium + helium + helium.

There are thousands of other possible reaction arrows. Fusion reactions generate atoms all the way up to iron. Atoms beyond iron are mostly formed when stars explode. The exploding stars fuse atoms all the way up to uranium and sometimes beyond. Many heavy atoms are also produced during the merger of neutron stars.

---

[15]Stars (GD 169; GSE 426). Fusion (SG 12–13; UR 52; AT 592–8).

Nuclear reactions preserve mass: the mass on the right side equals the mass on the left. The mass includes any mass converted into energy. Hence the nuclear reactions resemble equations involving mass numbers. The mass number of an atom is its total number of protons and neutrons. Since every atom has an exact mass number, the atoms are *natural models of numbers*. The mass number of an atom is written as a superscript before its symbol. Since hydrogen has mass 1, and its symbol is H, hydrogen is written as ^{1}H. An oxygen atom with eight protons and eight neutrons is symbolized as ^{16}O. The nuclear reaction equations involve something like addition:

$$^1H + {}^1H \rightarrow {}^2He \qquad\qquad 1 + 1 \rightarrow 2$$
$$^1H + {}^2H \rightarrow {}^3He \qquad\qquad 1 + 2 \rightarrow 3$$
$$^3He + {}^3He \rightarrow {}^4He + {}^1H + {}^1H \qquad 3 + 3 \rightarrow 4 + 1{+}1$$
$$^{16}O + {}^{16}O \rightarrow {}^{24}Mg + {}^4He + {}^4He \qquad 16 + 16 \rightarrow 24 + 4{+}4.$$

But the last two equations aren't exactly additions—they are transformations. What kind of transformations? The atoms are letters in an alphabet. Stringing atoms together makes a word. Thus H is a letter, He is a letter, H + H is a word, and He + H + H is a word. As letters get changed into letters, words get changed into words. Each atomic reaction arrow transforms an old string of letters into a new string of letters.

Every atom has a path from itself to itself, and the probabilities of these self-paths determine the *stabilities* of types of atoms.[16] If the path from an atom to itself is extremely probable, then it is *stable*. But if its self-path is improbable, it is *unstable*. Unstable atoms decay into stable atoms. Magnesium-23 decays in seconds into sodium-23 plus a positron (which we'll ignore). The arrows look like this:

$$^{23}Mg \rightarrow {}^{23}Na \qquad \text{high probability;}$$
$$^{23}Mg \rightarrow {}^{23}Mg \qquad \text{low probability.}$$

Atomic decay illustrates the Dawkinsian principle of the *survival of the stable*. It illustrates the principle of *differential survival*—unstable atoms perish, stable atoms persist. More complex atoms are usually less stable. Atoms with odd proton numbers are less stable than atoms with even

---

[16]Atomic stabilities (AT 594–8). Survival of stable (SG 12–13; BW 44).

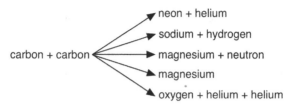

Fig. 2 Some nuclear reaction arrows

numbers. Hence some atoms are *fitter* than others, and the laws of physics exercise a *selective* effect. Dawkins says that atomic differential survival is the earliest form of natural selection. The physical laws select some possible atoms for actualization, while others remain merely possible.

A single atomic collision can often go down many paths. When stars fuse carbon, five reaction arrows gain almost all the probability.[17] Since these five are not all equally probable, probability is *non-randomly distributed* over them. Their probability distribution is *skewed*. Figure 2 shows these five arrows, with probability decreasing downwards. Obviously, arrows with higher probability are more frequently selected. Picture *resistance* as the inverse of probability. High probability arrows have low resistance. Flows of matter are more likely to follow arrows of lower resistance. This is *differential selection* of arrows. It is *arrow selection.* Of course, some explanation is needed here: why are the probabilities skewed? What is the skewer?

2.3   The Atomic Computers

Dawkins is obsessed with computers.[18] His *weasel program* illustrates the slow accumulation of complexity. His *biomorphs program* illustrates evolution. He used programs to simulate the evolution of spider webs

---

[17]De Loore and Doom (1992: 95–97).

[18]Biomorphs (BW ch. 3; EE; BCD 363–94). Spider webs (CMI ch. 2). Eyes (CMI ch. 5). Digital genetic code (BW 111–5; ROE 19). Cells compute (SG ch. 4; EP ch. 2). Brains compute (SG ch. 4; EP ch. 2; UR ch 12). Survival machines (SG *xxi*). Ecosystems compute (CMI 72, 326; ADC 12). Digital physics (SITS 80–5).

and eyes. He often points out the digital nature of the genetic code. He argues that cells are little computers running genetic programs, and that organisms are genetically programmed survival machines. The entire earthly ecosystem is a giant computer running an evolutionary algorithm. Brains too compute. He is interested in digital ideas in physics. Following Dawkins, computational ideas will play a central role in spiritual naturalism.

What defines natural computation? If some natural system runs an algorithm or program, then it is reasonable to say that it is a *computer*. It is arguable that the stars run algorithms. At first glance, it looks like they are doing arithmetical calculations with the mass numbers of atoms. The reaction arrow $H + H \rightarrow He$ looks like the calculation $1 + 1 = 2$. But a deeper look reveals something different. The arrow $O + O \rightarrow Mg + He + He$ shows that nuclear reactions transform old strings of atomic letters into new strings of atomic letters. If atoms are letters, then nuclear reactions are *string rewriting operations*, and every star is running a string rewriting program. Every star is a *massively parallel distributed string rewriting system*. The mathematician Emil Post showed that, in an exact sense, string rewriting really is computation.[19] Any string rewriting system is a computer. This is not a metaphor—on the contrary, *the stars are literally computing*. Nuclear reactions resemble Dawkins' weasel program.[20] It started out with a randomly generated string of letters. It ran through a Darwinian process of mutation and selection, until it reached its target string METHINKS IT IS LIKE A WEASEL. His weasel program, like the programs running in the stars, is a *string rewriting program*.

The stars are *celestial computers*. It does not follow that every physical system is computing. The stars compute because *and only because* they are physical models of some mathematical theories of computation. Those theories include the string rewriting systems of Emil Post, the machines of Alan Turing, and the lambda calculus of Alonzo Church, three theories which are all equivalent. As the planets revolve around their suns, they model physical equations; but since those equations are

---

[19] Post (1943).
[20] BW ch. 3.

not theories of computation, those planets are not computing their equations of motion. The computational powers of the stars do not depend on human interpretations. It isn't a human convention that chlorine has seventeen protons, or that two hydrogens fuse into helium. The system of nuclear reactions *naturally and objectively* computes.

For any computer, it is useful to distinguish between *hardware* and *software*. The hardware is invariant while the software can vary. For stellar computers, the hardware is the totality of possible nuclear reaction arrows, a totality fixed by the laws of physics. But the software is the *probabilities* assigned to these arrows. These probabilities vary from star to star, and they vary over the lifetime of any single star. The atomic software running on some star is its program or algorithm. An atomic algorithm is an assignment of probabilities to all possible nuclear reaction arrows.

Any algorithm can be classified as *divergent* or *convergent*. If an algorithm is convergent, then it has some set of final states (its *finalities*). When any convergent algorithm runs, its state grows increasingly similar to one of its finalities. The idea of convergence comes from the calculus: an algorithm converges to finality as a series converges to a limit. If an algorithm converges after running through only finitely many steps, then it halts. But algorithms can converge at infinity. For example, if an algorithm repeatedly divides its input number in half, then it converges to zero at infinity; it runs towards zero; zero is its finality. As another example, if an algorithm repeatedly doubles its input number, then it converges to infinity at infinity; it runs towards infinity, which is its finality. If an algorithm lacks finalities, then it is divergent. Divergent algorithms just wander around in their libraries of possibility.

Convergent algorithms *run towards, aim at,* or *strive for,* their finalities. To say that a convergent algorithm aims at its finalities does *not* imply that it has any mentality. Finalities need *not* be goals, and convergence does *not* require foresight, purpose, or teleology. Convergent algorithms strive *teleonomically* for their finalities.[21] To say that some computer strives for some finality just means that its algorithm is

---

[21]Teleonomic processes are analyzed in terms of dynamical systems. They run towards their ends like the mobile objects in dynamical systems are driven by their transition operators to their attractors in their phase spaces.

skewed or finely tuned to maximize the likelihood that it does reach that finality. Of course, as teleonomic algorithms gain complexity, they can also gain mentality. As swarms of atoms evolve into thinking organisms, teleonomy evolves into purposiveness. And if a convergent algorithm strives for some finality, it need not reach it. The doubling algorithm can run towards infinity without reaching it. Algorithms can fail.

Among convergent algorithms, some aim at finalities of greater complexity. They strive to transform their simpler inputs into more complex outputs. What should we call these complexity-increasing algorithms? Here we turn to Aristotle. Aristotle said that things have directionalities, they aim at more complex finalities, because they are running internal programs, which he called *entelechies*. Spiritual naturalism happily adopts this Aristotelian term: an entelechy is a convergent algorithm which teleonomically increases complexity. Computers that run entelechies need not be intelligent or purposive. They can be mindless automatons and utterly unconscious robots.

Any computer that runs an entelechy is a *crane*.[22] The stars are *atomic cranes*. A crane is a system of transformational arrows that tends to increase complexity. Cranes lift matter to greater heights of complexity. A crane is a big arrow made of little arrows, and it points to more complex finalities. It will be useful to have a word to refer to the lifting actions of cranes. The word *moil* is an older English word for labor. Thus cranes *moil* towards their finalities. Although moiling is convergent, it need not be goal-directed or teleological. Moiling is teleonomic striving. The finalities of cranes are their *ecstasies*. The term is used in the sense of *ek-stasis*, which means *to stand outside*. The limit of a series stands outside of the series and so is the ecstasy of the series. If any crane moils towards some finality, then that finality is its ecstasy.

2.4   The Evolution of Atomic Complexity

An atomic computer is any whole composed of atoms. Its atomic hardware is the totality of all possible nuclear reaction arrows, and this hardware is programmable. The programming of any atomic computer

---

[22]Dennett (1995). Dawkins (GD 99, 168, 185–8; AT 634, 688).

is its software. Atomic computers have two layers of software. The *first-order* software running on any atomic computer consists of the probabilities of its reaction arrows. These probabilities vary from thing to thing. The probabilities inside a uranium-powered nuclear reactor differ from those inside an apple. Those on the surface of the earth differ from those in the core of the sun.

The first-order programming of any thing can change over its lifetime. Our sun began as an immense but diffuse cloud of hydrogen. The probability of hydrogen staying hydrogen was very high, while its probability of fusing into helium was very low. As gravity pulled this cloud together, it formed a proto-star. When it ignited, its first-order probabilities changed. At the core of the sun, the probability of hydrogen staying hydrogen decreased drastically, while its probability of fusing into helium rose sharply. It began fusing hydrogen into helium. At the current time, in a star like our sun, the probability of helium fusion is very low. But as our sun turns into a red giant, that probability will grow. It will fuse helium into beryllium, carbon, and oxygen. As the sun ages, the laws of physics specify the changes in its first-order reaction probabilities. The factors that specify first-order changes are the *second-order* programming of the sun. The second-order programming *skews* the first-order programming so that our sun does not fuse atoms randomly. The probabilities of the first-order reaction arrows in our sun are skewed so that complexity increases—our sun runs an entelechy.

Three arguments now confirm that the stellar software is highly skewed away from randomness. The first argument starts with the fact that there exists a stable line of atoms rising through Segré space.[23] The atoms sitting in this *line of beta stability* have highly probable self-arrows. But a line is a highly non-random structure. The second argument observes that the stellar reaction probabilities facilitate the growth of complex atoms. If they were random, atomic complexity would not grow. Since it does grow, they are skewed far from randomness. The third argument observes that the stellar algorithms change in ways that *reinforce* the steady growth of complexity. The *changes* in the first-order stellar algorithms are skewed far from randomness.

---

[23]Jaffe and Taylor (2018: 310–312).

Hence the second-order stellar software is also highly skewed away from randomness. All these probabilities and their changes are skewed towards increasing complexity. The stars run entelechies, which are very improbable algorithms. And they run self-reinforcing entelechies, which are almost vanishingly rare in the total space of possible stellar algorithms.

All these probabilities and their changes are skewed in extremely improbable ways. Dawkins insists that improbable things cannot emerge from chance alone. Hence there must be some non-chancy explanations for these stellar algorithms. They must have been skewed by *non-random factors*. Since these factors skew the distribution of probabilities, they can be gathered together into the *skewer*. The skewer pushes probability around in the space of arrows. It twists the arrows so that they point upwards. The skewer drives stellar programs away from random software and towards entelechies. Since entelechies are very improbable, the skewer must be very powerful. *What is it?* Whatever it is, it is entirely natural, and it acts in every star. It emerges from the laws of physics, and they act everywhere in the universe. All the stars participate in the skew. All the stars are little complexity-increasing computers. Together, they make a big computer.

The big atomic computer consists of all the stars in the entire history of the universe. Its hardware consists of Segré space plus all possible nuclear reaction arrows. Its software consists of the first-order probabilities of those arrows, plus the second-order programming that drives them to change. The software guides the flow of energetic matter through Segré space, from the big bang until the end of time. If you could watch this flux from outside of Segré space, that is, from outside of the atomic library, and from the big bang to the end, then you would see a big *atomic arrow* rising like a flame up into the atomic library. The growth of this big atomic arrow emerges from the flow of matter through all the little first-order nuclear reaction arrows.

The skewer ensures that the big arrow of atomic complexity grows like a volcano. The height of a volcano grows over time. So the complexities of the most complex atoms increase over time. This does not mean that all atoms grow more complex—the universe still contains mostly hydrogen. The higher volcanic strata rest on the lower strata. So if some higher atomic complexities are populated, then all the lower complexities were or are populated. The volcano grows narrower as it grows higher: more complex atoms are rarer. It grows by *accumulation*. New rocks get piled on top of old rocks. If any rocks fall down the volcano, they were first carried to the top. Atoms can only lose the complexity which they previously gained. Simpler atoms can fuse into more complex atoms, and complex atoms can fission into simpler atoms. But fission depends on fusion, decay depends on growth. The atomic computer runs an entelechy. It is a crane which moils towards its ecstasy. The volcano only rises. Of course, it will eventually erode away; but *the volcano itself* does not perform that erosion. As the complex arrows disintegrate, the volcano and its arrow also disintegrate. The atomic arrow never *grows* downwards. It always grows upwards, and therefore points beyond itself, to possible atoms which will never be realized in our universe. It defines a self-surpassing process, which strives for an unattainable infinity. It moils towards that infinite ecstasy.

3   The Molecular Liturgy

3.1   The Library of All Possible Molecules

Since atoms join together to make molecules, the atomic liturgy is followed by the *molecular liturgy*, which deals with chemistry. The laws of chemistry in our universe define a *library of possible molecules*. It resembles the atomic library, except that the book s in this *molecular library* are abstract molecular forms. They are connect-the-dots networks, in which types of atoms are linked by types of chemical bonds. The books on the higher floors define molecular types with higher complexities.

The complexities of molecular types are defined combinatorially. Consider formic acid, whose network is shown in Fig. 3. Its set of atomic parts is {H, H, C, O, O}. There are lots of ways to arrange these five atoms into molecules. They can be arranged into carbon dioxide and dihydrogen: COO and HH. Or into water and carbon monoxide: HOH and CO. To find the Dawkinsian complexity of formic acid, you count the ways of rearranging its atoms and then you do some arithmetic.

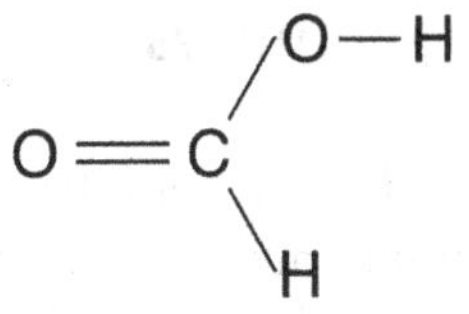

Fig. 3 Formic acid

Table 1 Some floors in the molecular library

Floor	Molecule	Floor	Molecule	Floor	Molecule
15	benzene	212	aspirin	14700	insulin
10	formic acid	119	dopamine	4030	vasopressin
8	boric acid	64	histamine	1870	oxytocin
6	ethylene glycol	31	acetic acid	527	LSD
2	formaldehyde	25	octane	335	adenosine
0	water, methane	18	carbon dioxide	245	tryptophan

Its complexity is the logarithm of its arbitrary multiplicity divided by its stable multiplicity. Formic acid turns out to be very complex, even though it has only a few atoms. Although chemists generally agree with Dawkins that molecular complexity needs to be defined combinatorially, they have developed their own detailed theories of chemical complexity.

The chemist Steven Bertz developed a combinatorial theory of molecular complexity, which has been applied to the PubChem molecular library.[24] Table 1 illustrates some molecules, along with their Bertz complexities. The Bertz complexities can be used to assign these molecules to their floors in the molecular library. But the complexity of any thing can also be defined as the number of bits of information it contains. Dawkins endorses informational ways of measuring complexity, and chemists like Thomas Bottcher have proposed ways to measure the information in any molecule. Both the combinatorial and informational approaches to molecular complexity are consistent with the Dawkinsian principle that complexity is improbability.

---

[24]Bertz (1981). PubChem <pubchem.ncbi.nlm.nih.gov>. Informational complexity (ADC 100–2, 210). Bottcher (2016).

3.2   The Molecular Computers

After the stars have produced some complex atoms, they start binding into molecules. The first simple molecules are just atomic pairs like H_2 or O_2. Or atoms decorated with hydrogen, like water (OH_2), ammonia (NH_3), and methane (CH_4). Molecular evolution starts with these molecular alphas. From the stars, we shift to the celestial bodies that run molecular evolution. We can focus on the planets (asteroids are little planets, moons are planets of planets). Molecules on planets enter into chemical reactions, defined, like nuclear reactions, by reaction arrows. For example, one methane molecule reacts with two oxygen molecules to form one carbon dioxide molecule and two water molecules. This combustion reaction looks like this:

methane + 2 oxygens → carbon dioxide + 2 waters.

This reaction is usually written in chemistry books as

$CH_4 + 2O_2 \rightarrow CO_2 + 2H_2O$.

But it can be written out in more detail in terms of the atoms in the molecules:

HHCHH + OO + OO → OCO + HOH + HOH.

The combustion of methane breaks down old molecular words and rearranges their atomic letters into new words. Of course, the atoms in molecules are arranged into networks, also known as *graphs*. Figure 3 showed the graph for formic acid. Thus *computational chemistry* uses graph rewriting.[25] But graph rewriting is just a kind of string rewriting. Consequently, the chemical reactions on any planet make big string rewriting system. Since string rewriting systems are computers, planets run molecular computations. However, since planetary chemical reactions are driven mainly by star-power, it's more accurate to say that an entire star-planet system is a molecular computer. Solar systems are *celestial computers* running both atomic and molecular algorithms. Once again, this does not imply that every natural system computes.

---

[25]Bournez et al. (2006).

The hardware of any molecular computer is the physically unchangeable totality of possible chemical reaction arrows. Its software is the changeable distribution of reaction probabilities to those arrows. Since molecules have variable stabilities, their self-arrows have varying probabilities. They have *differential survival* rates. The stable survive while the unstable perish. And if some molecular reaction can go down many alternative arrows, those arrows have different probabilities. Although all the planets in any solar system in the universe share the same hardware, their algorithms vary. The molecular software running on Earth differs from that on Venus or Mars. Our moon runs a simple molecular computation (it mostly preserves molecular types). An astonishingly rich molecular computation has been running on earth for billions of years.

From the perspective of molecular computation, our earth consists of the molecular library along with the flows of earthly matter through it. Matter flows through the little molecular arrows that run from groups of books to groups of books. It flows mostly through the most probable arrows. This is *arrow selection*. If you were able to watch this flux from outside of the molecular library, and from the origin of our earth to the present, then you would see something like an arrow rising up into the molecular library. The growth of this big *molecular arrow* emerges from the flow of matter through all the little molecular reaction arrows. The big arrow of molecular complexity grows like a volcano. So the complexities of the most complex molecules increase over time. This does not mean that all molecules grow more complex—the universe still contains plenty of water. But the height of this volcano grows over time. The higher molecular strata rest on the lower strata. The molecular arrow grows narrower as it grows higher. It grows by *accumulation*.[26] This volcanic logic confirms that our earth runs a molecular entelechy. The molecular arrow points to the ecstasy of this entelechy. It always points upwards. It points beyond itself to infinite molecular complexity.

---

[26] Detailed models of molecular evolution in space predict that molecular complexity is cumulative (Garrod et al. 2008). Work on the evolution of networks of linked objects also suggests that molecular complexity is cumulative (Johnston et al. 2011).

3.3   The Evolution of Molecular Complexity

Many other planets (and moons) in our solar system appear to run molecular entelechies.[27] Planets like Venus, Mars, Jupiter and Saturn have produced complex molecules. Moons like Enceladus, Titan, and Europa appear to run entelechies. Dawkins estimated that there are one-hundred billion billion planets in our universe. Current data suggests our universe contains about *two hundred billion trillion* planets—planets are common. Surveys indicate that the Milky Way contains tens of billions of earthlike planets in the habitable zones of sunlike stars. Many of these planets run molecular entelechies. Hence *molecular cranes* are common in our universe.

A molecular entelechy is a program which increases complexity. The arguments that applied to the atomic entelechies also apply to molecular entelechies. All the molecular entelechies in the universe are first-order and second-order programs skewed very far from randomness. Our best current estimates indicate that complex planetary chemistry is common. Look into the starry sky, and you are looking into a universe saturated with entelechies. Some deep non-random factors in the laws of physics skew molecular computation towards complexity everywhere. All across the universe, the skewer twists molecular matter into more complex shapes. The skewer acts directly on reaction probabilities and indirectly on flows of matter. What is it?

A Stoic would say the skewer is the Logos. It shapes the flow of the *pneuma*, the *pyr technikon*, the designing fire. This Stoic appeal to flowing fire-energy suggests that we turn to thermodynamics. On Earth, molecular evolution is driven mainly by flows of thermal energy from the sun. And on moons like Titan and Europa, it is driven by flows of heat generated by tidal forces. More generally, it looks like molecular cranes are heat engines. They are *far from thermodynamic equilibrium*. They dissipate their heat into cold dark space. This dissipation drives material flows

---

[27]Dawkins estimates planets (BW 142–6, 164–6; CMI 283; GD 165; GSE 421). There are at least two trillion galaxies in the observable universe (Conselice et al. 2016). The Milky Way contains between two and four hundred billion stars and at least one hundred billion planets. Multiplication yields at least two hundred billion trillion planets in the universe. Planets in habitable zones (Petigura et al. 2013).

on planets to self-organize. This suggests that *thermodynamic forces* skew molecular programs into molecular entelechies. They push the probabilities away from randomness and towards those that increase complexity. Moreover, complexity is linked with information, which is linked with entropy, and entropy is a thermodynamic quantity. Somehow, thermodynamics drives self-organization. But we can still ask: *why* do the laws work this way? To answer this question, we will eventually turn from Stoicism to Platonism.

3.4 The Replicator

At some point, the molecular crane running on Earth produced a molecule that could make copies of itself.[28] It produced a *replicator*. Dawkins is obsessed with replicators. Perhaps the first replicators were peptides (chains of amino acids). Or maybe they were RNA. But earthly replication converged onto DNA. A *strand* of DNA is a string of four chemical letters. Two strands of DNA bind to make a *helix*. Suppose some DNA helix consists of the two strands X_1Y_1. This strand replicates in two phases. During the first phase, the two strands X_1 and Y_1 separate. During the second phase, each strand binds with a copy of its old partner. Thus X_1 binds with Y_2 and Y_1 binds with X_2. The result is two new helixes X_1Y_2 and X_2Y_1. So DNA replication is a molecular arrow:

$$X_1Y_1 \rightarrow X_1Y_2 + X_2Y_1.$$

The first replicator crosses two important philosophical thresholds.[29] The first is the *information threshold*. Each strand of DNA is a series of letters (or bases) in a molecular alphabet. This alphabet contains the four letters A, C, G, and T. Since there are four letters, each can be expressed using two binary digits. For example, A is 00, C is 01, G is 10, and T is 11. Thus DNA stores information in digital form. The digital nature of DNA permits it to self-replicate. This threshold is Platonic, because evolution now turns away from the particular sequence of DNA atoms

---

[28] SG ch. 2; EP ch. 5; BW 128–37.

[29] Digital DNA (BW 115–20; ROE 11–20; UR 89–97; AT 603; SITS 214). Digital self-replication (BW 112, 115; ROE 19).

to the information encoded in them. Only the information gets copied when the DNA replicates.

The second philosophical threshold is *reflexivity threshold*. Like the information threshold, this is a Platonic threshold. The Platonists say our universe has a duty to maximize reflexivity. If they are right, then all the things in the universe are striving to do this duty. The stars strive to increase atomic complexity. By increasing it, they are contributing to the universal duty to maximize reflexivity. But they are not explicitly maximizing it. The planets strive to increase molecular complexity. Like the stars, they are helping to maximize reflexivity without explicitly doing it. Everything changes with the emergence of the first replicators. They are the first self-acting objects—they *self-replicate*. And so, with the first replicators, the universe begins to do its Platonic duty *explicitly*.[30] The replicators start to maximize reflexivity. As they do their duty, reflexivity elaborates and concentrates itself: the replicators evolve into complexes which become self-regulating, self-representing, self-legislating, and so on. Do they always do their duty? An *ideal replicator* does its duty without fail. The one becomes two; the two become four; and so it goes. An ideal replicator evolves into an infinitely self-reflecting computer. It evolves into a Plotinian divine mind. Every replicator has the property of *possibly* producing infinitely many self-copies and *possibly* evolving into a divine mind. It has these properties even if it fails to actualize them. And how did the one become two? Doesn't logic first demand that the zero becomes one? We will have to answer these questions. We will give arguments for this Platonism in due course.

Although the first replicators may have been naked, they soon surrounded themselves with cellular machinery.[31] The first living cells appeared on earth. These were the biological alphas. Dawkins often discusses the origins of life. He talks about the hypothesis that life first appeared in deep sea hydrothermal vents, or in fissures deep in the crust of the earth. He discusses the hypothesis that the first cells were built

---

[30]SITS 42–4.

[31]Origins of life (BW ch. 6; CMI 283–6; ROE 135–51; GD 162–8; GSE 419–21; AT 642–65). Hydrothermal vents (AT 662–3). Fissures (AT 663–4). RNA cells (BW 130–4; GSE 419–21; AT 657–62). Life needs luck (BW 141–6; CMI 283–5; GD 164–9). But not much luck (CMI 283; GD 166). Jeremy England (2014).

from RNA. The emergence of life involves some luck. But how much? Dawkins says he does not think life needs much luck. He thinks that life, including intelligent life, is common in the universe. The physicist Jeremy England has used thermodynamic ideas to argue that the emergence of life involves very little luck indeed. On the contrary, he argues, life is an almost inevitable consequence of self-organization. He says, "You start with a random clump of atoms, and if you shine light on it for long enough, it should not be so surprising that you get a plant." If thermodynamic forces drive physical systems to self-organize, then perhaps they also drive the emergence of life.

Of course, we don't know how life emerged.[32] Any theory about the origin of life has to be purely speculative. Worse, if science requires observational confirmation, it can be objected that we can't have any *scientific* theories of the origin of life. But Dawkins replies that we can speculate *responsibly* about the origins of life. And by using computers to simulate the early earth, we can virtually travel back in time. Dawkins argues that simulation extends the scientific method. It enables us to empirically test otherwise untestable theories. England has been using computer simulations to help refine his theories. The main lesson here is that we need not *guess* at the origins of life—we have a valid scientific method, using computer simulations, to explore those origins. Our ignorance should inspire research, not defeat.

4   The Biological Liturgy

4.1   The Library of All Possible Organisms

And so life begins on earth. Since the study of life is biology, the next liturgy is the *biological liturgy*. It starts with the study of the actual evolution of earthly life. The first organisms, like the first atoms and molecules, are simple alphas. They are primitive single-celled organisms.

---

[32]Speculate responsibly (BW 147). Simulation (MR 16). England's theories simulated (Horowitz and England 2017; Kachman et al. 2017).

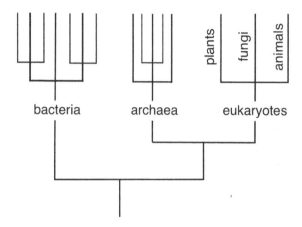

Fig. 4 A highly simplified tree of life

Over time, they evolve into more complex forms of life. The actual evolution of life fills out the phylogenetic tree of life.[33] Figure 4 shows an *extremely* simplified tree of life. Of course, these *actual organisms* are not the only *possible organisms*. Just as there are abstract spaces of possible atoms and possible molecules, so there is an abstract space of possible organisms.

Dawkins often talks about this abstract space, which we will call the *biological library*.[34] It resembles the atomic and molecular libraries. He says it is a mathematical structure, a multi-dimensional space in which the points are books. Each book contains all the information needed to define some possible type of organism. The biological library contains the forms of all possible organisms on all planets in the entire universe. Of course, various objections can be raised against this library. Nevertheless, since Dawkins refers to it, the Sanctuary includes this library. Its books are the Platonic forms of possible organisms. An old approach to these forms treats them as *biological essences*. But Dawkins objects to the old doctrine of biological essences. Do his objections defeat Platonism? Biological essentialism is an obsolete version of Platonism.

---

[33] Soltis and Soltis (2019).

[34] Biological library (BW ch. 3; CMI ch. 6; ADC ch. 2.2; AT 676). Is mathematical (CMI 200). Against essences (GSE 21–7). Better Platonism (Wagner 2014).

New versions of Platonism are consistent with modern biology. When the term *Platonism* is used here, it means Platonism in some modern scientific sense.

Dawkins frequently says that organisms have varying degrees of complexity.[35] But what is biological complexity? Dawkins defines it combinatorially. The combinatorial complexity of some organism-type is the logarithm of its arbitrary multiplicity divided by its stable multiplicity. And he defines it informationally. So the complexity of an organism is the length of the shortest or most compressed description of the organism. Other biologists say the complexity of an organism is its number of different cell types, or the percentage of its DNA which does not code for proteins. All these ways agree well with each other. Does biological complexity seem too vague? Dawkins thinks it is a real quantity. It plays a crucial role in his arguments for evolution and against creationism. Hence we agree with Dawkins that biological complexity can be mathematically defined and scientifically measured. It can be used to sort all possible organisms into ranks, which are the floors in the biological library. All organisms on the same floor have the same complexity, while more complex organisms are on higher floors.

4.2   The Biological Computer

Dawkins argues that cells are computers running genetic programs.[36] He extends these computational ideas into reproduction. When a parent organism fissions asexually into two offspring, the genes of the parent get copied (perhaps with mutations) into each offspring. This is a string rewriting operation:

parent $\rightarrow$ offspring + offspring.

---

[35]Biological complexity (BW 6–9; AT 688; SITS 122). Defined informationally (ADC 100–1, 210). Plays a central role in evolutionary arguments (BW, CMI, ROE; GSE; AT), and against creationism (GD ch. 4). Organism complexity is number of cell types (Bower 1988). Or percentage of DNA which does not code for proteins (Taft et al. 2007). These ways are mutually consistent.

[36]Cells compute (BW 115–20). Biosphere computes (CMI 72, 326).

When two organisms sexually reproduce, their gene strings fuse into the genetic strings of some offspring. This fusion involves genetic string rewriting:

male + female → offspring.

Both asexual fission and sexual fusion involve genetic string rewriting operations. Biological evolution really is a massively parallel distributed string rewriting system. But string rewriting is computation (Sect. 2.3). Since biological evolution is a physical model of a mathematical theory of computation, it computes. Dawkins says the entire earthly biosphere is a massively parallel distributed information-processing system. This does not mean that everything computes.

The planet earth is a celestial computer. It has atomic, molecular, and biological hardware. Its biological hardware is the totality of all possible reproductive arrows. These arrows link books with books in the biological library. Dawkins says the earthly biocomputer runs an evolutionary algorithm.[37] So our earth runs biological software. Its biological software is the non-random distribution of probabilities to the reproductive arrows. These probabilities constitute its evolutionary algorithm. They ensure that different types of organisms have *differential survival* rates. And they ensure that organism types have *differential production* rates. These differences entail that the flows of matter through the biological library are skewed very far from randomness. But the fact that DNA codes information ensures that more than just matter flows through the biological library. Information also flows through this library.

Our earthly biocomputer consists of the biological library along with the flow of living earthly matter through it. Matter flows through the little reproductive arrows that run from books to books. Biologist Thomas Ray describes this flow aesthetically: matter flows beautifully through this abstract space.[38] If you were to watch the flow of living earthly matter from outside of the biological library, and from the origin of life to the present, then you would see something like the phylogenetic tree of life growing up into the abstract space of the biological library.

---

[37]ADC 12; CMI 72, 326.
[38]Ray (1992: 181–182).

This tree supports an arrow rising up from its roots to its leaves. This big *biological arrow* emerges from the flow of matter through all the little reproductive arrows. Here the arrow of biological complexity resembles the atomic and molecular arrows. The earthly biocomputer lifts living matter up into the sky of complexity. It runs a biological entelechy, it is a biological crane.

Biological evolution builds a volcano, which Dawkins calls Mount Improbable. This biological volcano mostly resembles the atomic and molecular volcanoes. It grows by *accumulation,* and it obeys Dennett's *principle of the accumulation of design.*[39] Dennett's principle states that "since each new designed thing that appears must have a large design investment in its etiology somewhere, the cheapest hypothesis will always be that the design is largely copied from earlier designs, which are copied from earlier designs, and so forth."[40] Of course, if complexity goes up, then it can also go down. The concept of accumulation implies only that any complexity that gets lost is complexity that was previously gained. Cave fish gained their eyes before they lost them.

The biological volcano differs from the atomic and molecular volcanoes. It rises very high and it rises relatively rapidly. The *Argument from Replication* explains this rise: (1) Atomic and molecular evolution do not generate much complexity. (2) Biological evolution generates enormous complexities. (3) But the crucial difference in biological evolution is replication. (4) By inference to the best explanation, it follows that the generation of high complexities requires replication. This is the *replication principle.* (5) The replication principle does not depend on the details of the replicators. It does not require DNA. It does not even depend on the laws of physics in our universe. It is based not on physics, but on information theory, and it is justified by computer science. (6) Consequently, if *any possible object* is extremely complex, then it has been produced by some evolutionary process involving replication. For complexity to really increase, you need more than just flows of matter—you need flows of information.

---

[39]BW; CMI; GD ch. 4; GSE; AT.
[40]Dennett (1995: 72).

The biological volcano rises from the simple organisms at the dawn of life to the most complex organisms that will ever exist in our universe. They are probably not humans. We may transform ourselves, through genetic engineering and other technologies, into transhuman s (Sect. 3 in Chapter 9). Dawkins believes in superhuman aliens.[41] Anywhere that life exists in the universe, there exists a tree of life that points to infinity. An *ideal* tree of life branches to infinity. It generates infinitely many lineages that evolve to infinite levels of biological complexity. It moils towards that infinite ecstasy. Of course, this infinite ecstasy is only an ideal limit. Although the trees of life in our universe strive for infinity, they will almost certainly never reach it—entropy is grief.

4.3   The Evolution of Biological Complexity

Dawkins often talks about life on other planets.[42] He believes the universe contains intelligent aliens. And he argues for a conditional: *if* there is complex life on other planets, *then* it has evolved via Darwinian natural selection. He thus argues for *universal Darwinism*. If there is life on other planets, then it is climbing up its own biological Mount Improbable. It is working its way up through its own regions of the biological library. Other planetary ecosystems are planetary biocomputers running their own biological algorithms. They have their own biological volcanoes and arrows. If there is life on some other planet, then that planetary biocomputer is a part of the cosmic biocomputer. The cosmic biocomputer contains all the life in the universe.

Pick an arbitrary celestial computer running some biological software.[43] It is a planet powered by a sun. It is far from thermodynamic equilibrium. Energy flows across it in a well-defined direction: from its hot bright sun into the cold dark abyss. This teleonomic flow powers molecular evolution. Very complex molecules evolve. From these, the

---

[41]WHO 12; GD 98.

[42]Aliens (ROE 151–61; IA 96–7; UR 117; GD 98–9). Universal Darwinism (BW 288; ROE 151–8; ADC ch. 2.2; SITS 119–50, 191–2).

[43]First life needs luck (BW 141–6; CMI 283–5; GD 164–9). Life emerged quickly (Bell et al. 2015; Nutman et al. 2016).

first organisms appear. They emerge as pre-biotic chemistry slowly but insistently climbs higher on the molecular Mount Improbable. Dawkins says the emergence of the first living things requires some luck. But not much. So, at least relative to their chemical backgrounds, these simple initial organisms are highly probable. This high probability is supported by the previously mentioned thermodynamic ideas. It is also supported by increasing evidence that life emerged very quickly on Earth.

Every organism sprouts a quiver of arrows.[44] Each arrow specifies a change from the organism to some new version of itself. These arrows can be classified as *downgrades, equigrades,* or *upgrades.* Downgrades decrease complexity, equigrades preserve it, and upgrades increase it. Without skew, downgrades and equigrades are the rule, while upgrades are the exceptions. A random distribution of probability to these arrows makes them all equally likely. If probability were distributed that way, life would quickly die out. Complex organisms are too improbable to come into being, or remain in being, by chance alone. Thus biological evolution involves non-random factors, which pump living matter to higher levels of complexity. Mutation is not random. Natural selection is not random. As complexity increases, selective pressures tend to prevent it from decreasing. Selection works like the *ratchet* on a jack: the ratchet allows the jack to easily move up, but makes it hard for it to move down. Natural selection is a biological ratchet. Some lineages of organisms therefore start out simple and gradually accumulate complexity. They are little cranes, little arrows. Since the simple starting organisms are highly probable, and each simple change is highly probable, these cranes are highly probable. The little cranes add up. Celestial biocomputer are planet-sized cranes. The little cranes lift matter upwards on Mount Improbable. As life evolves on any planet, a biological arrow rises like a flame into the abstract biological library above that planet.

The teleonomic flow of energy across some celestial biocomputer (some planet, some moon) drives the evolution of biological complexity. Every celestial biocomputer runs an evolutionary algorithm, a Darwinian entelechy. It runs a program which slowly accumulates biological

---

[44]Too improbable for chance (BW 43; GD 136–48). Non-random factors (GSE 416). Non-random mutation (BW 305–12). Ratchet (CMI 26, 133–6; GD 99).

complexity. But a biological program is a distribution of probabilities to reproductive arrows. Dawkins stresses over and over again that if these probabilities were randomly distributed, then life would remain simple. It has not. Hence these probabilities are non-random. Biological entelechies (like their atomic and molecular counterparts) are skewed extremely far from randomness. Here is a *first-order question*: why does living matter flow through these reproductive arrows rather than those? The first-order answer is that the probabilities on these arrows are higher than on those. High probability is low resistance. Living matter flows through the paths of least resistance. This is biological arrow selection. But here is a *second-order question*: why are the first-order probabilities different? Not chance. It isn't *accidental* that life climbs Mount Improbable. Dawkins devotes enormous effort to explain this skew, and his explanations are correct.[45] But are they ultimate?

It is arguable that his biological explanations are merely proximate: the skew emerges from much deeper principles. Some have argued that the skew emerges from very deep thermodynamic principles (Sect. 1 in Chapter 3). Thermodynamics reveals deep directionality. Energy flows through celestial computers in the single direction from hot to cold, entropy is always increasing. But entropy is closely associated with information. So perhaps the skew emerges from even deeper informational principles. The skewer is the ultimate reason for the skew. Reasons need not lie in any mind. Why is the Pythagorean theorem true? Its proof is not built from thoughts. A Stoic argues that the skewer is the Logos, which is the rational ordering of nature. Fated by the well-ordered reasons in the Logos, the *pneuma* steers all things. The *pneuma* is the *pyr technikon*, the designing fire. It is the second-order force which skews the first-order probabilities. A Platonist argues that the skewer comes from deeper principles about duty. Deontic axioms entail that the probabilities *ought* to be skewed. Our universe strives to do what it ought to do, and its strivings often succeed; therefore, its probabilities are skewed.

---

[45]Explaining the skew (SG, EP, ROE, BW, CMI).

4.4 Bodies Filled with Mirrors

The nucleus of almost every cell contains a compressed representation of its body.[46] The cellular part represents its bodily whole. To say a whole is *self-representative* means that it contains some parts that represent the whole. Hence multicellular organisms are self-representative. Any nucleus that represents the whole body *reflects* that whole. It is like a little mirror in which a genetically compressed image of the whole body appears. Metaphorically speaking, every nucleus is like a little *eye* in the body which *sees* the entire body. Part of the body sees the whole body. To use some Dawkinsian language, the nucleus stands outside of the body, because it contains a model of the body. By producing self-representative organisms, evolution has made things which genetically see themselves. Of course, this metaphor should not be taken too literally.

As organisms gain complexity, they develop brains. Brains represent some things through maps. A human brain contains two neural maps of its own body.[47] The first is its *sensory homunculus* while the second is its *motor homunculus*. The spatial structure of each homunculus corresponds to the spatial structure of the body. The shape of the cortical map of the hand resembles the shape of the hand. The maps preserve shapes but not sizes. Body-parts with greater densities of neurons are bigger in their respective maps. For example, since hands have much higher densities of neurons than feet, the size of the homuncular hand is much bigger than that of the homuncular foot. Each homuncular map is a mirror of the body inside of the body. A part of the body represents or reflects its whole. Each homuncular map metaphorically sees the entire body. It is like an eye in the brain. Of course, while these maps do not literally see the body, they are aware of the body. Neural maps are mental maps.

As organisms gain complexity, so do their brains.[48] More complex brains build more complex models of their environments. Their mental models often represent the world like maps represent their territories:

---

[46]UR 312–3.
[47]AT 284–6.
[48]More complex brains (UR ch. 11). Isomorphic mental models (Johnson-Laird 1983: 419; Kosslyn 1994; Cummins 1996). Mental number line (Dehaene et al. 2003; Nieder 2005). Get outside of universe (UR 312).

they are *isomorphic* to their objects. Human brains contain visual models that represent visible objects by isomorphism. They contain a mental number line isomorphic to the Platonic number line. Our mental models are mirrors. And we can build models of the entire universe. Human brains are parts of the universe that reflect the universe. If you can make a model of your house inside your brain, then you can look at your house from an external perspective. Dawkins applies this logic to our models of the universe. If we can make models of the universe inside of our brains, then we can look at the universe as if from the outside. Of course, this does not mean that we have immaterial minds that magically escape from the physical universe. It just means that parts can build models of their wholes.

Human brains are parts of the universe that can mentally stand outside of the universe. Now, the Greek term *ek-stasis* means to stand outside. The Greek *ek-stasis* is the root of the English word *ecstasy*. The term *ecstasy*, as used here, means that a thing stands outside of itself. Of course, since a thing cannot spatially stand outside of itself, ecstasy refers to the logical condition in which some part reflects or mirrors its whole. A thing is *in ecstasy* when a part of that thing reflects its whole. The universe is in ecstasy when we get outside of it by putting models of it inside of our skulls. This self-modeling is the ecstasy of the universe. Obviously, this ecstasy does not refer to some orgasmic bliss—the universe does not have any emotions. And what goes for the universe also goes for you. You can get outside of your self by putting a model of yourself inside of your skull. Your self-modeling is your ecstasy. More accurate self-consciousness is more ecstatic. Perfect self-knowledge is maximal ecstasy.

As parts of the universe gain complexity, they represent more of the universe at greater levels of accuracy. The evolution of complexity from atoms to brains supports the evolution of reflection. It supports the evolution of cosmic self-reflection. Evolution produces bigger and better mirrors. Metaphorically speaking, these mirrors are all eyes; they are parts that see the whole. The Platonists took these metaphors literally. Thus Plotinus portrays the universe as an organism striving to *maximize vision*.[49] And the Plotinian divine mind, which does maximize vision,

---

[49] Plotinus (*Enneads*, 3.8).

is an infinitely self-conscious mind.[50] So a Platonist will argue that the best explanation for this growth of cosmic self-seeing is that the universe strives to maximize vision. Of course, the old Platonists were wrong to take these metaphors literally. But Platonism can be naturalized.

A naturalistic Platonism argues that the best explanation for this growth of cosmic self-mirroring is that the universe moils to *maximize self-reflection*. It moils to *maximize reflexivity*. For the Platonist, maximal self-reflection is the ecstasy of the universe. But why does the universe moil to maximize reflexivity? The Platonic answer involves duty: we ought to climb up out of the Platonic cave; we ought to climb up the divided line; we ought to climb up the great chain of being. Plotinus extended this duty to the whole universe: the universe ought to become like the perfectly self-reflecting divine mind. The Platonic answer involves the logic of value. This logic includes the logic of duty, which is deontic logic. But is this Platonic answer consistent with Dawkins? Dawkins says his rational magisterium includes rational moral philosophy; but that requires foundations in the logic of value; hence his rational magisterium includes deontic logic. It is consistent with Dawkinsian principles to say that the universe has a duty to maximize reflexivity. However, we need to know whether his Einsteinian religion (our spiritual naturalism) should affirm this duty. For that, we need an argument (Sect. 5 in Chapter 3).

4.5   The Parable of the Cell

Any evolutionary metaphysics says that *value* grows slowly, from zero through all the positive degrees. The *intrinsic value* of any thing is the value it has in itself, the value of its being. Degrees of intrinsic value rise gradually, from the simplest particles through all complex things. Cells have small positive degrees of intrinsic value. Any evolutionary metaphysics likewise says that *agency* grows slowly, from zero through all the positive degrees. Cells have small positive degrees of agency. They are primitive little agents whose *policies* or *maxims* are encoded in their

---

[50]Plotinus (*Enneads*, 5.3).

genes. Most cells can reproduce, they are fertile. Any fertile cell can increase the amount of biological value by reproducing. If a cell has one unit of biological value, then by reproducing (through fission) it makes two units. Further reproduction makes four units, then eight, and so it goes.

Agents (like cells) follow the rules or axioms of deontic logic. These say what they ought to do (but not necessarily what they actually do). Our deontic logic includes the axioms of standard deontic logic, to which we add two distinctive axioms of our own. The first distinctive axiom of our deontic logic states that if some agent can maximize value, then it ought to maximize it. Cells have both agency and value. So if it is possible for some cell to increase value, then it ought to increase it. Consider the first living cell on earth. Call it *Alpha*. Alpha has enough resources to reproduce. It is well-nourished and full of energy. It has the power to divide into two offspring. Since Alpha can do it, it ought to do it. It has a duty to reproduce. Its duty is *not* a moral duty—Alpha is not a person. Of course, the first axiom entails more than mere reproduction. It entails that Alpha ought to maximize value. It ought to reproduce to infinity. It can do this by passing its duty down to its offspring. It ought to be the root of an infinitely ramified binary tree of eternally dividing cells. But does Alpha do its duty?

The second axiom of our deontic logic states that any rational agent strives to do its duty. The laws of nature are rationally organized. Every thing that participates in those laws participates in their rationality. Since Alpha participates in the rational laws of nature, Alpha strives to do its duty. It strives to fission into two offspring, and to pass its duty down into them. Its striving is purely teleonomic. At first, Alpha succeeds in doing its duty. It fissions, its offspring fission, and so it goes. This fissioning continues until its offspring run out of resources, or start to destructively compete with each other. So the population of cells on earth reaches some finite maximum. Perhaps some of those cells travel to other planets in our solar system. But the sun eventually runs out of energy. Perhaps the cells travel further into the galaxy. It hardly matters. Entropy increases until everything ends in desolation. The descendents of Alpha never even get close to infinity. Since it does not reproduce to infinity, Alpha fails to do its duty.

Our first deontic axiom refers to something that Alpha *can* do. It refers to a property of Alpha which involves possibility. Since the logic of possibility is known as *modal logic*, it refers to a *modal property* of the cell. We are optimists about Alpha: we affirm that it *can* maximize value. It *can* reproduce to infinity. To say that it *can* do this means only that Alpha *possibly* does it.[51] Alpha *possibly* has descendents that always reproduce. Alpha *possibly* inhabits a universe whose laws permit cellular reproduction to infinity. Alpha *possibly* sits at the root of an infinitely ramified tree of biological complexity. Hence Alpha *can* maximize value. Of course, Alpha *actually* fails to realize its best possibility. It fails to do its duty. Hence the duty of Alpha points beyond its actuality. Following the interpretation of possibility given by the philosopher David Lewis, the possibilities of Alpha involve other better cells inhabiting other better universes. They involve *counterparts* of Alpha.[52] These counterparts are better possible versions of Alpha, in other parallel universes. For any number n, there exists some n-th universe in which the n-th counterpart of Alpha reproduces to the n-th generation of descendents. To say that the cell Alpha can maximize value means that there exists some *ideal cell* which does maximize it. But what is this ideal cell? The ideal cell exceeds cellularity. It is an unsurpassable genealogical tree of surpassable cells.

5   Planetary Replicators

Any replicator aims at infinity in three ways. It aims at *spatial infinity* as the number of organisms perpetually doubles to fill a larger volume of space. It aims at *temporal infinity* as every earlier generation of parent organisms is surpassed by its later generation of organisms. It aims at *functional infinity* as simpler organisms evolve into more complex organisms. These three infinities coincide to make the ecstasy of life. If life were to evolve without any constraints, then it would evolve into an infinitely complex ecology of infinitely complex organisms filling the entire universe for all time. Clearly, this is the *ideal* destiny of life. Life

---

[51]This is a possibility *of the thing* (in Latin, *de re*); it is a *de re* possibility.
[52]Lewis (1986).

moils towards this infinite ecstasy. If life always wins, then it will reach this ecstasy. However, ideals need not be actual. Many ideals are actually unattainable—they remain merely utopian possibilities.

Life already covers our earth.[53] If it keeps moiling towards its ecstasy, then it will expand beyond earth. Dawkins very briefly describes the expansion of self-replicating patterns of information beyond our earth. He says life on earth may cross the space-travel threshold. By crossing that threshold, the self-replicators which evolved on one planet begin to colonize other planets. It may even be that earth was colonized by alien replicators. These space-faring replicators would be like seeds that spread from planet to planet. But these seeds might not be carbon-based. Dawkins says they might be robots, and mentions that we have already sent many robots into deep space.

Where the Dawkinsian *River out of Eden* ends, the futurists begin.[54] The futurist Ray Kurzweil argues that life has a natural tendency to transcend itself. He argues that earthly life will first abandon its organic substrate. We will transform ourselves into robots made of silicon and metal. We will then spread out from earth to colonize the solar system. We will become patterns of pure information, living digital lives in ever greater computers. The futurist Anders Sandberg describes three kinds of great computers. He begins with *Zeus*. Zeus emerges from the conversion of an entire planet into a computer. But Zeus is surpassed by *Uranos*. Uranos emerges from the conversion of an entire solar system into a gigantic computer network. The network surrounds the central star like a spherical cage around a lamp. Now Uranos is surpassed by *Chronos*. Chronos emerges from the conversion of a neutron star into a computer. It is a quantum-mechanical machine. Sandberg argues that these computers are physically possible. They exist in some of the possible technological futures of our universe. They are godlike. However, since they evolved, Dawkins would say they are not gods.

Kurzweil says all these godlike machines will emerge as earthly life spreads out into the galaxy.[55] We will convert our whole galaxy into a

---

[53] Self-replicating patterns (ROE 160–1). Alien replicators (BCD 215; AT 633–5).
[54] Futurists (Moravec 1988; Tipler 1995; Kurzweil 2005). Great computers (Sandberg 1999). Godlike not gods (GD 98–9).
[55] Kurzweil (2005: 350–352). To infinity (2005: 389). Wakes up (2005: 21, 375, 390).

galactic computer. And from the Milky Way, intelligent life will spread out ever farther into the universe. Eventually, the entire universe will become a recursively self-improving cosmic machine. Evolution will produce an endless series of ever more perfect computers. They will climb ever higher on Mount Improbable. Although they will always be only finitely perfect, they will perpetually grow towards infinite perfection. Kurzweil says intelligence is essentially self-surpassing and self-transcending. As it expands, more and more of the universe *wakes up*. As it wakes up, it becomes ever more self-conscious, ever more self-mirroring. Its parts *reflect* more of the universe back to itself. They see the whole more deeply, accurately, and completely. Reflexivity becomes maximized.

Kurzweil argues for an infinite progression of ever greater finite computers. This progression can be defined by two laws. The *initial law* states that there exists some initial computer. Perhaps this was the ENIAC machine built during World War II.[56] It has some very low finite degree of complexity. It is slow and its memory is tiny. It is a little finite Turing machine. The *successor law* states that if there exists any computer with some complexity, then there exists some greater computer with twice the complexity. It has twice as much memory and twice the speed. But these are still just finite Turing machines. As they become more powerful, they become more reflexive. They become ever more self-conscious. They simulate more and more of their own histories and possible futures. They rise through all the finite degrees of self-representation. They simulate ever more of the universe at ever greater levels of detail. They become more ecstatic. Yet all these godlike machines remain only finite.

The limit of this progression of godlike machines is an infinite computer.[57] It is an infinitely powerful universal Turing machine. It has infinite memory and it runs at infinite speed, which means it can compress infinitely many operations into any finite time. It inhabits the first infinite level of Mount Improbable. The physicist Frank Tipler tried

---

[56]ENIAC (Dyson 2012).

[57]Infinite Turing machine (Copeland 1998; Steinhart 2003). Tipler (1995). Dawkins against Teilhard (UR 184–6; ADC 196–8; GD 183–4).

to argue that our universe will converge to an infinite computer at the end of time. He referred to it as the *omega point*. The term *omega point* originates with Teilhard de Chardin. While Dawkins is highly critical of Teilhard, his specific criticisms do not apply to all omega point theories. Tipler just says the omega point refers to an infinite physical machine, which is natural rather than supernatural. It is natural, in the sense used by Dawkins, because it emerges from an evolutionary process.

At the Tiplerian omega point, all the matter in the universe has been aroused from its slumber—the whole universe is wide awake. But when the omega point emerges, a new kind of temporality emerges with it. Since it operates at infinite speed, the omega point experiences an ever-lasting aeon of conscious life. The omega point is perfectly self-conscious. It therefore perfectly reflects its own image back to itself; it is perfectly self-mirroring. Its exact self-consciousness is its ecstasy. If the omega point is a quantum mechanical machine, then perhaps it encodes its thoughts holographically in networks of entangled quantum bits. An infinite holographic quantum computer would very closely resemble the Plotinian divine mind. Tipler says the omega point is all-powerful, all-knowing, and all-good. He refers to it as God. Of course, since it emerges from an evolutionary process, Dawkins would say it is merely godlike.

The Tiplerian vision suffers from monotheistic blindness. If it is possible for there to be one infinite machine, then it is possible for there to be infinitely many of these infinite machines. More accurately, the omega point is an infinite network of infinite computers. Every computer is like an infinite star in an infinite sky. Of course, it is almost certain that our universe will never reach any omega point. The omega point is only a vision of an ideal destination for intelligent self-surpassing. If the omega point is an ideal, then life always *aims at* it, regardless of whether life will ever reach it. Even as life dies out, it still aims at the omega point. More-over, *hope* can get by with an infinitesimal degree of probability. As long as the probability of reaching the omega point is not strictly zero, life can always hope. Ideals are possibilities. It is possible that there are other universes in which life climbs through all finite degrees of complexity. It is possible that life in other universes reaches its infinite omega point.

References

Bell, E. et al. (2015). Potentially biogenic carbon preserved in a 4.1 billion-year-old zircon. *Proceedings of the National Academy of Sciences, 112*(47), 14518–14521.

Bertz, S. (1981). The first general index of molecular complexity. *Journal of the American Chemical Society, 103*, 3599–3601.

Bottcher, T. (2016). An additive definition of molecular complexity. *Journal of Chemical Information and Modeling, 56*, 462–470.

Bournez, O., Ibanescu, L., & Kirchner, H. (2006). From chemical rules to term rewriting. *Electronic Notes in Computer Science, 147*, 113–134.

Bower, J. (1988). *The Evolution of Complexity by Means of Natural Selection.* Princeton, NJ: Princeton University Press.

Conselice, C., et al. (2016). The evolution of galaxy number density at $z<8$ and its implications. *The Astrophysics Journal, 830*(2), 83.

Copeland, B. J. (1998). Even Turing machines can compute uncomputable functions. In C. Calude, J. Casti, & M. Dinneen (Eds.), *Unconventional Models of Computation* (pp. 150–164). New York: Springer-Verlag.

Cummins, R. (1996). *Representations, Targets, and Attitudes.* Cambridge, MA: The MIT Press.

Dehaene, S., Piazza, M., Pinel, P., & Cohen, L. (2003). Three parietal circuits for number processing. *Cognitive Neuropsychology, 20*(3/4/5/6), 487–506.

De Loore, C., & Doom, C. (1992). *Structure and Evolution of Single and Binary Stars.* Boston: Kluwer.

Dennett, D. (1995). *Darwin's Dangerous Idea: Evolution and the Meanings of Life.* New York: Simon & Schuster.

Dyson, G. (2012). *Turing's Cathedral: The Origins of the Digital Universe.* New York: Vintage Press.

England, J. (2014). A new physics theory of life (interview with N. Wolchover). *Quanta Magazine.* Online at www.quantamagazine.org/a-new-thermodynamics-theory-of-the-origin-of-life-20140122/.

Garrod, R., Widicus Weaver, S., & Herbst, E. (2008). Complex chemistry in star-forming regions: An expanded gas-grain warm-up chemical model. *The Astrophysical Journal, 682*, 283–302.

Horowitz, J., & England, J. (2017). Spontaneous fine-tuning to environment in many-species chemical reaction networks. *Proceedings of the National Academy of Sciences, 114*(29), 7565–7570.

Jaffe, R., & Taylor, W. (2018). *The Physics of Energy.* New York: Cambridge University Press.

Johnson-Laird, P. N. (1983). *Mental Models*. Cambridge, MA: Harvard University Press.

Johnston, I., Ahnert, S., Doye, J., & Louis, A. (2011). Evolutionary dynamics in a simple model of self-assembly. *Physical Review, 83*(6), 066105.

Kachman, T., Owen, J., & England, J. (2017). Self-organized resonance during search of a diverse chemical space. *Physical Review Letters, 119*(3), 038001.

Kosslyn, S. (1994). *Image and Brain: The Resolution of the Imagery Debate*. Cambridge, MA: MIT Press.

Kurzweil, R. (2005). *The Singularity Is Near: When Humans Transcend Biology*. New York: Viking.

Lewis, D. (1986). *On the Plurality of Worlds*. Cambridge, MA: Blackwell.

Magill, J., & Galy, J. (2005). *Radioactivity Radionuclides Radiation*. New York: Springer.

Moravec, H. (1988). *Mind Children: The Future of Robot and Human Intelligence*. Cambridge, MA: Harvard University Press.

Nieder, A. (2005). Counting on neurons: The neurobiology of numerical competence. *Nature Reviews: Neuroscience, 6* (March), 177–190.

Nutman, A., et al. (2016). Rapid emergence of life shown by discovery of 3,700-million-year-old microbial structures. *Nature, 537*, 535–538.

Petigura, E., et al. (2013). Prevalence of Earth-size planets orbiting Sun-like stars. *PNAS, 110*(48), 19273–19278.

Post, E. (1943). Formal reductions of the general combinatorial decision problem. *American Journal of Mathematics, 65*(2), 197–215.

Ray, T. (1992). An approach to the synthesis of life. In C. Langton, C. Taylor, J. Farmer, & S. Rasmussen, *Artificial Life II* (Vol. 10, pp. 371–408). SFI Studies in the Sciences of Complexity. Reading, MA: Addison-Wesley.

Sandberg, A. (1999). The physics of information processing superobjects: Daily life among the Jupiter brains. *Journal of Evolution and Technology, 5*(1), 1–34.

Soltis, D., & Soltis, P. (2019). *The Great Tree of Life*. Cambridge, MA: Academic Press.

Steinhart, E. (2003). Supermachines and superminds. *Minds and Machines, 13*, 155–186.

Taft, R., Pheasant, M., & Mattick, J. (2007). The relationship between non-protein-coding DNA and eukaryotic complexity. *BioEssays, 29*(3), 288–299.

Tipler, F. (1995). *The Physics of Immortality: Modern Cosmology, God and the Resurrection of the Dead*. New York: Anchor Books.

Wagner, A. (2014). *Arrival of the Fittest: How Nature Innovates*. New York: Penguin.

3

Reflexivity

1   The Thermodynamic Liturgy

1.1   Entropy Is Not Disorder

The *thermodynamic liturgy* emerges from the previous liturgies. It starts with *entropy*, which Dawkins often portrays as disorder.[1] Since there are more ways to be disordered than ordered, any system naturally wanders towards disorder. This statistical point allegedly motivates the *second law of thermodynamics*, which states (roughly) that the entropy of a closed system generally stays the same or increases. Dawkins often gives this statistical explanation for the second law. However, *entropy is not disorder*.[2] Nor do natural systems necessarily wander. So Dawkins gets both entropy and the second law incorrect. Since the growth of order (and complexity) are central to his arguments about evolution, it is important to get these concepts right. This means doing a little science. Like the previous liturgies, the thermodynamic liturgy is a series of scientific exercises—but these are also spiritual exercises.

---

[1]ADC 84–5; AT 397.
[2]Wright (1970), Wald (2006), and Martyushev (2013).

© The Author(s) 2020
E. Steinhart, *Believing in Dawkins*,
https://doi.org/10.1007/978-3-030-43052-8_3

There are three ways to think about entropy (and the second law). The first way defines it in terms of *free energy*. The energy available to cause physical change is free. When a battery is fully charged, it is filled with free energy; as it runs down, its free energy is spent, and it becomes exhausted. So, as it runs down, its entropy increases. Thus entropy is the inverse of free energy. A second way to think about entropy defines it in terms of *energy concentration and dispersion*.[3] Concentrated energy is free. When a stick of dynamite is at rest, it is filled with concentrated energy; but when it explodes, the energy is dispersed; again, its energy becomes exhausted. So, as it explodes, its entropy increases. Thus entropy measures *energy dispersal*.[4] Dawkins sometimes correctly identifies entropy with exhaustion. And he correctly states that the second law implies that energy in a closed system becomes more dispersed.

On these energetic definitions, you can think of entropy as the *flatness* of an energy landscape (or probability distribution). Think of an energy landscape with a single extremely high and steep mountain. Energy is intensely concentrated in the height of *Mount Energy*. Since this landscape is far from flat, its entropy is very low. The second law, in this landscape analogy, says that the flatness of the landscape is always increasing. And here, in this landscape analogy, there is a force that increases entropy. This force, of course, is gravity. Gravity erodes Mount Energy. As the mountain erodes, the surrounding landscape fills up with its detritus. Boulders roll off and sit on the plains; they dissolve into sand. The final state is a flat sandy wasteland, in which the grains of sand no longer have any gravitational energy to roll anywhere. When the landscape is utterly flat, its energy is thoroughly dispersed, its entropy is maximal.

The third way is the most technical. It involves the distinction between *microstates* and *macrostates*. If you start with some set of small parts that can be arranged into larger wholes, then every possible way of arranging them into an arbitrary whole is a microstate. The parts could be the molecules in a gas or the smallest mechanical parts of an airliner.

---

[3]Kotz et al. (2009: ch. 19.2), Catling (2013: 6), Ebbing and Gammon (2017: ch. 18.2), and Tzafestas (2018: 128).
[4]GSE 413–16.

Since Dawkins often uses airliners to illustrate complexity (Sect. 1.2 in Chapter 2), we can start with a field covered with the smallest mechanical parts of an airliner.

Every possible way of arranging those airliner-parts is a microstate. The number of possible ways of arranging parts is the *multiplicity* of microstates. It is the *micro-multiplicity*. If there are ten billion possible ways of arranging the airliner parts, then the micro-multiplicity of the type *airliner* is ten billion. Since micro-multiplicities can be very large, it is often more useful to work with their logarithms. The logarithm of the micro-multiplicity of some system can be identified with the *available complexity* of that system. The field strewn with Boeing 747 parts has a high available complexity; after all, it can be organized into an extremely complex airliner.

Every *class* of microstates is a macrostate. Suppose all the microstates are classified as either *airliners* or *junk*. The airliner macrostate is the class of all airliner microstates, while the junk macrostate is the class of all junk microstates. Here the macrostates correspond to the types *airliner* and *junk*. If the macrostates correspond to types, then to be a member • of some macrostate is to be an instance of the corresponding type. To be a member of the airliner class is to be an instance of the *airliner* type. To say that a microstate *satisfies* a macrostate means that it is an instance of the macrostate type. Thus any microstate satisfies either the type *airliner* or the type *junk*.

The number of microstates that satisfy some macrostate is the multiplicity of the macrostate.[5] It is the *macro-multiplicity*. Suppose only ten ways of arranging the parts actually make airliners. On that assumption, the macro-multiplicity of the type *airliner* is ten. The macro-multiplicity of the type *junk* is ten billion minus ten, which is pretty close to ten billion. Now the third definition of entropy states that *the entropy of a macrostate is the logarithm of its multiplicity.* Since the logarithm of a number increases as the number increases, higher macro-multiplicities mean higher entropies, while lower macro-multiplicities mean lower

---

[5]Boltzmann defined entropy as $S = k \log W$, where W is the macro-multiplicity of the state S. His constant k can be set to 1. Macro-multiplicity (GSE 416).

entropies. Thus common macrostates (like junk) have high entropies, while rare macrostates (like airliners) have low entropies.

If entropy is defined in terms of multiplicities, then it is closely associated with the Dawkinsian combinatorial definition of complexity. A little arithmetic (see the note) shows that *the complexity of any type is the logarithm of its micro-multiplicity minus the logarithm of its macro-multiplicity*.[6] The logarithm of the micro-multiplicity is the available complexity of some set of parts. And the logarithm of the macro-multiplicity of some type made from those parts is just its entropy. So our equation says that *the complexity of a type is its available complexity minus its entropy.*

What, then, is entropy? *Entropy is wasted complexity.* Hence *the complexity of some type is its available complexity minus its wasted complexity.* More complex types have both higher available complexity and less wasted complexity. Their available complexity is actualized rather than wasted. And so, if you see the airliner parts just jumbled together by a windstorm, you can be sad that so much available complexity was wasted. But if you see them assembled into an airliner, then you are seeing the wasted complexity eliminated. You are seeing waste being *pumped out* of the field.

1.2   Stars Pumping Entropy into the Abyss

We now have three ways of defining entropy (in terms of free energy, energy dispersal, and microstates). Do these ways work well together? Some of the physics here is not yet settled. However, it is reasonable to assume that these ways are mutually coherent, and that they can be applied to our universe as a whole. At its start, our universe is an tiny dot with an extremely high temperature and density.[7] Since this dot is so tightly compressed, it is plausible that its micro-multiplicity is very small.

---

[6]The complexity of some type is the logarithm of its arbitrary multiplicity divided by its stable multiplicity (Sect. 1.2 in Chapter 2). This is just the logarithm of its micro-multiplicity divided by its macro-multiplicity. But the logarithm of x divided by y is the logarithm of x minus the logarithm of y. So the complexity of any type is the logarithm of its micro-multiplicity minus the logarithm of its macro-multiplicity.

[7]Silk (2001: 113).

But since every microstate satisfies the type *universe*, the macrostate of that type contains a very small number of microstates. So, if entropy involves multiplicities, then *the entropy of the initial universe is very low.* Likewise its available complexity is low, its wasted complexity is low, so its complexity is low. The universe starts simple.

The energy in this dot was extremely concentrated. If entropy is energy dispersal, then at its birth the universe has *extremely low entropy.*[8] The big bang is an explosion of fiery energy. Now the universe resembles Mount Energy, its energy expressed through gravity. Gravity starts pulling particles together into ever denser clumps. Gravity works on Mount Energy, pulling it down and flattening it out. Entropy increases. The final state of the universe will be energetically very flat. A black hole is almost totally flat, its gravitational energy exhausted. So the universe will end in a wasteland of black holes, which will evaporate into physical desolation. At its end, the energy of the universe will be maximally dispersed, so its entropy will be maximal.

One consequence of the Dawkinsian combinatorial concept of complexity is that the type *universe* is always simple. It is simple because every way to permute the parts of the universe preserves the type *universe*. The universe starts out simple because its available and wasted complexities are both low. But as the universe grows, its available complexity increases enormously. Likewise its wasted complexity, which is its entropy, also increases enormously. If available and wasted complexity both increase together at the same rate, then the universe always remains simple. Nevertheless, the thesis that both of these quantities are increasing gives rise to *directionality.* The low entropy at the start of the universe, plus the second law of thermodynamics, gives rise to the thermodynamic *arrow of time.* All the arrows of complexity ride on this arrow of time. But how can complexity emerge if all the available complexity of the universe is wasted? How can complex things evolve in a universe which remains eternally simple?

Complex things can emerge in our simple universe because the available and wasted complexities of our universe are not evenly distributed. Our universe as a whole can be extremely simple while containing some

---

[8]Penrose (1979), Greene (2005: 173–174), and Wald (2006).

small yet very complex parts. And planets like our earth are very small compared with the cosmos. Of course, while this unevenness allows complexity to emerge, it does not explain how it emerges. Complex things emerge in our universe because it contains great engines that pump entropy out of some parts of the universe and into others. Some parts of our universe have high available complexities. These are the solar systems in galaxies. They contain engines that pump out the wasted complexity. They contain *entropy pumps*.[9] These entropy pumps are *stars*. As the entropy gets pumped out, the complexity increases.

As our solar system condensed from clouds of cosmic dust, its available complexity increased rapidly. And its wasted complexity increased rapidly too. It had high entropy. But once the sun ignited, it began bathing the embryonic solar system with free energy.[10] It began pumping entropy out of the solar system and into space. Ultimately, it pumps that entropy into the abysmal black hole at the center of the Milky Way. As this entropy gets pumped out, actual complexity increases. Our earth was born. During its early stages, our earth was literally a hot mess. It had enormous available complexity, but almost all of that available complexity was wasted. The entropy of the early earth was very high. But free energy from the sun (and from the nuclear reactions in the core of the earth) was flowing through the matter near the surface of the earth.

As these energies pass through the earth, they drive a pump which moves entropy away from the surface of the earth and into space.[11] Hence the entropy of the surface of the earth goes down. Entropy is wasted complexity. As wasted complexity gets pumped out of the earth, more and more of the available complexity turns into actual complexity. Earthly matter self-organizes into increasingly complex structures. Thus life evolves on the surface of the earth. Wasted complexity is negative, but pumping wasted complexity out is negating the negative, and negating the negative makes a positive. Pumping entropy *out* pumps complexity *in*. Dawkins identifies complexity with improbability. So the solar free energy that pumps entropy away from the earth pumps improbability

---

[9]GSE 414–16.
[10]BW 94; GSE 412–6; ADC 84–5.
[11]Improbability pump (GSE 416). Blessed (UR 5).

into the structures that evolve on the earth. Thus Dawkins says the free energy from the sun drives an *improbability pump*. He says we are "hugely blessed" to live in a universe whose laws concentrate complexity into material structures.

Figure 1 illustrates the falling line of free energy, the rising line of entropy, and the arc of complexity. This Figure illustrates relations among concepts—it isn't intended as an exact physical diagram. The *arc of complexity* rises, peaks, and falls. Of course, this arc traces only the complexities of the most complex things in the universe. The universe itself always remains simple. The ascending part of this complexity arc is a thermodynamic arrow of complexity that points upwards on Mount Improbable. Mount Improbable rises as Mount Energy falls. This thermodynamic arrow drives all the other arrows of complexity (see Chapter 2): it drives the atomic, molecular, biological, and other arrows. During this ascending part, the universe is filled with cranes lifting things up to higher levels of Mount Improbable. And, during this ascending part of the arc of complexity, both *complexity and entropy are increasing together*. Eventually, however, there won't be enough free energy to push things up any higher, or to keep them at their high levels. The stars will burn out and Mount Improbable will erode away. All complexity will disintegrate, all value will fade away. Our universe will end in desolation,

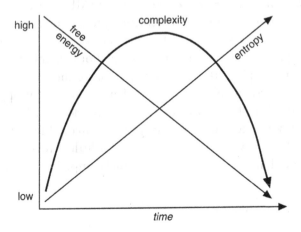

Fig. 1 Free energy, complexity, and entropy

bereaved of all of its valuable machines. These ideas motivate a more spiritual interpretation of entropy: *entropy is grief*. The atheist Iris Murdoch said that grief reveals the *void*.[12] For spiritual naturalists, this void is the *abyss*.

1.3   Maximizing Entropy Production Rates

How do entropy pumps like the sun drive the evolution of complexity on planets like the earth? As solar radiation heats the earth, it drives flows of material stuff. Any flow produces entropy at some rate. Flow links entropy production rates to self-organization. According to the physicist Rod Swenson, *ordered flow creates entropy faster than disordered flow*.[13] Consider the currents of air in our atmosphere. They carry energy from the surfaces heated by the sun to the cold emptiness of space. There are three main ways they can be structured: (A) turbulent currents poorly organized; (B) currents somewhat organized into convection cells; (C) currents highly organized into cyclones. More highly organized currents transfer heat-energy more rapidly. They minimize energy potentials more efficiently. They help energy disperse itself more rapidly. Hence they produce entropy more rapidly. As currents shift from turbulent winds to highly organized cyclones, their entropy production rates increase.

The *maximum entropy production principle* (MEPP) states that physical systems tend to maximize their entropy production rates.[14] Swenson puts it like this: "A system will select the path or assembly of paths out of available paths that minimizes the potential or maximizes the entropy at the fastest rate given the constraints."[15] Since the MEPP states that every physical system *selects* the paths that maximize entropy production rates, the MEPP is a principle of natural selection. Every physical system, at any moment in time, can go down many future paths. As it follows the big arrow of time, each path is a little arrow directed from past to future. These arrows enter into virtual competition – they struggle

---

[12]Murdoch (1992: chs. 18 and 19).

[13]Swenson (2006: 318).

[14]MEPP (Martyushev and Seleznev 2006).

[15]Swenson (2009: 334).

against each other.[16] The arrows that produce entropy at the fastest rate win the competition. They are the fittest paths. This is *arrow selection*. It is differential survival based on fitness. And since thermodynamical principles are among the deepest physical principles, the MEPP is among the most fundamental principles of natural selection.

The MEPP means that if a process *can* shift into a faster entropy production mode, then it almost certainly *will*. Consider an air stream that starts in a chaotic mode. If it can become a convection cell, it almost certainly will; if it can become a cyclone, it almost certainly will. Hence processes tend to evolve in ways that increase their entropy production rates. But why say they merely *tend*? The MEPP acts teleonomically: it *drives* systems to produce entropy faster. A great deal of evidence supports the MEPP.[17] It has been successfully employed in physics, chemistry, biology, ecology, and economics. It seems to act at all times and places in the universe. The driving power of the MEPP has been confirmed by many examples of self-organization. It drives many examples of *biochemical and biological* self-organization.[18] Dawkins points out that thermodynamic principles play roles in protein folding.[19] Nevertheless, the MEPP is not fully confirmed as a physical fact—it does not yet have the status of a settled law of physics. More study must be done, and the MEPP may be modified. Jeremy England has argued for thermodynamic principles which are somewhat different from the MEPP but which play the same role.[20] It seems likely that *something like* the MEPP will turn out to be a universal and close to ultimate law of physics. So

---

[16]Rescher states that "in the virtual competition for existence among alternatives it is the comparatively best that is bound to prevail" (2010: 33–34). The MEPP says the path that produces entropy the fastest will usually prevail.

[17]Steinhart (2018) surveys evidence for the MEPP.

[18]The MEPP has been confirmed in many examples of biochemical and biological self-organization. The MEPP accurately predicts the evolution of beta-lactamase enzymes (Dobovisek et al. 2011) and the evolution of the enzyme ATP synthase (Dewar et al. 2006). Proton pumps in photosynthesis operate produce entropy at rates very close to the maximum (Juretic and Zupanovic 2003). It correctly predicts bacterial metabolism (Unrean and Srienc 2012). Replicator systems evolve towards states in which entropy production is maximized (Martin and Horvath 2013). A model based on MEPP does well at predicting evolutionary trends (Skene 2015).

[19]GSE 236.

[20]England (2013, 2014, 2015).

the law that converts entropy into complexity is the MEPP or something like it. From now on, I'll just use the term "MEPP" to whatever principle converts entropy into complexity.

If the MEPP is correct, then we can make a *Thermodynamic Argument*: (1) Ordered flow produces entropy faster than disordered flow. (2) The MEPP asserts that physical systems tend to maximize their entropy production rates. If they can produce entropy faster, they almost certainly will. (3) Therefore, physical systems tend to increase their orderliness. If they can become more orderly, they almost certainly will. If physical flow can self-organize, then it almost certainly will. If it can gain complexity, it almost certainly will. Dawkins says the universe could have easily stayed internally simple.[21] The Thermodynamic Argument refutes that thesis. Since processes evolve according to the MEPP, they at least have a tendency to self-organize. But saying they merely have a tendency is too passive. They have a real directionality pointing from simple to complex. This directionality is the arrow behind every other arrow of complexity in our universe. The MEPP generates a *thermodynamic force*, which *drives* the self-organization of matter.[22] It drives the algorithmically patterned upthrust of complexity. If cranes are complexity-increasing computers, then the MEPP powers every crane.

The Thermodynamic Argument can be visualized using Mount Improbable. Every physical system, at some moment of time, sits in the middle of a landscape. It sits on its *central point*. This central point is surrounded by a ring of points. Each point in this ring is some immediate future variation of that physical system—it is some immediate way in which the system can evolve. So each point on the ring is a variant point. An arrow runs from the central point to each variant point. The arrow has some probability. The MEPP entails that arrows with higher entropy production rates are more probable. Hence probability is non-randomly distributed over these arrows. It is skewed towards maximizing the rate of entropy production. These probabilities can be thought of as resistances: higher probability is lower resistance, while lower probability is higher resistance. Flows of matter are more likely to follow arrows of lower

---

[21]AT 699.
[22]Dewar (2006).

resistance. So the matter flowing through some physical system flows through the arrows which maximize entropy production rates. This is *arrow selection*. From this selection, there emerges teleonomic direction: flows of matter increase their entropy production rates. Since ordered flow produces entropy faster, those flows generate order. The MEPP drives them to self-organize. So simplicity evolves into complexity.

1.4 Thermodynamic Forces Drive Self-Organization

On one definition, entropy measures energy dispersal. Low entropy is highly concentrated energy. But the energy in a high entropy system is highly dispersed. The second law of thermodynamics now says, roughly, that energy tends to disperse itself over time. Energy tends to distribute itself more uniformly. As energy disperses itself, thermodynamic forces can arise. Since these are often associated with increasing entropy, they are often said to be *entropic forces*.[23] An entropic force drives a system to maximize its entropy. Entropic forces include *elastic forces*. When you stretch a rubber band, the force which pulls it back into its unstretched shape is entropic. Entropic forces include *depletion forces*. When you mix small particles with large particles, the mixture evolves until the amount of open space in the mixture is minimized. Entropic forces drive this evolution. Entropic forces help drive protein folding.

It turns out that entropic forces can drive disordered systems to order. However, if entropy is disorder, then how is this possible? The answer is that *entropy is not disorder*. An easy way to illustrate how an entropic force can increase order is the *tornado in a bottle*.[24] Start with two empty bottles, about four liters in size. Fill one about halfway full with water. Now join the two bottles using a length of tubing about four centimeters long. When thus joined, the empty bottle is on top. Flip the whole thing over, so that the bottle with water is on top. Since gravity pulls on the water in the top bottle, you've just concentrated some gravitational free

---

[23]Steinhart (2018).
[24]Schneider and Kay (1994).

energy. Since energy is highly concentrated right after you flip the bottles over, the whole system has low initial entropy.

The second law says that the entropy of this two-bottle system will increase to its maximum. As entropy increases, the energy concentrated in the top bottle disperses. It disperses by flowing into the bottom bottle. Thus dispersed, it can no longer do any work. At first, after you've flipped the bottles, the water will most likely drain out in a turbulent and disorderly way. It chaotically gurgles and glubs. But the MEPP entails that entropic forces are acting on this draining water. As they act on this draining water, its disorderly gurgling will very likely turn into an orderly vortex. Entropic forces drive the emergence of this orderly vortex. Why? Because this orderly vortex disperses the energy faster than the disorderly gurgling. The orderly vortex emerges because it is the fastest way to maximize the entropy—it maximizes the entropy production rate. And so entropic forces drive the flow to self-organize into a vortex. The flowing water can follow two dynamical arrows: the turbulent arrow or the vortex arrow. Entropic forces push the flow towards the vortex arrow. This is arrow selection.

More generally, it has been argued that entropic forces drive the evolution of complexity by maximizing entropy production rates in open systems.[25] Many specific examples of entropy-driven self-organization have been studied. Depletion forces are entropic forces. They drive mixtures of molecules to become more orderly. They drive the self-organization of large molecular structures. They drive the self-organization of cellular structures like cytoskeletons and chromosomes. It has been argued that the maximization of thermodynamic quantities drives ecological competition. If that is right, then entropic forces drive the emergence of biological *arete*.

The thermodynamic liturgy ends with a general picture: All the definitions of entropy cohere. All complexity in the universe emerges from the operation of entropy pumps. It emerges from the action of something similar to the maximum entropy production principle. Entropic forces

---

[25]Entropic forces drive self-organization (Steinhart 2018). They drive ecological competition (Martin and Horvath 2013; Yen et al. 2014; Chapman et al. 2016).

drive the self-organization of matter. Dawkins often says that evolution does not violate thermodynamics.[26] But the thermodynamic liturgy says that thermodynamics actually *drives* the evolution of complexity. Of course, since there are many open questions in thermodynamics, the general picture of this liturgy is not entirely clear. We need to be skeptical about it, and we need to keep our minds open for further refinements. Nevertheless, we can be confident that the general picture is sufficiently accurate for us to proceed. Moreover, while the Sanctuary for Spiritual Naturalists depends on this general picture, it does not depend on its details.

2   The Physical Liturgy

2.1   Lessons from the Earlier Liturgies

We have worked through four physical liturgies (atomic, molecular, biological, and thermodynamic). They teach several lessons in the general physical liturgy. The *first lesson* is that the universe contains natural models of computation. They are systems of arrows which model the string rewriting theory of computation developed by Emil Post. Atomic arrows define nuclear reactions, molecular arrows define chemical reactions, and biological arrows define reproductive reactions. Here are some examples:

hydrogen + hydrogen → helium;
methane + oxygen → carbon dioxide + water;
parent → offspring + offspring;
male + female → offspring.

These arrow systems are concentrated in celestial computers. Every celestial computer is a solar system. It contains some star pumping entropy into the abyss. It probably also contains dust clouds, planets, moons, asteroids, and so on.

---

[26]BW 94; ADC 84–5; GSE 413–6; AT 397; SITS 337.

The *second lesson* is that these celestial computers run complexity-increasing programs. These programs are assignments of probabilities to their reaction arrows. The atomic program running in interstellar space gives very low probabilities to complexity-increasing fusion reactions. But the atomic program running in the core of the sun gives a high probability to the reaction that fuses simple hydrogen into more complex helium. Other stars carry these fusion reactions even higher into the sky of atomic complexity. The molecular programs running in most regions of space give very low probabilities to complexity-increasing reactions, while the molecular programs running on the surfaces of illuminated planets give much greater probabilities to complexity-increasing chemical reactions. Similar remarks hold for biology.

The *third lesson* is that random chance does not suffice for the increase in complexity. If complexity increases, then probabilities are not randomly distributed across arrows; they are distributed in non-random ways which make the increase of complexity more likely than its decrease or equivalence. Say the *skew* of a program is the degree to which its probabilities deviate from random. Programs that produce small increases in complexity are skewed a little. But programs that produce large increases are greatly skewed. They require chains of complexity-increasing reactions. Any program greatly skewed to complexity-creation is an *entelechy*. Any computer that runs an entelechy is a crane. Our solar system is a celestial crane.

The *fourth lesson* is that programs are not skewed towards complexity by chance. Random drift preserves randomness. If a program gets skewed towards complexity generation, that skew was caused by some non-random skewer. The thermodynamic liturgy teaches us that the skew comes from entropy pumps. As the stars pump entropy out of their solar systems into empty space (or into black holes), this pumping skews the probabilities of the celestial computers towards complexity generation. The maximum entropy production principle (the MEPP) defines the skew. As stars pump entropy out of their solar systems, the MEPP redistributes the probabilities. It rewrites chaotic programs into entelechies. The MEPP is the physical skewer. Of course, since we can still seek an explanation for the MEPP, it cannot be the *ultimate* skewer.

2.2 The Library of All Possible Physical Things

The physical liturgy abstracts from the earlier and more specific liturgies. All the things discussed in those earlier liturgies are composed of particles like quarks and electrons. Since all physical things in our universe are composed of particles, it is possible to define a single *library of all possible physical things*.[27] This general library is the *physical library*. Of course, here possibility is restricted to the physical possibilities in our universe. All the books in the atomic, molecular, and biological libraries are in the physical library. All these books define networks of simple particles (mainly quarks and electrons). The Dawkinsian combinatorial definition of complexity (Sect. 1 in Chapter 2) ranks these books by complexity. Complexity corresponds to improbability.

Complexity is usually thought to involve information. Dawkins briefly considers an informational definition of complexity: the complexity of any thing is the length in bits of its most concise description.[28] One informational definition of complexity stands out for its fidelity to evolutionary principles. This definition identifies complexity with a mathematical quantity known as *logical depth*.[29] The logical depth of any thing is the amount of computational work needed to produce its structure. Logical depth obeys a *slow-growth law*, which states that logically deep things cannot easily be produced by chance.[30] On the contrary, they result from long processes in which depth slowly accumulates. Machta writes that "depth is sensitive to embedded computation and can only be large for systems that carry out computationally complex information processing."[31] If complexity is depth, then things gain complexity by accumulating computations nested in computations. Since complexity increases with self-nesting computation, it increases *reflexivity*. More complex things are more reflexive, and this is confirmed by the growth

---

[27] BW ch. 3; CMI ch. 6; ADC ch. 2.2; AT 676.

[28] ADC 100–2, 210.

[29] Bennett (1988, 1990).

[30] The slow-growth law states that "deep objects cannot be quickly produced from shallow ones by any deterministic process, nor with much probability by a probabilistic process, but can be produced slowly" (Bennett 1988: 1).

[31] Machta (2011: 037111–037116).

of complexity on earth. But the computations which increase complexity are cranes. So things gain complexity by nesting cranes in cranes. Logical depth can be used to rank the books in the physical library. However, logical depth is hard to measure. Fortunately, there are measures of physical complexity which can serve as proxies for logical depth.[32] These informational approaches to complexity, like the combinatorial approach, entail that complex things are improbable.

2.3   The Great Chain of Being

The idea of sorting possible physical things into complexity ranks resembles an old idea known as the *great chain of being*.[33] The classical great chain was a way of sorting types of things into ranks based on their degrees of perfection. It was first articulated by the ancient Platonists and Stoics. It was taken up by medieval Christians. The classical chain contained only a few major ranks. These contained the natures of minerals, plants, animals, humans, angels, and God. They subdivided the first major ranks into many minor ranks. So, for instance, there were different choirs of angels, with some choirs more powerful than others. Of course, God alone occupied the top rank. The great chain influenced scientists from Aristotle to modern times. When Linnaeus drew up his classification of species, he was influenced by the great chain.

The classical great chain involved many errors. Its concept of perfection (or excellence, or greatness) was not clear. Its rankings of living things did not correspond to any real natural rankings. The classical great chain appeared to imply that the natures of organisms were eternally fixed; it thus appeared to invalidate the evolutionary change of one species into another. And the classical great chain contained only modern species. It did not contain any fossil species. So, when it was interpreted in terms of evolution, it mistakenly implied that some modern species

---

[32]One proxy for the logical depth of some thing is the time it takes to decompress a compressed description of it (Zenil and Delahaye 2010). Another proxy is the free energy rate density of Chaisson (2006). This is the amount of free energy passing through one gram of matter of the thing in one second.

[33]For the great chain, see Lovejoy (1936). For the Stoic great chain, see Cicero (*On the Nature of the Gods*, II.33–5).

evolved into other modern species. And the classical great chain incorporated the tragic cultural biases of ancient and medieval societies. It mistakenly sorted humans into antiquated social ranks. It wrongly sorted humans into higher and lower races. Finally, as far as naturalists are concerned, it contained things which do not exist, such as choirs of angels and God.

For all these reasons, Dawkins correctly rejects the classical great chain of being.[34] But when he sorts organisms into ranks based on complexity, he is following the logic of the great chain in a more modern and scientific way. The elevations on his Mount Improbable resemble the ranks of the great chain. There are simpler things on lower elevations, and more complex things on higher elevations. Of course, these elevations are defined using modern science, and they include fossil species as well as modern species. And the modernized great chain rejects both the classism and the racism of the old great chain. It likewise does not imply that the natures of things are fixed. Species can evolve into new species. Nevertheless, all the Dawkinsian libraries resemble the classical great chain in their orderly structure. The physical library, sorted into floors based on complexity, resembles the classical great chain.

Although the details of the classical great chain are incorrect, it contains an important insight: it orders the natures of things by *value*. Anselm writes that "things are not all of equal value, but differ by degrees. For the nature of a horse is better than that of a tree, and that of a human more excellent than that of a horse."[35] The value that appears in the Anselmian ranking is *not* any kind of ethical or moral value. Moral value applies only to interactions among moral agents. It applies only to the actions or characters of persons (which may be either human or nonhuman). A virtuous person is ethically better than a vicious person; but a horse is not ethically better than a tree.

Since the Anselmian ranking involves only abstract natures, it does not depend on any minds. Those abstract natures are the types or forms of things, and they exist even if no things exist. Likewise their properties and relations exist even if no things exist. So the nature of a human is

---

[34]ADC 208; GSE 155–9.
[35]Anselm (*Monologion*, ch. 4).

more valuable than the nature of a horse even if there are no humans and no horses. The ranking exists even if there are no minds. Hence the value in the Anselmian ranking is *objective* rather than subjective. It is the value that each nature or form has *in itself* rather than the value it has for something else. The value that something has in itself is referred to as its *intrinsic value*, while the value it has for other things is its *extrinsic value*. But what is intrinsic value?

2.4   Complexity Is Intrinsically Valuable

A long tradition in Western philosophy says that being is goodness: more being is more goodness. Hence the intrinsic value of any existing object is its quantity of being. Since every existing object has some positive quantity of being, intrinsic value is always positive. It cannot be negative or zero. One approach to intrinsic value was developed by Leibniz. He argued that the intrinsic value of any thing is its degree of perfection; its degree of perfection is its quantity of being.[36] The quantity of being in any thing is its amount of inner harmony. Harmony is proportional to both order and variety. It is minimal when there is all order and no variety (regularity). It is maximal when order and variety are both maximal. It returns to minimality when there is all variety and no order (randomness). But any quantity which varies with order and variety in that way is complexity. So Leibniz thinks of complexity as intrinsically valuable.

More recently, Daniel Dennett discusses ranking types of organisms by intrinsic value.[37] He says it is worse to kill a condor than to kill a cow; worse to kill a cow than a clam; worse to kill a redwood than to kill an equal mass of algae. It is worse to kill a condor than a cow because "the loss to our actual store of design would be so much greater if the condors went extinct." A condor is more valuable than a cow because it contains a greater store of design than a cow. Of course, by a store of design, Dennett just means accumulated complexity. Dennett says Bach

---

[36]Leibniz (1697), Leibniz (*Monadology*, sec. 58), Rutherford (1995: 13, 23, 35), and Rescher 1979: 28–31).
[37]Dennett (1995: 511–513).

is precious because he contained "an utterly idiosyncratic structure of cranes, made of cranes, made of cranes, made of cranes." Thus Dennett thinks of complexity as intrinsically valuable.

Dawkins often compares the values of organisms.[38] He defines value as adaptive fitness, which he defines in terms of combined adaptive features. As organisms get fitter and fitter, they also tend to get more and more complex. Of course, while organisms can gain fitness by *losing* complexity (as when cave fish lost their eyesight), Dawkins stresses the tendency of life to gain adaptive fitness by gaining complexity. Hence the adaptive fitness of an organism usually follows the complexity of its system of combined adaptive features. As species climb Mount Improbable, they get intrinsically better in terms of adaptive fitness—but height on that mountain is complexity. Across all his books, Dawkins persistently *praises* the evolution of complexity. Hence Dawkins also thinks that complexity is intrinsically valuable. Many other writers argue for the same point.[39] Consequently, we affirm that *complexity is intrinsically valuable*.

This concept of value provides some clarity about agency. It is plausible to say that an *agent* is any self-powered thing that uses a *maxim* to change some value. A maxim is some rule that directs the agent to teleonomically strive to increase or decrease its value. A maxim can be expressed as an imperative in a single sentence or as a policy in hundreds of pages of symbols. More generally, a maxim is any convergent algorithm skewed to increase (or decrease) its value. An agent is therefore any self-powered computer that can run some maxim. Stars are self-powered computers running maxims encoded in their nuclear reaction probabilities. A thing can be self-powered by surfing on the energy gradients generated by some other self-powered thing. Planets are self-powered only by surfing on the gradients of their stars. Planets run maxims encoded in their chemical reaction probabilities. Replicators run maxims encoded in their digital coding. Cells run maxims encoded in their genes. Organisms run maxims encoded in their genes or their brains. Stars, planets, replicators, cells, and organisms are all agents. At

---

[38]Values of organisms (CMI ch. 3; ADC ch. 5.4; AT 681–9). Adaptive fitness (ADC 208).
[39]Intrinsic value is complexity (Steinhart 2014: secs. 72–74).

some point in biological evolution (long before humans), animals evolve into moral agents running moral algorithms. Humans are moral agents self-consciously running moral algorithms. Artificial computers (like self-driving cars) run maxims encoded in their software. These artillects increasingly teach themselves to act morally. They are self-legislating agents. Our universe is saturated with agency. And, wherever there are agents, no matter their kinds, the axioms of deontic logic apply.

The first axiom of our deontic logic states that if any agent can maximize value, then it ought to maximize it. So if any agent can maximize complexity, then it has a duty to maximize it. Of course, duties can come into conflict, agents can fail to do their duties, and duties may be poorly or inaccurately defined. Nevertheless, agents have duties. Our first axiom does not entail that everything has duties. If something is not an agent, or if it is an agent which cannot increase value, then it need not have any duties. Here agency is far deeper and far more general than moral agency. Agents need not have personality or mentality. If some computer can run an entelechy, then it is an agent. If it can run an entelechy, then it ought to run it. Celestial computers are agents that can run atomic, molecular, and biological entelechies. Hence they ought to run those entelechies.

The second axiom of our deontic logic states that if any agent is rational, then it strives to do what it ought to do. Rational agents strive to do their duties. Rationality does not require personality or mentality. How rational is our universe? All the cranes in our universe regulate themselves in accordance with the MEPP. Since maximizing entropy production rates maximizes the production of complexity, the MEPP is a *maximum value production principle* (an MVPP). All the cranes in our universe work very hard to do their duties. The best explanation for their intensive striving is that our universe has a high degree of rationality. So, by inference to the best explanation, our universe has a high degree of rationality. Its rationality is not made of thoughts.

2.5   Evolution Increases Reflexivity

The liturgies in Chapter 2 showed that, as evolution increases complexity, it also increases *reflexivity*. The reflexivity of any whole is the degree to which its parts carry information about the whole. Sagan said "The cosmos is within us. We are made of star-stuff. We are a way for the universe to know itself."[40] His words illustrate reflexivity: humans are parts of the universe, and the scientific and mathematical theories in our brains approximately reflect the structure of the universe back to itself. Our brains are like mirrors: through this mirroring, the universe knows itself.

The history of the universe provides evidence for increasing reflexivity.[41] Reflexivity increases in both atomic and molecular evolution. But reflexivity really takes off with biological evolution. Every cell in your body contains a genetic description of your entire body. Dawkins says the genes of every species make a *Genetic Book of the Dead*, which describes the ancient worlds in which it evolved. Animal brains contain models of their bodies and environments. Brains are mirrors in which their surrounding structures are reflected. Some brains literally contain neural images of their bodies. These images are your cerebral homunculi. Your cerebral homunculi are approximate images of your body inside of your body. And human brains can be mirrors of the entire universe. Our brains can contain physical theories which describe the whole universe. Dawkins recognizes that parts of the universe can become conscious of the whole: our brains can run models of the universe. Our brains are parts which can simulate the whole.

Reflexivity rises to enormous heights through technological evolution. Using many technologies, we build detailed phylogenetic charts of the evolution of life. Through structural correspondence, those charts reflect billions of years of biological evolution. We make maps of the whole universe, such as the map of the microwave background left over from the big bang. Computer technology greatly increases reflexivity. We build ever more accurate computer simulations of the universe at all scales.[42] At the smaller scales, our computers simulate quantum mechanics. At

---

[40]Carl Sagan (1980), *Cosmos*, Episode 1, "The Shores of the Cosmic Ocean."

[41]Genes model past environments (UR ch. 10). Brains model current environments (UR ch. 11). Brains model the universe (UR 312).

[42]Computers simulate quantum mechanics (Gattringer and Lang 2009). Millennium Simulation (Springel et al. 2005). Illustris Simulation (Vogelsberger et al. 2014).

the larger scales, they run cosmic simulations like the Millennium Simulation or the Illustris Simulation. These computer models reflect the universe from the big bang to the present at the galactic level of detail. They are approximate instances of the universe in the universe. They are parts of the universe which approximately instantiate the form of the universe.

Reflexivity is self-representation: a part of the whole represents the whole through structural correspondence. Since our universe contains parts which represent the whole, it is a self-representative system. However, so far, our universe is only a *finitely self-representative system*. To say that a system is finitely self-representative means that all of its internal self-representations are only partial and approximate. If any part of our universe represents the whole, then its representation is only partial and approximate. It is not perfectly accurate. Its level of detail is not complete. Yet the growth of reflexivity is easily extrapolated. If reflexivity continues to increase, then it rises through a series of ever greater *finite degrees of self-representation*, just like the endless series of ever greater finite numbers. Each of these finite degrees of self-representation corresponds to a *finite physical reflection principle*. Such reflection principles assert that there are some parts which reflect the whole to some partial degree of precision.

As finitely self-representative systems rise towards their infinite limit, they contain parts which ever more accurately represent their wholes. The limit of this increasing accuracy is *exact* self-representation. The limit of the series is a whole which contains a perfect self-image. This perfection leads directly to infinity: the perfect self-image in some whole contains a perfect self-image, which contains a perfect self-image, and so it goes. It's like a ideal mirror reflecting itself back to itself. The philosopher Josiah Royce described such *infinitely self-representative systems* using the example of a perfect map of England in England.[43] Any such map contains an infinitely nested series of self-images. More precisely, to say that a system is infinitely self-representative means that some proper part of that system *exactly* represents the whole. If our universe were to contain some exact computer simulation of itself, then that computer

---

[43]Steinhart (2012).

would contain a computer, and so on. Beyond the endless series of finite degrees of self-representation, there exists an infinite degree. A whole (such as a universe) satisfies an *infinite physical reflection principle* if it is infinitely self-representative.

Many writers have described the extrapolation from finite self-representation to infinite self-representation.[44] Hegel said the universe was making progress towards an absolute mind. His absolute is perfectly self-conscious, and its perfect self-consciousness is infinitely self-reflective. Peirce argued that the universe will evolve into a perfectly self-representative mind.[45] Teilhard de Chardin argued that evolution would run to an infinitely self-conscious omega point. Freeman Dyson showed how evolution could make progress to an infinite omega point. Tipler argued that the infinite omega point is an infinite computing machine at the end of time. Kurzweil argued that the universe will rise perpetually towards this infinite computer but will never reach it.

It is extremely doubtful that our universe will endlessly increase its reflexivity. As its gravitational free energy runs out, its entropy pumps will grind to a halt. All complexity will disintegrate back into simplicity. And so it is *almost* certain that our universe will rise only to the n-th degree of finite self-representation. It will *almost* certainly never become infinitely self-representative in the limit. It is *almost* certain that it will fail to do its duty to maximize reflexivity. *Almost*. Yet our cosmic fate is not completely sealed until our universe ends. An infinitesimal probability is not zero. As long as it continues, our universe has *possible futures* which rise to its omega point. Although our universe has no mentality, these futures can be logically classified as *hopeful* or *aspirational*. The study of possible cosmic futures is *physical eschatology*.[46] Dawkins puts

---

[44]Dyson (1985), Tipler (1995), and Kurzweil (2005).

[45]Peirce (1965) presents an evolutionary cosmology (6.33). He says the universe began in a state of chaos (1.409, 6.214, 6.215, 6.33, 8.317). Through the self-negation of its nothingness, this chaos starts to self-organize (6.217–20). Through continued self-relation, self-reinforcing regularities emerge (1.409, 6.490, 8.317). Time and space emerge (1.411–16, 6.214, 8.318). Laws of nature emerge (1.412, 6.13, 7.513–15). The flow of cosmic energy can produce a branching tree of universes (1.412). The streams of cosmic activity converge to an infinite omega point (1.409, 6.33, 8.317). The process of cosmic evolution is driven by the imperative to maximize reflexivity.

[46]Cirkovic (2003).

eschatology into his Einsteinian religion.[47] He acknowledges that the universe has many possible futures. He thinks it will probably expand forever, but it might collapse back in on itself. Entropy will probably increase to some maximum, but it might reset. Time might begin again. Our universe might give birth to new universes with new laws.

3   The Naturalized Organic Design Arguments

3.1   Designed Versus Designoid

All our liturgies suggest that nature designs things.[48] From atoms to artifacts, nature does design. But is *design* the right word? Dawkins often talks about design. He says design requires higher-order cognitive powers. Thus design involves "creative volition." When a human potter makes a pot, it requires creative imagination. He says humans design things through conscious foresight. Before she makes a watch, a watch-maker has a mental representation of her *plan* for building the watch. Her plan includes the mental representation of the completed watch as a *goal*.

Dawkins contrasts designed things with *designoid* things.[49] Although designoid things may look designed, in reality they are not. Designoid things manifest the *illusion of design*. They are produced by *cumulative finding*. One way to produce an artifact (such as a knife) is through finding: you search through a vast number of accidentally shaped stones until you find one with a sharp edge. Finding can be iterated. Dawkins thinks of cumulative finding as a blind search process. By contrast, he says that designed objects are not produced through cumulative finding – they are not found at all.

---

[47]SITS 272.

[48]Creative volition (CMI 16–17). Imagination (CMI 19). Foresight (BW 5).

[49]Designoid versus design (CMI ch. 1). Illusion of design (CMI 7, 25). Cumulative finding (CMI 28). Knives (CMI 11). Design is not finding (CMI 28).

Since evolution produces things by cumulative finding, it produces designoid things rather than designed things.[50] Dawkins gives three classes of designoid things. The first class of designoid things includes organs. The pots of pitcher plants are designoid, as are the traps of the Venus fly trap. The second class includes the artifacts produced by non-human animals. These designoid artifacts include: the houses of caddis flies, the webs of spiders, the houses of wasps and bees, the nests and bowers of birds, the mounds of termites, and so on. The third class includes the harmonies of ecosystems.

However, during his early writings, Dawkins often says evolution designs things. The contrast between designed and designoid does not appear. He does not talk about the illusion of design. The phrase does not occur in *The Selfish Gene* or *The Extended Phenotype*. But as Dawkins begins to contrast Darwinian evolution with Christian fundamentalism, he introduces the illusion of design. The phrase first appears once in *The Blind Watchmaker*, and it occurs twice in *Climbing Mount Improbable*.[51] It does not occur in *River Out of Eden*. During his later writings, he comes to recognize that the contrast between designed and designoid is false.

There are several objections to the Dawkinsian contrast between the designed and the designoid.[52] The first is that it leads Dawkins to make false claims about design. He thus makes the spectacularly false claim that design is not cumulative. The entire history of technology illustrates the accumulation of design. Technology *obviously* follows Dennett's principle of the accumulation of design. The second objection runs like this: if technology developed through foresight, then it would develop smoothly and rapidly; it does not; therefore, it does not develop through foresight. Thirty-five centuries ago, lenses were used as magnifiers. Devices for holding things on our heads are at least as old. So why did it take *twenty-nine centuries* to develop eyeglasses?

---

[50]Designoid organs (ADC 225–6; GD 24, 139, 143, 168, 188; GSE 21, 334, 371; AT 633). Like pots of pitcher plants (CMI 12) and Venus fly traps (CMI 14). Designoid non-human artifacts (CMI 6–18). Designoid ecosystems (ADC 225–6).

[51]BW 21; CMI 7, 25.

[52]Design is not cumulative (GD 169). But in fact it accumulates (Basalla 1988; Temkin and Eldredge 2007; Brey 2008). Lenses (Enoch 1998). Eyeglasses (Rubin 1986).

The third objection is that Dawkins draws a false contrast between the designoid and the designed. Dawkins tries to draw a bright line between cumulative finding and human creative volition. Hence he tries to draw a bright line between the designoid and the designed. However, Dawkins constantly stresses the gradual evolution of all complex qualities. It follows that there is a gradual slope of intelligence which rises from the primitive potter wasp all the way up to *Homo sapiens*. There is no principled place to put the line between cumulative finding and human creative volition. Hence there is a continuum of technical skill from the potter wasp all the way up to *Homo sapiens*. There is no principled place to draw the line between designoid and designed. Were the artifacts of Neanderthals designed or merely designoid? How about *Homo erectus*? Were the stone tools at Olduvai designed or merely designoid?

3.2   Mental Evolution Designs Artifacts

The fourth objection is that technologies initially grow more through blind trial and error rather than through foresight.[53] Modern research indicates that early human technologies (like bows and arrows) grew through cumulative finding. Even advanced technologies seem to emerge and grow in this way. The history of technology after technology confirms that invention is mostly cumulative finding. Dawkins says that watchmakers use foresight. However, the history of timekeeping devices shows that they were discovered mostly through trial and error. An analysis of over one thousand patents by Thomas Edison shows that he proceeded primarily through blind variation and selective retention. One history of airplanes describes their early development as "flailing in the dark." The same holds for computers. If you were to apply Dawkins' own conception of the designed-designoid distinction to the first instances of any technology, you would have to say they were more designoid than designed. But that would be a strange use of language. Better to say design is cumulative finding.

---

[53]Technologies that grow through cumulative finding include early artifacts (Derex et al. 2019); watches (Bruton 1979); inventions by Thomas Edison (Simonton 2015); airplanes (Anderson 2002); and computers (Essinger 2004; Dyson 2012).

The fifth objection against the designed-designoid contrast comes from the study of creativity. Invention begins with some *indefinite* goal: make a device that keeps time; make an artifact that flies; make a machine that calculates. These indefinite goals are neither blueprints nor plans. They are not mental pictures of perfected artifacts. The indefinite goal opens a landscape of possibilities in which height is functional success. The aircraft pioneers were exploring the aeronautical Mount Improbable. Inventors explore the space of possible artifacts by running evolutionary algorithms on mental models of them. These evolutionary algorithms use variation and selection.[54] Sometimes the variation is blind: it comes from random guessing or lucky accidents. Sometimes the variation occurs through reasoning by analogy. The Wright brothers made extensive use of bicycle-airplane analogies and ship-airplane analogies.[55] This use of analogy resembles horizontal gene transfer in organisms. It implements Dennett's principle of the accumulation of design by copying designs across branches in the technological phylogenetic tree.[56] So the Wright Flyer accumulated its design both vertically from earlier aircraft and horizontally from bicycles and ships. All this accumulation was just designoid. It was cumulative finding. Of course, the Wrights could make increasingly well-educated guesses about the functionalities of their inventions. But they could not know *in advance* that their wings or propellers would work. They could not use Dawkinsian foresight. They had to proceed through experimental tests.

Even when goals are specified with some precision, evolutionary means are used to reach them. When designers have goals specified in advance, they proceed towards them using goal-directed evolution. When humans breed animals to serve goals defined in advance, those goals are merely abstractly defined. Thus chickens are bred to lay more eggs. But this abstract functional goal is not a detailed blueprint of any champion egg-layer. To reach these abstract goals, human breeders use blind variation and selective retention. We design animals through evolution by artificial selection. Goal-directed evolution also occurs in nature. During *affinity*

[54]Dennett (2004) and Simonton (2010).
[55]Johnson-Laird (2005).
[56]Temkin and Eldredge (2007).

maturation, the human immune system designs antibodies to precisely fit targeted antigens. Dawkins knows that the immune system builds up its database of antibodies by goal-directed evolution.[57] Dawkins himself developed a goal-directed evolutionary algorithm. Through cumulative finding, his weasel program *designed* the sentence METHINKS IT IS LIKE A WEASEL.

Dawkins suggests that creative intelligence emerges from something like Darwinism running in our brains.[58] He says that thoughts are populations of active neurons which compete with each other like organisms. Thus natural selection runs in our brains. Design is a kind of evolution running on populations of ideas. So it is paradoxical that Dawkins ever opposed design to evolution. The objections to the design-designoid contrast show that design just is cumulative finding. Reflecting on the history of the airplane, Dawkins concludes that design is cumulative finding. If engineers gain foresight, that is just because they evolve eyes.

3.3   All Design Work Is Evolutionary

Dawkins, in his early writings, often says biological evolution does design-work.[59] He *frequently* and *explicitly* says that evolution *designs* bodies, organs, and behaviors. He says new organisms inherit ancestral designs encoded in genetic programs. Things designed by evolution include the alarm calls of birds, feathered wings, worker ants without germ cells, eyes, the bodies of fish, and so on.

Dawkins, during his later writings, returns to his earlier idea that *evolution designs things*.[60] The shapes of insect houses were designed by natural selection. Human inventors design artifacts by running Darwinian algorithms in their brains. Human technological design is a

---

[57]GSE 407.

[58]Darwinism in our brains (UR 8). Design is cumulative finding (AT 688).

[59]Evolution designs things (EP 59–71). Things designed by evolution include: ancestral bodies (SG 26, 261); alarm calls of birds (SG 170); feathered wings (EP 68); worker ants (EP 128); eyes (EP 261; ROE 78); bodies of fish (ROE 93). Electronic searches of his texts will reveal dozens of other examples.

[60]Evolution designs things (BCD 323). Brains run Darwinian design algorithms (IA 104; 2015: 15). Airplanes (AT 688).

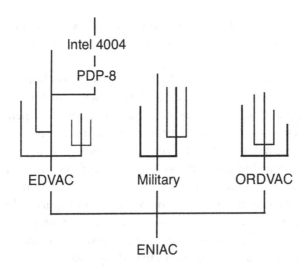

Fig. 2 A highly simplified phylogenetic tree of computers

kind of Darwinian evolution. Dawkins illustrates this with the evolution of airplanes: the Wright brothers built a barely functional plane; the process from their planes to Stealth bombers was a process of gradual accumulation of complexity. Later planes inherited most of their design from earlier planes. Just as biological species form phylogenetic trees, so technological species form such trees. They have been traced out for musical instruments, computers, and other machines.[61] Figure 2 shows a highly simplified phylogenetic tree for computers. While most of these computers went extinct, the Intel 4004 line continues to evolve.

Thus Dawkins returns to the more accurate thesis that living things were *designed* by evolution. Dennett puts this nicely: "A *designed* thing, then, is either a living thing or a part of a living thing, or the artifact of a living thing."[62] Technology evolves.[63] Kevin Kelly argues that technology is the seventh kingdom of life.[64] Technological evolution runs on top of biological evolution. The *technological liturgy* follows the other liturgies:

---

[61]Musical instruments (Temkin and Eldredge 2007). Computers (Mataxis 1962: 63).
[62]Dennett (1995: 69; his italics).
[63]Basalla (1988), Dyson (1997), Temkin and Eldredge (2007), and Brey (2008).
[64]Kelly (2010).

it includes the library of all possible technologies, the technological arrow and volcano, and the other themes of other liturgies. We need not repeat them here.

On this view, earthly animals (both human and non-human) do *technological design work*. The internet and spider webs are equally products of technological design. This technological design work depends on the *biological design work* done to bring animals into existence. Animal bodies and ecosystems are the products of biological design. All this technological and biological design work depends on the fact that the surface of the earth is far from thermodynamic equilibrium. It depends on the flow of energy from the hot sun into cold space. This biological design work depends on *molecular design work*. Complex molecules are the products of design work done on the surfaces of planets and in clouds of celestial dust. This molecular design work also depends on the flow of energy from hot to cold. These three design levels depend on thermodynamic principles. It seems likely that they depend on the maximum entropy production principle (the MEPP) or something very much like it. Some thermodynamic principles (like the MEPP) skew the probabilities of change away from randomness and towards design.

A solar system is a celestial computer. If its star is fusing elements, then it is doing *atomic design work*. Heavy atoms are the products of stellar design. By fusing elements, the star pumps entropy out of its surrounding celestial computer. This transforms its program into an entelechy. It transforms the computer into a crane. By pumping entropy out of their solar systems into empty space or black holes, the stars drive all design work. But this story is not ultimate. Why does all this design work get done? It gets done because the skewer skews the probabilities. The skewer is the ultimate designer. Should we say the skewer is God? On the contrary, the skewer serves as a naturalistic *replacement* for God in the old design arguments.

3.4   Evolution Plays the Role of Watchmaker

The *Organic Design Argument* reasons from organisms to God. Dawkins often deals with this argument.[65] It runs like this: (1) Living things from organs to ecosystems exhibit complexity, harmony, and other highly improbable features. (2) Since these features are improbable, they cannot be explained by chance. (3) The alternative to chance is design. (4) So, there exists some designer of living things. (5) Any designer of life is either natural or supernatural. (6) No natural designer of life exists. (7) Hence there exists some supernatural designer of life. This supernatural designer is God.[66]

It is crucial to cover this argument step by step. Dawkins agrees with the first two steps. When he distinguishes between design and the illusion of design, he emphatically rejects the third step.[67] This leads directly to his spectacular error that design is not cumulative. And all the objections to the distinction between designed and designoid imply that Dawkins *should accept* the third step. Hence he *should accept* the fourth step. Of course, the texts in the last section show that he often *does accept* this step. Along with the self-consistent version of Dawkins, spiritual naturalists affirm that there exists some designer of living things. Step five is just logic. The real action enters at step six. The Dawkinsian designer must be natural. Step seven now just identifies this natural designer with evolution: evolution designs organs and bodies and ecosystems.[68] So the last steps of the Natural Organic Design Argument say: (6) evolution is a natural designer of life. (7) Since there is a natural designer, there is no need to invoke any other designer. Evolution suffices for the design of all living systems.

The argument remains haunted. Theists often say the alternative to chance is not merely design—it is *intelligent* design. A clever theist may accept all the empirical data of evolution. They may agree that evolution designs living systems. But there is a ghost in the machine: God

---

[65] BW 4–6; GD ch. 4; etc.
[66] GD 96–99.
[67] Rejecting step three (GD 145) leads to a spectacular error (GD 169).
[68] EP 449; SITS 120–1; etc.

intelligently skews the probabilities of the reproductive arrows. If there were no intelligent skew, evolution would not climb Mount Improbable. Dawkins says he cannot disprove this intelligent skew.[69] He should not have conceded this point – it opens a door through which theistic hijackers will crawl.

Dawkins should have argued that evolution contains all the intelligence needed to skew the probabilities towards increasing complexity, value, and so on.[70] Dawkins says evolution computes and has memory. And he explicitly states that evolution makes value-laden progress. On his own terms, evolution looks very much like a mind. Some biologists have gone further and argued that evolution learns. Perhaps some adventurous biologists would even argue that evolution is intelligent, but spiritual naturalists need not go that far. To close the open door, we need only this: if the skew seems to require mentality, then evolution is sufficiently *mindlike* to produce that skew. Nevertheless, lacking sufficient informational integration, evolution is almost certainly not a mind. It is neither any person nor even personal. It is not the Lovelockian Gaia.[71]

The Natural Organic Design Argument can be extended to include all forms of complexity. So the *Natural Design Argument* goes like this: (1) Atoms, molecules, organisms, and technologies all exhibit highly improbable complexity. (2) Since this complexity is improbable, it cannot be explained by chance. (3) The alternative to chance is design. (4) So, there exists some designer of all complex things. (5) The proximate designer of any complex thing is some celestial computer running an entelechy; it is some crane. Perhaps these cranes have some mindlikeness. More deeply, the ultimate designer is the skewer itself. It skews the probabilities of change away from random drift and towards design. Nietzsche thought of the universe as a work of art without an artist; it is

---

[69] BW 316–7.

[70] Evolution computes (CMI 72, 326; ADC 12) and has memory (UR 257; GSE 405–8). It increases value (AT 681–9). Evolution learns (Partner et al. 2008; Watson and Szathmary 2016; Watson et al. 2016; Kouvaris et al. 2017).

[71] Dawkins rejects the Lovelockian Gaia (EP 357–61; CMI 268; UR 222–4; SITS 153). Perhaps a more scientific pagan might devise a better Gaia (EP 360; ADC 173).

a self-designing work of art.[72] Thanks to the skewer, our universe seems to satisfy this Nietzschean description.

4   On Stoic Axiology

4.1   Violent Beauty Shining Brightly

All design work involves non-random maximization; it therefore involves directed motion along some ranking; it therefore involves values.[73] Dawkins affirms this. He proposes a thought experiment: we imagine "that living creatures were made by a Divine Engineer and try to work out, by reverse engineering, what the Engineer was trying to maximize." The quantity the Engineer strives to maximize is their *optimality function*. So what is the optimality function? What value is the Divine Engineer trying to optimize? The abundance of suffering entails that the Divine Engineer is not a utilitarian, and is not trying to maximize happiness or pleasure.

Dawkins answers that life on earth maximizes genetic survival and replication.[74] So the Divine Engineer's optimality function is genetic replication. It follows the policy *Maximize genetic replication!* Maximizing genetic replication entails maximizing many interesting quantities. Dawkins hints that maximizing genetic replication is merely a means to the end of maximizing other values. Writing about the violent competition between gazelles and cheetahs, he seems to say that the Divine Engineers are really interested in maximizing *dramatic beauty*. He wonders whether the designer "enjoys the spectator sport and is forever upping the ante on both sides to increase the thrill of the chase." He suggests that this is sadistic, but that's just a rhetorical flourish. Dawkins stresses that evolution is neither cruel nor kind. It is utterly indifferent to pleasure and pain, to happiness and misery. It is not a utilitarian.

---

[72]Nietzsche (*The Will to Power*, sec. 796).

[73]Divine Engineer (ROE 104–5) is not utilitarian (ROE 103–4, 131–2; GSE 390–5).

[74]Maximize genetic replication (ROE 131; GSE 392). Maximizing dramatic beauty (SG 78; ROE 119–20; UR 219–20). Thrill of the chase (GSE 384; GD 161). Neither cruel nor kind (ROE 95–6, 131; ADC 8–9). Deeper than its utility (UR 5–6, 21–4).

However, indifference to utility does not imply indifference to value. And since utilitarianism is not the only theory of good and evil, it does not imply indifference to good and evil. Of course, there are moral contexts in which utility *ought* to be maximized. But that very fact shows that utilitarianism presupposes duty. Utility is not the deepest value. Dawkins says the value of science is deeper than its utility. He is ultimately not utilitarian.[75]

Any consideration of alternatives to utilitarianism leads quickly to Stoicism. To build the Sanctuary for Spiritual Naturalists, it will be useful to briefly look at some of the Stoic themes found in Dawkins. The Stoics were not utilitarians. They did not think that happiness was the ultimate positive value, or that misery was the ultimate negative value. Of course, they thought that happiness was *preferable* to misery and misery was *dispreferable* to happiness. But preference and dispreference were psychological rather than ethical comparisons. So the Stoics did not think that happiness was good and misery was evil. For the Stoics, *virtue* is good and *vice* is evil. Virtue is the competence revealed in struggle; vice is the incompetence revealed in struggle. The Stoics regarded the excellence revealed in struggle and competition as far superior to happiness. They recommended indifference to personal suffering. Far more important, on their view, was the struggle for personal virtue—and a virtuous life is a beautiful life. The Stoic heroes, like Hercules and Socrates, chose virtuous suffering and death over happiness.

The Stoics argued that the universe contains an objectively existing rational power, which they called the Logos. Dawkins often strongly affirms that the universe has an objective rationality.[76] On this point, he is Stoic. Here Dawkins goes beyond science. By affirming the rationality of the universe, Dawkins does metaphysics. Of course, it does not follow that Dawkins agrees with any old Stoic Logos doctrines. Those old doctrines conflict in many ways with modern science. By affirming the rationality of the universe, Dawkins merely situates himself at the modern end of the Stoic tradition. His rational magisterium includes a

---

[75]Dawkins *proximately* endorses utilitarianism (e.g. ADC ch. 1.3; SITS 301–8). But his *ultimate* axiology assumes Stoic and Platonic values deeper than utility.
[76]UR *xi*, 151; TL 73–4; SSSF.

modernized Logos doctrine. This new Logos is just the rational order of nature. It is just the system of logically organized natural laws. It lacks all mentality and personality. Of course, the Stoic Logos designs organisms—it is the *pyr technikon*, the designing fire. Does this contradict evolution? We already argued that evolution designs things. Since the Logos is just the unified system of natural laws, evolution emerges from it in a purely natural way. And since the *pyr technikon* is fire-energy, the Stoic theory can be modernized via the thermodynamic liturgy.

Stoicism helps to make sense of cosmic value. The Stoics derived their theory of value (their *axiology*) from the rationality of the universe. The Logos drives the universe to maximize its values. But the Logos is not utilitarian. It does not aim to maximize happiness or minimize suffering. As the Logos regulates earthly biological evolution, it drives it to maximize the *dramatic intensity* of biological competition. It drives it to maximize the opportunities for the manifestation of *arete* in competitive struggles. But struggles in which *arete* is maximized are also struggles in which beauty is maximized. Hence the Logos drives it to maximize the *dramatic beauty* of biological competition. If it displays beauty, then there is some sense in which the biological competition is a spectacle. So the Logos drives the earth to maximize the dramatic beauty of the biological spectacle. Since Dawkins affirms universal Darwinism, this idea generalizes: the Logos drives the universe to maximize the dramatic beauty of the cosmic spectacle. This beauty is the ultimate good, and ugliness is the ultimate evil. Happiness does not add to this goodness, and suffering does not detract from it. Nietzsche expresses a Stoic idea when he says that nature is "*holy enough* to justify a monstrous amount of suffering."[77] Of course, Nietzsche remains an atheist.

4.2   The Providential Ordering of Nature

The Logos is the rational and providential ordering of nature. When naturalized and modernized, the Logos is just the basic laws of physics. The surprising usefulness of mathematics in fundamental physics helps

---

[77] Nietzsche, *The Will to Power*, sec. 1052.

justify the thesis that the laws of physics are rational. But what about the *providential* aspects of the Logos? Our previous reasoning indicates that the Logos maximizes dramatic intensity and dramatic beauty. It maximizes values that emerge through competitive struggle of life on earth. For the Greeks, the *agon* was the field of competitive struggle (the jungle, the battlefield, the Olympic games, the *agora*). The excellence that emerges in the *agon* is *arete*. At least in earthly biology, the Logos providentially maximizes *arete*. Universal Darwinism implies that all possible life maximizes *arete*. We should look more closely at *arete*.

It is plausible that natural selection maximizes *arete*.[78] Dawkins describes exactly how this happens. Dawkins presents Darwinian evolution as having two stages. The first stage is mutation, which rarely leads to improvements. The second stage is natural selection. Natural selection is *differential survival*—the survival of the fittest, the survival of the stable. He says "Natural selection, the second stage in the Darwinian process, is a non-random force, pushing towards improvement." Among the wolves, as among all organisms, selection prefers "the fleetest of foot, the canniest of wit, the sharpest of sense and tooth." Selection prefers the elite genes. Dawkins describes the evolution of the eye as climbing steadily upwards on Mount Improbable. But height on this mountain is optical excellence. So natural selection only preserves mutations that improve sight. Dawkins does not shy away from using value-laden language: natural selection makes organisms *better*. It *improves* them. He says "Evolutionary change is, to a far greater extent than chance alone would expect, *improvement*." Natural selection explains why "living things are so good at doing what they do."

Dawkins argues that competitive excellence emerges most intensely through *arms races* between competing types of organisms.[79] Dawkins often discusses arms races. As they compete with each other, predator and prey both grow greater in excellence, parasite and host both grow

---

[78]Maximize *arete* (ADC ch. 5.4; CMI ch. 3; ROE ch. 4). Mutation (CMI 80–5). Survival of the stable (SG 12). Pushing towards improvement (CMI 85; ADC ch. 5.4). Wolves with elite genes (CMI 86). Eyes improve (CMI 163). Natural selection *improves* organisms (BW 305, his italics; AT 681–9). Good at what they do (CMI 90).

[79]Arms races (EP ch. 4; BW ch. 7; ADC ch. 5.4; AT 683–9; etc.). Teeth and toxins (AT 685). Evolution of evolvability (ADC ch. 5.4; EE). Adaptive complexes (ADC 206).

greater in excellence. Thus "saber teeth get sharper and longer as hides get tougher. Toxins get nastier as biochemical tricks for neutralising them improve." Over longer evolutionary time-scales, the *evolution of evolvability* leads to greater functional potentials. It opens up new spaces for functional excellence to emerge in new arms races. Hence Dawkins declares that evolution does make progress. He defines it very precisely as the "tendency for lineages to improve cumulatively their adaptive fit to their particular way of life, by increasing the numbers of features which combine together in adaptive complexes." As evolution makes progress, the non-random forces of natural selection always entail maximization of functional competence and minimization of functional incompetence. Natural selection necessarily optimizes the virtue revealed in competitive struggle—it optimizes *arete*.

Suppose *arete* is the highest value. If that is right, then it is *for the best* that this cheetah violently kills that gazelle; it is *for the best* that those escaping gazelles make that cheetah starve to death. Every struggle is for the best because it allows the competitors to manifest their virtues. The struggle is the *agon* in which the competitors strive to express their virtues. The cheetah manifests its virtue when it kills the gazelle; the gazelle manifests its virtue when it starves the cheetah. Of course, to say that the struggle is for the best *does not* imply that it is equally for the best *in each case*. It is for the best that the cheetah and gazelle compete; if the gazelle loses, then it is for the best for the cheetah but not for the gazelle; if the cheetah loses, then it is for the best for the gazelle but not for the cheetah. Since the struggle between cheetahs and gazelles involves racing, they act for the best when they maximize their speeds. Moving with speed, grace, and agility are their virtues. Moreover, to say that the struggle is for the best does not imply that either side wins. Some struggles end in stalemate. Some *agons* end with the defeat of both sides. As they race, *both* cheetah *and* gazelle may die from exhaustion. Finally, some *agons* are decided not by virtue but almost entirely by luck.

If providence means maximizing *arete* in the evolutionary *agon*, then a Dawkinsian can say the Logos is providential. This *arete* shines out brightly in the evolutionary war of all against all. Against this conception of providence, it may be objected that the maximization of *arete* entails social Darwinism. According to its traditional definition, social

Darwinism urges humans to seek short-term gains through ethically and politically unconstrained aggression. It urges humans to engage in a Hobbesian war of all against all; it urges us to move towards the Hobbesian state of nature. Dawkins rightly insists that social Darwinism is wrong.[80] He campaigns against social Darwinism by saying we can rebel against the blind aggression of our selfish genes.[81] Thus we get an argument: (1) maximizing *arete* entails social Darwinism; (2) but social Darwinism is morally wrong; (3) therefore, maximizing *arete* is wrong. But is this argument sound? Hopefully, we can all agree that social Darwinism is wrong. But does it follow that it is wrong to maximize *arete*? Are there any replies to this argument?

The reply is that social Darwinism was incorrectly defined. As traditionally defined, social Darwinism wrongly urges humans to act *as if they were non-human animals*. A bird that correctly derives its values from Darwinism strives to fly better. Since humans are rational social animals, we act for the best when we maximize our virtues of rationality and sociality. A human that correctly derives its values from Darwinism strives to be *more rational* and *more social*. If social Darwinism were correctly defined, it would urge us to act rationally and socially. Thus maximizing *arete* does not entail social Darwinism as traditionally defined. The traditional definition was wrong. If social Darwinism were correctly defined, it would emphasize human cooperation and altruism. It would emphasize long-range self-sustaining policies over short-sighted self-defeating policies. It would involve maximizing justice and fairness. It would affirm the Kantian categorical imperative (Sect. 3 in Chapter 8). Dawkins shows that it is through the maximization of *arete* that cooperation and altruism emerge among all animals.[82] As we maximize our virtues of rationality and sociality, it will often but not always turn out that happiness goes up and misery goes down. Utilitarianism will often but not always emerge as a side-effect or by-product of Stoicism. So the argument from social Darwinism does not refute the Stoic thesis that we ought to maximize our *human arete*. *Arete* transcends utility.

---

[80]SG *xiv*; GD ch. 6; GSE 62; AT 458–62, etc.
[81]SG 201, 331; ADC 9–11; GD 246; SITS 39–40.
[82]SG ch. 12; GD ch. 6.

4.3   Naturalizing the Designing Fire

The first three physical liturgies dealt with three types of self-organization (atomic, molecular, and biological). The fourth physical liturgy used thermodynamic ideas to explain those types of self-organization. According to that liturgy, all self-organization emerges from something like the maximum entropy production principle (the MEPP). The MEPP skews probability towards reaction arrows on which entropy production rates are maximized; but ordered flows produce entropy faster than disordered flows; therefore, the MEPP skews probability towards arrows that maximize order production. This argument suggests that the MEPP is the skewer.

Advocates of the MEPP argue that it powers all the self-organization in the entire universe. However, critics of the MEPP will correctly argue that it is not fully settled science. It is well-justified, but there are many thermodynamic mysteries still waiting to be solved. Both advocates and critics of the MEPP can agree that thermodynamic principles play central roles in the evolution of complexity. And they can agree that other factors are at work. The programs running on the celestial computers have to satisfy many constraints (such as conservation laws). So it seems more careful to say that the skewer includes the MEPP and any other non-random factors that bias the celestial algorithms towards complexity-creation. The skewer designates all the factors that bias programming away from drunken randomness and towards teleonomic upwardness. It acts directly on the probabilities distributed over reaction arrows, skewing them so that some paths offer less resistance to flow than others. The skewer makes it more probable that matter will flow from simplicity to complexity. By directly acting on probabilities, it indirectly acts on matter. So it looks like the skewer generates a *force*.

If the skewer contains thermodynamic principles, then it generates thermodynamic forces.[83] Because he misunderstands entropy, Dawkins rejects thermodynamic forces. On this point, he is just plain wrong. Physics affirms existence of thermodynamic forces. It affirms that they

---

[83]Misunderstands entropy (ADC 84–5; AT 397). Objects to vital forces (ROE 18; ADC chs. 1.6, 3.3; SITS 4, 213).

drive the self-organization of matter. Physics affirms that the skewer generates some thermodynamic force. Of course, that force it is *not* mental or vital; it is not some will to complexity. Dawkins correctly objects to vitalistic or New Age pseudo-forces. Any force emerging from the skewer is purely physical. And since both thermodynamics and complexity are closely linked with information, the *skewing force* may ultimately emerge from purely informational principles at the roots of physics.[84] After all, if the skewing force rewrites celestial programs, then it is a *computational force*. By rewriting programs, it rewrites the flows of matter.

Skewing force? It's a terrible name. Perhaps we can find a better name by exploring its associations. The skewing force is closely associated with thermodynamics and thus with information. The Stoic *pneuma*, as the *pyr technikon*, the designing fire, was also closely associated with thermodynamics and information.[85] The English word for the *pneuma* is *spirit*. Should we use the word *spirit* to refer to the skew? Unfortunately, the word *spirit* is almost certainly too corrupted to reclaim. Some other word is needed to refer to the skewing force. The technologist Kevin Kelly suggests *exotropy*.[86] But the prefix *exo-* suggests moving from rather than moving towards.

A better word is needed.[87] Dawkins does not shy away from coining new words. He famously coined the word "meme". He also coined "designoid" and "theorum". So I will introduce a new word here. Since complexity is intrinsically valuable, and since the Greek syllable *axio* indicates value, this new word will contain the syllable *axio*. Since the Greek *trope* means growth, it will contain *trope*. Putting these together, I will use the term *axiotropy* to refer to the entirely physical order-creating or complexity-generating force that emerges from the skewer. Axiotropy acts directly on probabilities and indirectly on material flows. If the skewer is just the MEPP, then axiotropy is just the

---

[84] Some physicists speculate that gravity is an entropic force (Verlinde 2016). Or that all fundamental forces are entropic (Dil and Yumak 2018). Spiritual naturalism does not depend on these speculations.

[85] Steinhart (2018).

[86] Kelly (2010: 63).

[87] Memes (SG ch. 11); designoid (CMI ch. 1); theorum (GSE ch. 1).

entropic force emerging from the MEPP. It acts on material things at strengths in the piconewton range over distances of nanometers.[88] If the skewer involves other principles, then axiotropy includes them. Nevertheless, it remains a purely physical force, generated from purely physical principles, including informational principles.

Axiotropy drives the evolution of complexity.[89] If any physical systems moil towards greater complexity, they are driven by axiotropy. Axiotropy drives the low-entropy radiation at the big bang to become particles; then simple atoms; then stars. Axiotropy drives the moiling in the atomic, molecular, and biological computers. It drives planets to boil with life. Axiotropy drives life to contend with life; it drives the emergence of *arete* in the *agon*. If this Stoic picture is right, then axiotropy makes it more likely that these natural computers do their duties. Something is a virtue (or is virtuous) if it makes it more likely that an agent will do its duty. Axiotropy is virtuous.

5   On Platonic Axiology

5.1   The Beatific Vision of Nature

Dawkins often adopts a Stoic position on the values in nature.[90] He often endorses *amor fati*, the love of fate. He writes that nature, like the Stoic Zeus, is *serene* in its indifference to suffering, and that there is a *grandeur* in this serenity (Sect. 2.1 in Chapter 9). If evolution is indifferent to pleasures and pains, it may still be providential in other ways. Dawkins says he finds it to have "an inspiring, if grim and austere,

---

[88]Many entropic forces acting on molecules and molecular assemblies have strengths of a few kT per nanometer, thus producing forces in the piconewton range (Marenduzzo et al. 2006). So axiotropy acts with similar strengths. It changes the microstates of systems. But changes in microstates scale up to become changes in macrostates.

[89]Anscombe says that machinery "should or ought to be oiled, in that running without oil is bad for it, or it runs badly without oil" (1958: 6). Universes are machines that can run well or badly. They ought to have axiotropy, in that they run badly without it.

[90]*Amor fati* (ADC 12–3; GD 20, 403–5; AK 188). Serenity (GSE 401). Austere poetry (EP 258). Indifference is more beautiful (UR *xi*, 41–2, 118). Valuable spectacle (UR 2–6; GD 404–20). Vision makes life worth living (UR *x*, 313). Scientific truth is too beautiful (1995c; see UR 114–21). Aesthetic argument (FH 99).

poetry of its own." But he does not rest in this austerity. When Dawkins reflects more deeply on cosmic value, he shifts to *beauty*. He says that a universe which is indifferent to humanity is *more beautiful* than one which answers to human emotions. Dawkins thinks the vision of the cosmic spectacle has great value, even ultimate value. Seeing the cosmic spectacle makes life worth living—it provides our lives with meaning and purpose. Seeing the deep mathematical structure of the universe "is truly one of the things that makes life worth living." Dawkins says "Scientific truth is too beautiful to be sacrificed" for the money or pleasure gained from astrology. Astrology is aesthetically offensive, and religion is aesthetically offensive too. But why does scientific truth have the value of *beauty*? Why is it *more beautiful* than astrological or religious falsity? These appeals to beauty (and to mathematical beauty) are Platonic. If Dawkins is assuming a Platonic scale of values, these appeals make sense; if not, they don't. On the basis of charity at least, it is plausible to say that Dawkins is motivated by Platonism. It will be useful to briefly look at some of the Platonic themes found in Dawkins.

The shift from *amor fati* to beauty marks a shift from Stoic axiology to Platonic axiology. Beauty is a Platonic value. Plotinus argued that the goodness of the universe consists in its *aesthetic value*.[91] Since beauty shines out most intensely in competition, maximizing beauty entails maximizing competitive excellence, that is, *arete*. But maximizing *arete* entails suffering. Great dramatic works of art often involve great suffering. So the duty to maximize vision entails the duty to maximize the dramatic beauty of the cosmic spectacle. If beauty did not increase with vision, then there would be no reflections of the Good for sentient life to see. Nietzsche wrote that: "it is only as an aesthetic phenomenon that existence and the world are eternally justified."[92]

This *beatific vision of nature* replaces the old beatific vision of God.[93] We gain this beatific vision through science. Thus Dawkins argues that pseudoscience blinds us to the beauty of the cosmic spectacle. After discussing the stars and galaxies, Dawkins writes (again) that astronomy

---

[91] Plotinus (*Enneads*, 2.3.18, 3.2.15–18, 3.6.2).

[92] Nietzsche (*The Birth of Tragedy*, sec. 5).

[93] Astrology is ugly (1995c; UR 118). Crystals (ADC 46). Religions impoverish us (AT 700). True reverence (AT 700). Vision is a blessing (UR 5).

is beautiful but astrology is ugly. He says the scientific theory of crystals is "more illuminating and more uplifting" than its New Age perversions. Pseudoscience and religion blind us to the beauty of the cosmic spectacle. Religions "miserably fail to do justice to the sublime grandeur of the real world. They represent a narrowing-down from reality, an impoverishment of what the real world has to offer." Just as Plato chastised the false poetry of Homer and Hesiod, so Dawkins spends much time in *Unweaving the Rainbow* chastising modern poets for their disdain for science. He says that once we understand the universe properly, that is, through science, we will be "moved to celebrate the universe" and to show "true reverence" for it. Dawkins says that to accurately see the universe is a blessing. True vision reveals its beauty. This vision has great value. For Dawkins, as for many New Atheists, truth is a supreme value. But why? The Platonists answer that truth reveals *the Good*.

Plato says we have a duty to change from ignorance to knowledge. He illustrates this duty with his myth of the cave and his divided line.[94] According to those parables, we initially find ourselves in the dim light of the cave. We can barely see. But we ought to realize our highest possible state of vision. We ought to see the Good. We therefore have a duty to escape from the cave. We have a duty to climb up the divided line until our vision is strong enough to look directly at the sun. Of course, the sun is the symbol of the Good. Should atheists avoid talking about the Good? Iris Murdoch was an atheist who sought to replace God with the Good. But what about Dawkins? He is driven by the Platonic duty to ascend from the cave. The cave is religious superstition, the upper world is scientific enlightenment. For Plato, the ascent ends with the mystical vision of the Good. Although I will deal with atheistic mysticism later (Sect. 4 in Chapter 7), it is worth stating here that Dawkins seems to have had mystical experiences.

For Dawkins, we have a duty to climb out of the dark cave of religious superstition and up into the scientific light of day. We *ought* to become rational. This duty requires some rational explanation. For if we cannot rationally derive values from existence, then there is no reason to rise from religion to science. Dawkins will be just as irrational as the most

---

94 Plato (*Republic*, 508b–520a).

savage fundamentalist. If the duty to rise from ugly falsehoods to beautiful truths is not an objective obligation, then the Dawkinsian claim that science is superior to religion is merely subjective opinion. The scientific enterprise collapses into relativism. Moreover, the entire therapeutic project of the New Atheism collapses into incoherence. This risk alarms Dawkins. He emphatically rejects relativism (Sect. 5 in Chapter 1). We have an objective moral duty to rise from religion to reason. The objectivity of this duty cannot be grounded in human psychology—it must have some deeper ground.

5.2　On the Duty of the Universe

According to Plato, you ought to climb up out of the cave in order to see the Good. You ought to become as much like the divine mind as possible, you ought to become godlike (Sect. 3 in Chapter 9).[95] Plotinus extended these Platonic ideas. He argued that the Platonic obligations apply to the universe itself. Since the universe strives for the Good, it has a cosmic duty. Its duty is to maximize its vision of the Good. Thus Plotinus pictures the universe as striving to looking at the Good.[96] He pictures the entire universe as climbing up the divided line to become as much like the divine mind as possible.

For Plotinus, the duty of the universe is to maximize vision. This duty can be expressed in evolutionary terms. Early organisms have primitive senses. Plants see the sun in the weak sense that they sense its radiation. Animals develop sophisticated organs of perception, including eyes. Humans have eyes that can see physical things, plus brains that can see abstract patterns. For the old Platonists (and Stoics), the universe is literally a rational animal. Hence it was appropriate for them to say that the duty of the universe is to become like the divine mind by maximizing its vision of itself. The universe maximizes its vision of itself by containing parts that ever more perfectly reflect the whole. For example, as humans or aliens do science, and thereby produce ever better models of the universe inside the universe, the universe is maximizing its vision. Doing

---

[95] Plato (*Theaetetus*, 176a5–b2).
[96] Plotinus (*Enneads*, 3.8).

science has cosmic significance. It provides life with cosmic meaning. The suffering of all life in the universe is redeemed by this holy project. To participate in this scientific project of cosmic self-knowledge makes life worth living.[97] However, at least two objections can be raised against this Platonic theory of cosmic obligation.

The first objection is that the universe lacks vision. Although *we* have eyes, the universe does not. The universe is utterly blind. The Saganesque slogan that we are the universe coming to know itself commits a mereological mistake. We are actually just *proper parts* of the universe coming to know the universe. If we are maximizing our vision of the universe, then we are giving our own lives some human meaning; but this human meaning is hardly cosmic. This objection can be defeated by reformulating the Platonic duty in terms of information: the duty of the universe is to maximize the information which its parts carry about the whole. To maximize this self-information is to *maximize reflexivity*. Blind things can maximize reflexivity. If something (like the universe) maximizes reflexivity, then its parts evolve through self-reaction, self-organization, self-regulation, self-replication, self-representation, self-awareness, self-legislation, and so on. This process is self-starting, self-bootstrapping, self-sustaining, and self-amplifying. The physical liturgies (Chapter 2) show that the universe has been doing a surprisingly good job at increasing reflexivity. Extrapolation suggests that it will keep working. The first objection fails. The universe does not need eyes in order to moil towards perfect self-representation, exact self-mirroring. Of course, it can have the duty to maximize reflexivity even if fails to fully do its duty.

The second objection runs like this: only *persons* have duties; but the universe is not a person; hence the universe has no duties. The objection is defeated by showing that the universe is an agent. A celestial computer is an agent whose maxims are encoded in the probabilities of all its transformational arrows. The whole composed of all the celestial computers in the universe is just the *sidereal computer*. It runs the *sidereal entelechy*,

---

[97]UR 312.

which strives to maximize complexity.[98] Hence it is the *sidereal crane*. Since complexity is intrinsically valuable, it is an axiological agent. Moreover, the best explanation for the fact that it runs its entelechy is that it is rational. Of course, this rationality is logical and structural rather than mental. So, by inference to the best explanation, the sidereal crane is a rational agent. Hence the sidereal crane is a *rational axiological agent.* Although the sidereal crane is really only a proper part of the universe, it does no harm to identify it with the universe. By this (not-so-strict) identity, the universe is a *rational axiological agent.* The second objection fails. The old Platonists and Stoics were almost right: the universe is *sufficiently like* a rational animal that it can have duties. It does not need to be a person to have duties. Of course, since it is not an intelligent social animal, its duties are not moral. But the first axiom of our deontic logic applies to our universe: if it can maximize value, then it ought to maximize it. Likewise the second axiom of our deontic logical applies to it: the universe strives to do its duty.

The first and second objections raise a deeper question. The first objection talked about reflexivity, while the second talked about complexity. Of course, these two go hand in hand. But which comes first? Three lines of reasoning indicate that reflexivity is prior to complexity. The first line concerns rationality. Any self-reflexive system is also a self-consistent system. But self-consistency is a logical value. Since it is rational to maximize logical values, it is rational to maximize reflexivity. Maximizing reflexivity is more rational than maximizing complexity. So, if the rationality of the universe explains the fact that it maximizes complexity, then this explanation runs from reflexivity to complexity. Since the universe is rational, it strives to maximize reflexivity; as a side-effect, it maximizes complexity. The second line concerns duty. Kant used the *categorical imperative* to define duty. It says that agents ought to use maximally self-consistent maxims to regulate their actions (Sect. 2 in Chapter 8). They should adopt policies that remain self-consistent through universalization. This self-consistency is reflexivity. And agents should not

---

[98]Just as organisms run entelechies, so universes run entelechies. By this analogy, much of Foot's (2001: chs. 2 and 3) theory of natural goods and duties transfers to universes.

adopt policies that become self-defeating when universalized. This self-defeat is anti-reflexivity. So the Kantian logic of duty (his deontic logic) explains why agents have duties to maximize reflexivity and to minimize anti-reflexivity. By doing those duties, they maximize complexity. A third line looks at the stages of evolution. Evolution sequentially produces things that are self-acting, self-extending, self-regulating, self-replicating, self-moving, self-sensing, self-simulating, and self-legislating. It seems possible to maximize complexity without going through this series of increasingly intensive self-relations. Hence this series is much better explained by the hypothesis that the cosmic entelechy is maximizing reflexivity. So, by inference to the best explanation, the universe is maximizing reflexivity.

For any computer, the duty to maximize reflexivity becomes a duty to maximize self-computation. The universe *ought* to climb up through all finite degrees of computational complexity towards an infinite computer. The infinite computer will be infinitely self-reflecting; it will be an infinitely self-representative system, exactly like the Plotinian divine mind. If the universe has some duty, then all the agents in the universe participate in that duty. Since you are an agent in the universe, you ought to help the universe to wake up: you have an ethical duty to actualize the omega point. As the universe strives to become godlike, so all its agents ought to strive to become godlike. Since you are a rational social animal, you ought to strive to acquire godlike animality, godlike sociality, and godlike rationality. Participation in this cosmic project makes life worth living. It justifies or redeems all the suffering of all life. As you participate in creating the great cosmic work of art, your life gains cosmic meaning. Of course, it is almost certain that the universe will fail to do its duty. It almost certainly will exhaust itself at some small finite level of reflexivity. Nevertheless, the near certainty of failure does not excuse any agents from doing their duties. As long as there is any possibility that it might succeed, the universe ought to do its duty. And you should do your duty too.

5.3   From Reflexivity to the Good

Our Platonic axiology includes our deontic axioms. The first axiom of our deontic logic states that if any agent can maximize value, then it ought to maximize it. Agency includes computers that can run entelechies. Of course, many other things can be agents. But it does not follow that everything is an agent, nor does it follow that everything can maximize value. Many things lack duties. This first axiom derives an *ought* from a *can*. It derives a deontic property (an *ought*) from a modal property (a *can*). This derivation is purely logical. It merely defines part of the meaning of the term *ought*. The second axiom of our deontic logic states that rational agents strive to do their duties. So if a rational agent can maximize value, then it strives to maximize it.

Our cosmology depicts our universe as agent which can maximize value. From our first deontic axiom, it follows that it ought to maximize value. Our cosmology also depicts our universe as rational. From our second deontic axiom, it follows that it strives to maximize value. But it does not follow that its strivings succeed. On the contrary, it seems pretty clear that our universe fails to do its duty. While the *imperative* "Maximize reflexivity!" holds at our universe, the *declarative* "Reflexivity is maximized" does not. If the imperative holds while the declarative does not, then the meaning of the imperative points beyond our universe.[99] An imperative to maximize some value is made true by a series in which that value increases from its minimum to its maximum. So the cosmic imperative is made true by a series of universes in which reflexivity increases from its minimum to its maximum. Wouldn't Dawkins reject all this metaphysics? On the contrary, to account for the cosmic value of our universe (its apparent fine tuning for life), Dawkins often invokes a plurality of universes.[100] It is entirely consistent with his principles

---

[99] If some thing x has a duty or imperative to F, then x ought to F; this is a deontic *de re* property. It is axiomatic that ought implies can. So the deontic *de re* implies the modal *de re* property that x possibly Fs. If x possibly Fs, then there exists some possible y such that y is a version of x and y does F. Since the deontic property entails the modal property, and the modal points beyond x, the deontic points beyond x.

[100] EP 384; GD 173–6, 185; GSE 426; MR 165; AT 2–4; SITS 272.

to turn to some multiverse model for the explanation of cosmic value. Platonic axiology remains entirely within the rational magisterium.

Consider one model of the imperative to maximize reflexivity. Its minimum is zero, while its maximum is at least infinity. It is defined by three rules. The *initial rule* states that there exists an initial universe with no reflexivity. This universe is simple. The *successor rule* states that every universe in the series is surpassed by a universe with greater reflexivity. Each successor universe contains more complexity and intrinsic value. The *limit rule* states that the entire series is surpassed by a universe which has infinite reflexivity. The limit universe is infinitely complex and valuable. Evolution in this limit universe rises from a simple origin to some infinite omega point. Its internal evolution mirrors the evolution of the entire series of which it is the limit.

This infinite series of universes is a model of the imperative to maximize reflexivity. An *Argument for the Model* goes like this: (1) Our universe has the duty to maximize reflexivity. (2) If any universe has this duty, then there exists some model of this duty. (3) So there exists some model of this duty. (4) But the model of this duty is an infinite series of universes. (5) So this infinite series of universes exists. The argument goes further to specify the series: (6) Our universe only partly satisfies this duty. (7) If any universe only partly satisfies its duty, then it is only part of a model that fully satisfies this duty. (8) The parts of this model are its universes. (9) Consequently, the model of this duty contains our universe and others. Since our universe is neither simple nor infinitely reflexive, it is some finite successor universe. It surpasses some lesser universe in the series and it is surpassed by some greater universe in the series.

This Platonic line of reasoning points beyond our universe. Its plausibility is far from clear. But it is almost certainly the smallest model for the Dawkinsian theory of the value of vision, and thence for his axiology. This Platonic model incorporates Stoic insights into the origins of suffering. The Stoics argue that we have natural duties. And they constantly say that doing your duty entails suffering. They present case after case in which people suffer horribly for doing their duty. Likewise our universe does its duty; likewise it suffers for doing its duty. It suffers by containing suffering. But the Stoics argue that duty beats utility. It is axiologically more important to do your duty than to be happy. The same holds for universes. Our universe is skewed towards duty. The skew makes it more likely that our universe does its duty. Something is a virtue (or is virtuous) if it makes it more likely that an agent does its duty. The skew is virtuous.

This Platonism posits an infinite series of finitely reflexive universes, a series which converges in the limit to an infinitely reflexive universe. However, since every infinity is surpassed by greater infinities, every infinite universe is surpassed by even greater universes. So this Platonism ultimately posits an unsurpassable series of surpassable universes. Because it is unsurpassable, this series extends the type *universe* beyond itself. It *idealizes* the type universe. An unsurpassable series of surpassable universes is an *ideal universe*. This seems paradoxical. On the one hand, an ideal universe is a universe; on the other, a series of universes is not a universe. The paradox is resolved by noting that adjectives can over-rule their nouns: just as a glass eye is not an eye, so an ideal universe is not a universe. However, while a glass eye is less than an eye, an ideal universe is more than a universe. An ideal universe is the ecstasy of an unsurpassable series of surpassable universes; as the ecstasy of the series, it transcends the series. Only the entire series can represent this ideal. Ideal universes play crucial roles in deontic logic. At any ideal universe, the duty to maximize reflexivity is satisfied in a particular way. For the Platonist, every ideal object expresses a particular kind of transcendence. It thus participates in the universal transcendence of the Good.

Both Dawkins and Plato use an absolute axiological standard to morally evaluate the actions and characters of deities. Plato frequently judges the actions and characters of the Olympian deities, and finds them morally defective. To make those judgments, Plato needs some standard. His standard is the Good. Dawkins likewise judges the actions and character of the Abrahamic God, and finds it morally defective.[101] To make that judgment, he need some standard. His opposition to relativism entails that he believes he is using some objective standard. That standard is the Good. If there were a God (there isn't), the Good would

---

[101]GD 51, chs. 7–9.

be beyond it. The Good is not God.[102] The Good *opposes* God. Our Platonic atheism replaces God with modal and deontic logic.

References

Anderson, J. (2002). *The Airplane: A History of Its Technology*. Reston, VA: American Institute of Aeronautics and Astronautics.

Anscombe, G. E. M. (1958). Modern moral philosophy. *Philosophy, 33*(124), 1–19.

Basalla, G. (1988). *The Evolution of Technology*. New York: Cambridge University Press.

Benitez, E. (1995). The good or the demiurge: Causation and the unity of good in Plato. *Apeiron, 28,* 113–139.

Bennett, C. (1988). Logical depth and physical complexity. In R. Herken, *The Universal Turing Machine: A Half-Century Survey* (pp. 227–257). New York: Oxford University Press.

Bennett, C. (1990). How to define complexity in physics, and why. In W. Zurek (Ed.), *Complexity, Entropy, and the Physics of Information* (pp. 137–148). Reading, MA: Addison-Wesley.

Brey, X. (2008). Technological design as an evolutionary process. In P. Vermaas, P. Kroes, A. Light, & S. Moore (Eds.), *Philosophy and Design* (pp. 61–76). New York: Springer.

Bruton, E. (1979). *The History of Clocks and Watches*. New York: Rizzoli.

Catling, D. (2013). *Astrobiology: A Very Short Introduction*. New York: Oxford.

Chaisson, E. (2006). *The Epic of Evolution: The Seven Ages of our Cosmos*. New York: Columbia University Press.

Chapman, E., Childers, D., & Vallino, J. (2016). How the second law of thermodynamics has informed ecosystem ecology through its history. *BioSciences, 66,* 27–39.

Cirkovic, M. (2003). Physical eschatology. *American Journal of Physics, 71*(2), 122–133.

Dennett, D. (1995). *Darwin's Dangerous Idea: Evolution and the Meanings of Life*. New York: Simon & Schuster.

---

[102]For Plato, God is the Demiurge of the *Timaeus*, identified with the Socratic *Nous*. It is not the Good. See Murdoch (1992: 37–8, 343, 475–7); Benitez (1995). Plotinus identified God with the divine mind, which is below the Good.

Dennett, D. (2004). Could there be a Darwinian account of human creativity? In A. Moya & E. Font (Eds.), *Evolution: From Molecules to Ecosystems* (pp. 273–279). New York: Oxford University Press.

Derex, M., et al. (2019). Causal understanding is not necessary for the improvement of culturally evolving technology. *Nature Human Behavior.* Online at www.nature.com/articles/s41562-019-0567-9. Accessed 29 April 2019.

Dewar, R. (2006). Maximum entropy production and non-equilibrium statistical mechanics. In A. Kleidon & R. Lorenz (Eds.), *Non-Equilibrium Thermodynamics and the Production of Entropy* (pp. 41–55). New York: Springer.

Dewar, R., Juretic, D., & Zupanovic, P. (2006). The functional design of the rotary enzyme ATP synthase is consistent with maximum entropy production. *Chemical Physics Letters, 430,* 177–182.

Dil, E., & Yumak, T. (2018). Emergent entropic nature of fundamental interactions. Online at arxiv.org/abs/1702.04635. Accessed 19 April 2019.

Dobovisek, A., et al. (2011). Enzyme kinematics and the maximum entropy production principle. *Biophysical Chemistry, 154,* 49–55.

Dyson, F. (1985). *Infinite in All Directions.* New York: HarperCollins.

Dyson, G. (1997). *Darwin Among the Machines: The Evolution of Global Intelligence.* Reading, MA: Perseus Books.

Dyson, G. (2012). *Turing's Cathedral: The Origins of the Digital Universe.* New York: Vintage Press.

Ebbing, D., & Gammon, S. (2017). *General Chemistry* (Eleventh ed.). Boston, MA: Cengage Learning.

England, J. (2013). Statistical physics of self-replication. *Journal of Chemical Physics, 139*(121923), 1–8.

England, J. (2014). A new physics theory of life (interview with N. Wolchover). *Quanta Magazine.* Online at www.quantamagazine.org/a-new-thermodynamics-theory-of-the-origin-of-life-20140122/.

England, J. (2015). Dissipative adaptation in driven self-assembly. *Nature Nanotechnology, 10,* 919–923.

Enoch, J. (1998). The enigma of early lens use. *Technology and Culture, 39*(2), 273–291.

Essinger, J. (2004). *Jacquard's Web: How a Hand-Loom Led to the Birth of the Information Age.* New York: Oxford University Press.

Foot, P. (2001). *Natural Goodness.* New York: Oxford University Press.

Gattringer, C., & Lang, C. (2009). *Quantum Chromodynamics on the Lattice.* New York: Springer.

Greene, B. (2005). *The Fabric of the Cosmos*. New York: Vintage.

Johnson-Laird, P. (2005). Flying bicycles: How the Wright brothers invented the airplane. *Mind and Society, 4,* 27–48.

Juretic, D., & Zupanovic, P. (2003). Photosynthetic models with maximum entropy production in irreversible charge transfer steps. *Computational Biology and Chemistry, 27,* 541–553.

Kelly, K. (2010). *What Technology Wants*. New York: Viking.

Kotz, J., Treichel, P., & Townsend, J. (2009). *Chemistry & Chemical Reactivity* (Vol. 2). Belmont, CA: Thompson.

Kouvaris, K., et al. (2017). How evolution learns to generalise. *PLoS Computational Biology, 13*(4), e1005358.

Kurzweil, R. (2005). *The Singularity Is Near: When Humans Transcend Biology*. New York: Viking.

Leibniz, G. W. (1697). On the ultimate origination of the universe. In P. Schrecker & A. Schrecker (Eds.) (1988) *Leibniz: Monadology and Other Essays* (pp. 84–94). New York: Macmillan Publishing.

Lovejoy, A. (1936). *The Great Chain of Being*. Cambridge, MA: Harvard University Press.

Machta, J. (2011). Natural complexity, computational complexity, and depth. *Chaos, 21,* 0371111–0371118.

Marenduzzo, D., Finn, K., & Cook, P. (2006). The depletion attraction: An underappreciated force driving cellular organization. *Journal of Cell Biology, 175*(5), 681–686.

Martin, O., & Horvath, J. (2013). Biological evolution of replicator systems: Towards a quantitative approach. *Origins of Life and Evolution of Biospheres, 43,* 151–160.

Martyushev, L. (2013). Entropy and entropy production: Old misconceptions and new breakthroughs. *Entropy, 15,* 1152–1170.

Martyushev, L., & Seleznev, V. (2006). Maximum entropy production principle in physics, chemistry, and biology. *Physics Reports, 426,* 1–45.

Mataxis, T. (1962, September). Change of life, computer style. *Army, 13*(2), 61–67.

Murdoch, I. (1992). *Metaphysics as a Guide to Morals*. London: Chatto & Windus.

Partner, M., Kashtan, N., & Alon, U. (2008). Facilitated variation: How evolution learns from past environments to generalize to new environments. *PLoS Computational Biology, 4*(11), e1000206.

Peirce, C. S. (1965). Collected papers of Charles Sanders Peirce. In C. Hartshorne & P. Weiss (Eds.), *Cambridge*. MA: Harvard University Press.

Penrose, R. (1979). Singularities and time-asymmetry. In S. Hawking & W. Israel (Eds.), *General Relativity: An Einstein Centenary Survey* (pp. 581–638). New York: Cambridge University Press.

Rescher, N. (1979). *Leibniz: An Introduction to His Philosophy*. Totowa, NJ: Rowman & Littlefield.

Rescher, N. (2010). *Axiogenesis: An Essay in Metaphysical Optimalism*. New York: Lexington Books.

Rubin, M. (1986). Spectacles: Past, present, and future. *Survey of Opthamalogy, 30*(5), 321–327.

Rutherford, D. (1995). *Leibniz and the Rational Order of Nature*. New York: Cambridge University Press.

Schneider, E., & Kay, J. (1994). Life as a manifestation of the second law of thermodynamics. *Mathematical and Computer Modelling, 19*(6–8), 25–48.

Silk, J. (2001). *The Big Bang* (3rd ed.). New York: Henry Holt & Co.

Simonton, D. (2010). Creative thought as blind-variation and selective retention. *Physics of Life Reviews, 7*, 156–179.

Simonton, D. (2015). Thomas Edison's creative career. *Psychology of Aesthetics, Creativity, and the Arts, 9*(1), 2–14.

Skene, K. (2015). Life's a gas: A thermodynamic theory of biological evolution. *Entropy, 17*, 5522–5548.

Springel, V., et al. (2005, June 2). Simulations of the formation, evolution and clustering of galaxies and quasars. *Nature, 435*, 629–636.

Steinhart, E. (2012). Royce's model of the Absolute. *Transactions of the Charles S. Peirce Society, 48*(3), 356–384.

Steinhart, E. (2014). *Your Digital Afterlives: Computational Theories of Life After Death*. New York: Palgrave Macmillan.

Steinhart, E. (2018). Spirit. *Sophia, 56*(4), 557–571.

Swenson, R. (2006). Spontaneous order, autocatakinetic closure, and the development of space-time. *Annals of the New York Academy of Sciences, 901*, 311–319.

Swenson, R. (2009). The fourth law of thermodynamics or the law of maximum entropy production (LMEP). *Chemistry, 18*(1), 333–339.

Temkin, I., & Eldredge, N. (2007). Phylogenetics and material cultural evolution. *Current Anthropology, 48*(1), 146–153.

Tipler, F. (1995). *The Physics of Immortality: Modern Cosmology, God and the Resurrection of the Dead*. New York: Anchor Books.

Tzafestas, S. (2018). *Energy, Information, Feedback, Adaptation, and Self-Organization*. New York: Springer.

Unrean, P., & Srienc, F. (2012). Predicting the adaptive evolution of *Thermoanaerobacterium saccharolyticum. Journal of Biotechnology, 158,* 259–266.

Verlinde, E. (2016). Emergent gravity and the dark universe. *SciPost Physics, 2*(3.016), 1–41. https://doi.org/10.21468/scipostphys.2.3.016.

Vogelsberger, M., et al. (2014). Introducing the Illustris Project: Simulating the coevolution of dark and visible matter in the universe. *Monthly Notices of the Royal Astronomical Society, 444*(2), 1518–1547.

Wald, R. (2006). The arrow of time and the initial conditions of the universe. *Studies in History and Philosophy of Modern Physics, 37,* 394–398.

Watson, R., & Szathmary, E. (2016). How can evolution learn? *Trends in Ecology & Evolution, 31*(2), 147–157.

Watson, R., et al. (2016). Evolutionary connectionism. *Evolutionary Biology, 43,* 553–581.

Wright, P. (1970). Entropy and disorder. *Contemporary Physics, 11*(6), 581–588.

Yen, J., et al. (2014). Thermodynamic extremization principles and their relevance to ecology. *Austral Ecology, 39,* 619–632.

Zenil, H., & Delahaye, J.-P. (2010). On the algorithmic nature of the world. In G. Dodig-Crnkovic & M. Burgin (Eds.), *Information and Computation.* Singapore: World Scientific.

4

Actuality

1   The Cosmological Liturgy

1.1   Our Universe Is Complex

There are at least four ways to think about cosmic complexity.[1] The first way comes straight from Dawkins: the complexity of the type *universe* is its arbitrary multiplicity divided by its stable multiplicity. On the one hand, every way to rearrange the parts of any possible universe preserves the type *universe*. So the arbitrary multiplicity always equals the stable multiplicity. Hence all possible universes are utterly simple, hence equally complex. On the other hand, Dawkins says a universe containing only hydrogen is far simpler than one containing intelligent life. Moreover, he says our universe is extremely improbable. Since improbability is complexity, it is extremely complex. His combinatorial theory of complexity doesn't work well for universes.

The second way to think about cosmic complexity also comes from Dawkins.[2] He indicates that the complexity of a universe is the

---

[1] Complexities of universes (AT 699). Our improbable universe (GD 146, 185).

[2] Complexity of basic laws (GD 176). Most concise description (ADC 102–4, 210).

© The Author(s) 2020
E. Steinhart, *Believing in Dawkins*,
https://doi.org/10.1007/978-3-030-43052-8_4

complexity of its basic laws. The laws in our basic physical theories, such as quantum mechanics and relativity, are mathematical equations. They can all be spelled out in some string of symbols. Assuming that any possible universe can be described mathematically, the complexity of any universe is the length of its shortest mathematical description (its *most concise* description). Since the shortest mathematical description of our universe is almost certainly extremely long, it looks like our universe is extremely complex.

The third approach to the complexity of universes looks at their parts: the complexity of any universe is proportional to the complexity of its most complex part. This approach allows us to stratify possible universes into complexity ranks. Universes that just contain atomic evolution are on the bottom rank. Those that contain atomic and molecular evolution are on the next higher rank. Then universes containing atomic, molecular, and biological evolution. The fourth rank holds universes containing atomic, molecular, biological, and technological evolution. On this third approach, our universe is also extremely complex. It might be objected that many possible universes don't contain material things like those in our universe. They lack atoms and molecules. But just as any measure of biological complexity has to apply to all possible organisms, so also any measure of cosmic complexity has to apply to all possible universes. This objection is correct. To meet it, it is necessary to get more abstract.

The fourth approach to cosmic complexity does not require specific types of parts, such as material atoms. It focuses instead on their abstract features. More specifically, it focuses on their computational features. If any thing in any universe is not a computer, it can be replaced with some computer that simulates it. So the complexity of any universe is the complexity of its most complex computer. This approach to cosmic complexity links up nicely with our earlier discussions of reflexivity and self-representation. Computers carry information. Say the reflexivity of any part of a universe is the number of bits of information it carries about the whole. The reflexivity of any universe is the reflexivity of its most reflexive part. More reflexive universes have parts which more accurately reflect the universe back to itself. But more reflexive parts are more complex computers. So the complexity of any universe is the reflexivity of its most reflexive part, its degree of self-representation. Since our universe contains extremely complex computers—like human brains—it is extremely complex. It does not seem likely that our universe will evolve to any infinite omega point. However, if it did evolve to such an omega point, it would be infinitely complex.

1.2 Tuned for Very Fine Music

Among the four ways of thinking about cosmic complexity, the first one works only from perspectives inside of universes. It doesn't let us compare universes. So only the last three are viable. They all agree that our universe is extremely complex, and they all closely correlate its complexity with life. Our universe has many features which make life possible. First, it has laws. If it were lawless chaos, then it would be sterile. Second, its basic laws have certain mathematical forms. If they were even slightly different, our universe would be lifeless. Third, its basic laws contain special numbers, often called the *fundamental physical constants*. For example, the speed of light and the strength of gravity are constants. There are many others. If they were even slightly different, our universe would be sterile, so the range of basic features that makes life possible is very small. Our universe resembles a violin whose strings have been very carefully tuned to play the right notes. It appears to be *finely tuned* for life.[3]

Dawkins thinks this apparent fine-tuning for life is real.[4] And he thinks this feature of our universe is so extremely unusual that it needs an extremely unusual explanation. But is our universe really finely tuned *for life*? Consider the constants. It turns out that if they were slightly different, then there would be no *stars* in our universe. Of course, if there were no stars, there would be no life. But since stars precede life, you could argue that the universe is finely tuned *for stars*, and that life is merely an accidental by-product. One of the main problems with much

---

[3]Leslie (1989); Rees (2001).
[4]GD 173; FW 79.

of the work on fine-tuning is that it is very hard to see why the universe is finely tuned *for life* rather than *for something else*.

Fortunately, for the friends of fine-tuning, more help is on its way from physics. The laws of physics entail that our universe has *three significant features*: (1) it started in an extremely low entropy state; (2) it maximizes entropy production; and (3) ordered flow produces entropy faster than unordered flow. Chapters 2 and 3 used these features to argue that our universe is an enormous complexity-producing machine. Its entire structure is *skewed* very far away from random simplicity and towards the generation of ever-greater complexity. These thoughts suggest that our universe is not merely finely tuned for life—it is finely tuned for self-organization. It is finely tuned to produce as much complexity as possible by producing entropy as fast as possible. But if complexity is intrinsically valuable (Sect. 2.4 in Chapter 3), then our universe is finely tuned for the production of intrinsic value. More deeply, it is finely tuned for the production of reflexivity. If Platonism is right, then the universe has a duty to maximize reflexivity. Its fine tuning makes it more likely that it will do its duty. Any feature that makes it more likely that an agent will do its duty is a *virtue*. Hence the three significant features are virtues. Our universe has some very high degree of virtue.

Dawkins says this fine tuning requires an explanation.[5] He considers many hypotheses intended to explain the fine tuning.[6] Five hypotheses will be examined here: (1) The *God hypothesis* says our universe was designed and created by some supernatural intelligence. (2) The *simulation hypothesis* says our universe was designed and created by some natural intelligence. (3) The *fecund universe hypothesis* says that universes resemble self-reproducing organisms. This cosmic begetting produces our universe. (4) The *eternal inflation hypothesis* offers another approach

---

[5]The fine-tuning needs explaining (FH 79; GD 169–80). Explanatory hypotheses include: God (GD 171–2); simulation (GD 98, 186); fecund universes (GD 174–5, 185; AT 2–4); eternal inflation (GD 185); and several others.

[6]I will ignore three other cosmological hypotheses (GD 171–76). Dawkins dismisses the *uniqueness hypothesis*, which says our universe is the only possible universe. He dismisses the *lottery hypothesis*, which says our universe exists by chance. He briefly considers the *plenitude hypothesis*, which says every possible universe actually exists. But plenitude is hard to reconcile with his evolutionary principles.

to cosmic begetting. (5) The *biocosmic hypothesis* further develops the theme of cosmic self-reproduction. But this hypothesis includes the evolution of cosmic complexity. The study of these five hypotheses makes up the *cosmological liturgy*. This liturgy aims to answer the *first Leibnizian question:* Why is our universe the way that it is? After all, there are many other cosmic possibilities—our universe could be different.

1.3   The Oracle at Delphi

Many of the hypotheses in the cosmological liturgy point to structures beyond our universe. But can we get outside of our universe? Dawkins says that we can get outside of it by building models of our universe in our brains.[7] We can transcend our universe cognitively, through science. For spiritual naturalists, this cognitive escape from our universe is driven by the duty or imperative to maximize reflexivity. This duty drives our universe to evolve parts which can represent the whole. When we do get outside of our universe in this way, we see that it is orderly, rational, and beautiful.[8] Its deep structure is finely tuned for the creation of complexity, reflexivity, and value. For many people, these cosmic features are mysterious signs which urgently demand interpretation. Our universe looks like an oracle which points to things beyond itself. These things have included divine creators, like the Platonic demiurge or the Biblical God. Spiritual naturalists say those deities do not exist. Nevertheless, the oracular cosmic features, like luminous hands, still seem to point to things beyond our universe.

Perhaps this pointing is illusory. Dawkins says that if there is no real evidence for the existence of something, then there is no good reason to believe that it exists.[9] Can there be any real evidence for the existence of any things beyond our universe? Obviously, our senses and our scientific instruments can't help us to observe any things beyond our universe. Our senses and instruments only observe things inside our universe. So it doesn't look like there can be any evidence of things beyond our universe.

---

[7]UR 312; GD 405.
[8]UR 151; TL 73–4.
[9]ADC ch. 7.1; MR 15; SITS 309.

Should we deny such things? A *positivist* says that if we cannot verify something by observing it using our senses or scientific instruments, then it does not exist. More generally, if we cannot devise a scientific experiment to verify that something exists, then it does not exist. Positivists say that absence of evidence is always evidence of absence.

Consider an argument against chakras. It goes like this: (1) There is no evidence for chakras. More precisely, there is no evidence for the statement that chakras exist. (2) If there is no evidence for some statement, then it is false. (3) Therefore, it is false that chakras exist. The second premise is the *positivist axiom*, also known as the *verification principle*. But what is the evidence for the positivist axiom? The answer is: *none*. The positivist axiom is self-refuting.[10] So the positivist argument against chakras fails. Likewise positivist arguments against gods, souls, mathematical objects, and possible universes fail. More generally, positivism is a self-refuting doctrine. Of course, there are still good arguments against things like chakras: (1) There is no evidence for chakras. (2) If chakras exist, then there is evidence for them. (3) Therefore, they don't exist. To apply this argument, you first have to ask whether there *should be* evidence for some type of thing. If not, then absence of evidence can't be evidence of absence.

The philosopher Massimo Pigliucci rightly criticizes the New Atheists for being positivists.[11] So is Dawkins a positivist? He knows about the difficulties associated with positivism. For example, he points out that, in the far future, earth-bound cosmologists will have evidence only for the existence of the Milky Way. Our universe will eventually expand so much that all other galaxies will move into unobservability. After that, our universe will appear to every cosmologist in the Milky Way to be composed of exactly one galaxy. Other galaxies will be things beyond our observable universe. We will have no evidence of other galaxies—no scientific experiment would reveal their existence. But that would not imply that they do not exist. You can easily imagine a future cosmologist proposing that there are other galaxies. You can hear the future skeptics

---

[10]Ayer and Copleston (1949: 118).

[11]Pigliucci (2013) criticizes New Atheists. Dawkins opposes positivism (BW 316; UR 21; ADC ch. 1.2; GD 98, 185). Milky Way (AK 187–8).

howl with laughter: "Other galaxies! Woo woo!" But the skeptics would be wrong. There are plenty of truths about physical reality for which there may not be any evidence at all. Absence of evidence is *not* always evidence of absence.

Dawkins is not a positivist.[12] He does not restrict science to the issues that can be decided by experimental tests. He thinks there are scientific issues that lie beyond experimental decidability or empirical testability. For example, he believes the existence of God is a *scientific question* even if there is no *scientific test* which decides for or against God. This means that we can have a valuable *scientific discussion* about the existence of things beyond our universe even if we can't prove that that they exist by observation or experiment. But what would we be discussing? Here Dawkins has an astute answer: we would be discussing *scientific models*.

Dawkins points out that modeling is richer than experimentation. Models can include mathematical theories or computer simulations.[13] Dawkins says there are three ways that we learn about reality: (1) directly through our senses; (2) indirectly through extensions of our senses with scientific instruments; and (3) very indirectly by making mathematical or computational models that have observable consequences. For example, we might build a computer model that evolves universes as if they were organisms. If this model eventually evolves some universe like ours, then that resemblance provides evidence for the model. And since that model involves other universes, there is evidence for them, even though they are only very indirectly observable.

Now the question is this: are there any things beyond our universe which explain the order, rationality, and beauty of our universe? Although we cannot observe such things using our senses or scientific instruments, we can make models that include them. Those models can be tested by seeing whether they produce universes which have the order, rationality, and beauty of our universe. At this point, an enterprising theologian might try to argue that theology develops models in which God designs and creates our universe, so there is scientific evidence for God. But Dawkins points out that those models are *not* scientific.

---

[12]SSSF; Dawkins and Jollimore (2008).
[13]MR 16–18.

What is the difference between scientific and unscientific models? Here Dawkins has a brilliant answer: *the difference is mathematics*. Unscientific theories, such as theologies, lack mathematics.[14] They rely instead on magic.

This reliance of science on mathematics provides spiritual naturalists with a good way to define *naturalism*. Naturalism does not imply any ontological doctrine. It does not try to tell you what exists and what does not exist. It does not mean that you recognize only material things or that you think that our universe is the only thing that exists. *A naturalist is somebody who restricts his or her attention to those theories which can be expressed mathematically.* For the naturalist, the order, rationality, and beauty of our universe point to large-scale mathematical models. Those models may contain things beyond our universe. But they will be mathematical. And so we will be able to run computer simulations which test those models. Of course, it may take centuries to state some scientific theory with mathematical precision. More generally, then, a naturalist strives to mathematize his or her theories. A naturalist believes that mathematics reveals the structure of reality—even the structure of ultimate reality. Twenty five centuries of practical success justify this faith in mathematization.

2   On Cosmic Designers

2.1   The God Hypothesis

According to Dawkins, the term "God" is defined by the scriptures of the Abrahamic religions.[15] They portray God as a male person with maximally high social status. Since relations between God and humans resemble those between a king and his subjects, we should relate to God through love, obedience, worship, prayer, fear, and so on. As the king of kings, God has the highest social qualities: God sees and judges your

---

[14]AK 190; UR 63–4.

[15]God (TL 64; GD 33, 41, 52, 56–7, 84, ch. 4, 184; BCD 420). Theistic sense of God (TL 64–7; GD 15, 33, 41, 56–7; GD chs. 8 & 9). Avoid theological doublespeak (1996b; OTTR 399; TL 67; GD 33, 83–4; AK 190).

deeds; He has the power to punish and reward. The Abrahamic scriptures define the *theistic system of social practices* through which humans constitute the *theistic sense* of the term "God". Dawkins insists that he will only use the word "God" in the theistic sense. On this point, I will resolutely follow Dawkins. By insisting that God has only one meaning, we avoid *theological doublespeak*. On the one hand, sophisticated theologians define God in abstract senses when talking to well-educated people. On the other hand, they define God in the theistic sense when talking to their congregations. But God cannot be both the abstract ground of being and a person who hears your prayers.

One explanation for the fine tuning of our universe for complexity-creation or for life is the *God hypothesis*. The God hypothesis asserts that our universe was designed and created by the Abrahamic God. This God is a supernatural person. To say that this God is *supernatural* means that He did not evolve. Of course, there are many arguments for this God. But right now we are only focusing on His role as the intelligent designer and creator of our universe. The *Fine Tuning Argument* for God is a kind of cosmic design argument. It runs roughly like this: (1) Our universe is finely tuned for life. (2) The best explanation for this fine tuning for life is a *Fine Tuner* that values life. (3) But this Fine Tuner is God. (4) So, by inference to the best explanation, God exists.

Dawkins presents an *Infinite Regress Argument* against the God hypothesis.[16] It is motivated by a rejection of unexplained complexity. He also calls this regress argument the *Ultimate Boeing 747 Gambit*. It goes like this: (1) Abrahamic theism entails that God is complex. (2) Since God is complex, some explanation is required for God. (3) Since God is by definition supernatural, God did not evolve from simpler things. So God must have come from something even more complex: God came from Super-God. (4) Hence there will be an infinite regression of ever more complex gods. (5) More complex things are less probable. (6) So the infinite regression of ever more complex gods is an infinite regression of ever less probable gods. (7) Since the probability of a chain is the probability of its least probable link, the probability of this chain of

---

[16]Infinite regress (GD 136–8, 145–6). Unexplained complexity (BW 141; CMI 77; BCD 420).

gods drops to zero. (8) Therefore, the God hypothesis is almost certainly false.

Dawkins correctly argues that unexplained complexity cannot provide an ultimate origin for our universe (or for anything else). Nevertheless, it is hard to understand why he gives this Infinite Regress Argument, since it does not agree very well with his own principles. From the fact that beavers design dams, an infinite regression of ever more glorious beavers does not follow. When we encounter a designer, Dawkins tell us that *the designer evolved from simpler and more probable antecedents*. He explicitly rejects the idea that it takes a greater thing to make a lesser thing.[17] Hence it does not take a big smart Super-God to make a lesser dumber God. The use of supernaturalness to block the evolution of God looks artificial (it looks ad hoc). We're never told why supernatural things can't evolve. Dawkins does say that theologians have not argued that "God evolved to his awesome complexity by slow, gradual degrees."[18]

The difficulty here is that Dawkins is making an exception for God. If Dawkins can say that *supernatural* means complexity that did not evolve, then the theist can just say that *supernatural* means complexity that is *necessary*. So the Ultimate 747 doesn't even get off the ground. To avoid this loophole, and for greater self-consistency, Dawkins should have insisted on the universality of his evolutionary principles: *all* complex things depend on simpler antecedents. Consequently, if the God hypothesis is true, then God evolved from simpler and therefore more probable deities. We should infer a regression of ever simpler and more probable deities, which terminates in a simple and maximally probable deity. Dawkins recognizes this when he discusses the cosmological arguments.[19] He says that if the regression of causes terminates in some Alpha, then it would not be the Abrahamic God. While that God is complex, the Alpha is simple.

These considerations motivate a better argument against God, one that is more consistent with Dawkinsian principles. The *Argument from*

---

[17]GD 142.
[18]Dawkins (1995b: 49).
[19]GD 101–2.

the Failure to Evolve goes like this (1) Suppose that God is the designer-creator of our universe. (2) But that God is highly complex. (3) All complex things come into being by evolution from simpler antecedents. (4) So God evolved from simpler deities. (5) But God did not evolve from simpler deities. (6) Since this is a contradiction, the God hypothesis fails. Its failure leaves a strange pagan residue—consider the Greek deities. They had sex and made babies. They had descent with modification. Since they have divine blood (*ichor*) in their veins, they can have divine genes in their genitals. Pagan deities can evolve. Why not define a population of evolving pagan deities? Dawkins says that a population of evolving gods would be an intriguing hypothesis.[20]

Dawkins considers the possibility that our universe was produced by some natural superhuman designer.[21] He argues like this: (1) It is possible that our universe was designed and created by some superintelligent and thus superhuman agent. (2) If it was, then that agent is highly complex. (3) All highly complex agents appear only after long evolutionary processes. From which it follows that (4) the designer of our universe "must be the end product of some kind of cumulative escalator or crane, perhaps a version of Darwinism in another universe." Therefore (5) It is possible that our universe has a superhuman designer-creator which evolved. It exists at the end of some long evolutionary chain of cosmic designer-creators. Of course, Dawkins denies that our universe was designed—but *evolution* can design things.

2.2   The Simulation Hypothesis

The *simulation hypothesis*, to put it roughly, says we live in a video game.[22] More precisely, our universe is a software process generated by some program running on some hardware substrate. The hardware is some *cosmic simulator*, some great machine built inside of some larger universe. It was designed and created by some earlier civilization with

---

[20]Dawkins (1995b: 49).

[21]Superhuman designer (GD 186, see 97–9). Long process (GD 52). Darwinism in another universe (GD 186). Denies cosmic design (GD 186).

[22]Bostrom (2003). Dawkins on simulation (GD 98).

extreme technological powers. Perhaps the Divine Engineers of *River out of Eden* built this machine. The omega points of Kurzweil or Tipler could be simulating our universe. Dawkins explicitly does not reject this hypothesis.

The simulation hypothesis entails that the physics of our universe rests on computational foundations—our local physics is ultimately digital. Digital accounts of physics have been advocated by many physicists and computer scientists.[23] Although digital accounts of physics are far from certain, they are defensible. Computer scientists have defined infinitely complex computers that can manipulate infinitely many bits of information. These infinity machines are infinitely more powerful than classical Turing machines. Digital physics does not need to be discrete or finite. It also needs to be said that, if our whole universe is generated by the cosmic simulator, it cannot be one of the things in our universe. We cannot observe the cosmic simulator. If we want to justify its existence, then we need to use mathematical models.

One argument for the simulation hypothesis comes from reflexivity: (1) Our universe moils to maximize reflexivity. (2) As the reflexivity of our universe increases, more and more of its parts are engaged in ever more accurate simulations of the whole. (3) So, as the reflexivity of our universe increases, it is increasingly likely that we are living in some simulation. (4) Our universe therefore moils to make it more likely that we are living in some simulation. (5) But nothing in the universe can prevent the success of this cosmic moiling. (6) Therefore, it is highly probable that we are living in a simulation. Of course, it can be objected that the structure of the universe itself can prevent the success of this cosmic moiling. The universe might not have enough free energy for something like the MEPP to drive physical evolution to produce myriad self-simulations. Still, we needn't worry much about the simulation hypothesis.

Suppose we are living in a simulation constructed by some intelligent organisms. They might be aliens, or they might be our descendents.

---

[23] Digital physics (SITS 85). Many physicists argue that our universe computes (Deutsch 1985; Zeilinger 1999; Lloyd 2002; Fredkin 2003). Computer scientists argue the same (Schmidhuber 1997; Wolfram 2002).

Where did they come from? Dawkins says they must have evolved.[24] The Dawkinsian theory of complexity entails that they were produced by some very long evolutionary process in which complexity gradually accumulates. They owe their existence to a cumulatively ratcheting crane running in their own metaverse. So the simulation hypothesis just pushes the question back: where did the metaverse come from? And the same reasoning applies to any nesting of simulations in simulations. The very definition of *simulation* implies the existence of some bedrock reality, some bottom level universe. We need to explain this universe, and the simulation hypothesis cannot help. So the simulation hypothesis is not an ultimate explanation for our universe. It is therefore more relevant to look at the ways that Dawkins uses the simulation hypothesis in his atheism.

Dawkins uses the simulation hypothesis to distinguish between things that are gods and things that are merely *godlike*.[25] If we are living in a simulation, then it was created by some superhuman agents. They would be *superhuman but not supernatural*. Dawkins thinks our universe is probably filled with superhuman alien civilizations. He thinks they would deserve to be called *godlike*. Nevertheless, he says they would not be *gods*. Dawkins writes that the difference between gods and godlike aliens lies in their histories. If something is a god, then it is superhuman but it did not evolve. If something is godlike, then it is superhuman and it did evolve. Godlike beings include the omega points of Kurzweil and Tipler. If they exist at all, they emerged through long processes of gradual evolution—so they are not gods. (Oddly, this suggests that the Greek and Norse deities are not gods; but I won't pursue this here.) If our universe is a video game, then you could use the term "God" in an approximate but nevertheless incorrect way to refer to its designers. The game designers are godlike but not gods.

---

[24]GD 98–9.

[25]But not supernatural (GD 33–41, 52, 96–99, 186). Godlike but not gods (GD 98; see IA 96–7; BCD 202–3; AT 635). Evolved superhumans (GD 98; WHO 12).

2.3   The Argument from Cosmic Beauty

Dawkins recognizes that, from the beauty of human works of art, we infer the existence of human artists.[26] But this inference motivates the *Argument from Cosmic Beauty* to a cosmic artist. It goes something like this: (1) The universe resembles a beautiful work of human art. (2) Beautiful works of human art are created by human artists (Sect. 2 in Chapter 7). (3) But like effects have like causes. (4) So, reasoning by analogy, our beautiful universe was created by a cosmic artist. For theists, of course, this cosmic artist is God. But this argument also supports the simulation hypothesis: from the beauty of a video game, we infer some artistic game designers. Dawkins, strangely, does not criticize this argument. He instead criticizes an irrelevant inference from human artistic genius to God. We focus here on the Argument from Cosmic Beauty.

Dawkins rejoices in the extreme sensual beauty of our universe.[27] But science reveals even deeper beauties. While flowers look lovely to our senses, biology reveals their deeper loveliness. The starry sky looks beautiful to our eyes. But astronomy reveals that the large-scale structure of the universe has even deeper beauty. While crystals dazzle our eyes, physics reveals they have even deeper structural beauties. To the naked senses and untrained minds, the universe may appear beautiful. But that beauty is shallow and small. As science reveals the structure of the universe, it reveals even deeper and greater beauty—it reveals *mathematical beauty*. Through mathematics, we pass from shallow sensory beauty to deep *rational beauty*. Dawkins affirms "the elegance and beauty of an orderly universe in which clocks stop for reasons."

But what kind of beauty does our universe have? Following the Platonists and the Stoics, Hume argued that our universe is more like an organism than like a human work of art. So its cosmic beauty more closely resembles biological beauty. Accordingly, the Argument from

---

[26]Argument from beauty (GD 110). The cosmic artist is God (Swinburne 2012). From artistic genius to God (GD 110–2).

[27]Our beautiful universe (ADC 43). Rejoice in its beauty (UR *x*, 27, 312–3; FH 99; GD 99; GSE 18, 291; AT 700). Beautiful flowers (UR 41–2) and starry sky (UR 63, 81–2, 118). Beautiful crystals (ADC 43–6). To deepest beauty (UR 41), which is mathematical (UR 63; TL 75–6) and rational (UR 151; TL 73–4).

Cosmic Beauty goes like this: (1) Our universe resembles a beautiful organic body. (2) Beautiful bodies were shaped by evolution (Sect. 3 in Chapter 3). (3) But like effects have like causes. (4) So, reasoning by analogy, our beautiful universe was shaped by cosmic evolution. Here the Dawkinsian arguments from improbability to evolution apply with extra force. After all, beauty is even rarer than complexity. You can make things more complex without making them more beautiful.

Nevertheless, there are at least two objections to the analogy with biological beauty. The first objection arises because biological beauty serves reproductive ends. But the universe is not analogous to a peacock trying to use its beautiful tail to seduce a peahen. So the argument from cosmic beauty to cosmic evolution fails. The second objection comes from the lack of motivation. On the one hand, beauty is a kind of goodness; but all minds (including divine minds) are attracted to goodness; hence the beauty of the universe provides an explanation for its existence. On the other hand, blind evolution (including cosmic evolution) has no attraction to any sort of goodness; hence the beauty of the universe provides no explanation at all for its evolution.

Due to these objections, theists may still argue that the best explanation for cosmic beauty is that there exists some creative mind which values beauty and which therefore created a beautiful universe. They will explain the beauty of our universe like this: (1) There exists some maximally good divine mind. (2) Because this divine mind is maximally good, it creates only the best. (3) But beautiful cosmic forms are better than ugly cosmic forms. (4) Therefore, the divine mind selects the most beautiful cosmic form for creation. (5) Our universe is an image of this most beautiful cosmic form.

The most atheistic reply to these objections is Platonic. Since the divine mind is motivated by goodness, we can cut out the middleman by appealing directly to the goodness. From the fact that our beautiful universe exists, the Platonist infers the existence of a law that selects beautiful cosmic forms. Since cosmic forms are systems of physical laws, this Platonic law is a law-of-laws. The divine mind gets replaced with this mindless law-of-laws. The law-of-laws acts on all the cosmic forms in some great Platonic library. The Platonic library contains mathematical structures that exist eternally and necessarily. The law-of-laws is aesthetic:

for every cosmic form in the library, if that form is beautiful, then there exists a physical universe which models that form. For consistency with evolutionary principles, this law-of-laws must apply to *sequences*: for every *series* of cosmic forms in the library, if that series maximizes beauty, then there exist physical universes which model the forms in that series.

More deeply, we can argue that the evolution of beauty in our universe shows that it gets maximized as a by-product of the maximization of reflexivity. Thus, for every series of cosmic forms in the library, if that series maximizes reflexivity, then there are physical universes which model the forms in that series. But what is this law-of-laws? It is just the form of forms. For a Platonist, this form of forms is *the Good*. Of course, you might try to avoid all this metaphysics by denying that our universe is beautiful; or by arguing that its beauty is merely subjective. But Dawkins affirms that our universe has an objective or intrinsic beauty. Dawkinsian thought marries claims about objective cosmic beauty with claims about the evolution of complexity. It will be hard to save that marriage without appealing to something like the Platonic Good.

3   Almost Cosmological Hypotheses

3.1   The Eternal Inflation Hypothesis

Dawkins briefly considers the *cyclical universe hypothesis*.[28] It states that physical reality as a whole is an infinite series of universes. It runs back endlessly into the past, and it goes forward endlessly into the future. The Stoics argued for a cosmic cycle in which universes end in conflagrations. During each conflagration, only Zeus remains, in absolute desolation and loneliness, bereft of all the complexity of his beloved universe.[29] Yet Zeus remains undisturbed. He grieves serenely. At peace with his own destruction, he is purified by the flames. From his ashes, he will arise to rebuild the universe. Dawkinsian nature often resembles this Stoic Zeus: it is *serene* in its indifference to suffering, and there is a *grandeur* in its

---

[28]Steinhardt and Turok (2007); Dawkins (GD 174).
[29]Epictetus (*Discourses*, 3.13.4–7).

serenity (Sect. 2.1 in Chapter 9). More recently, some physicists say every universe starts with a bang and ends with a crunch. Each crunch triggers the next bang. Every universe has a parent and a child. Leibniz argued that even if this series of universes goes back forever, we still have to ask two questions. The *first Leibnizian question* asks: Why these universes rather than some others? The *second Leibnizian question* asks: Why is there any series rather than none? The eternity of the series does not remove the need to answer these questions.

The cosmologist Andrei Linde has proposed a physical explanation which at least tries to answer the first Leibnizian question.[30] Dawkins mentions Linde's theory, which Linde calls *chaotic eternal inflation*. It relies on an analogy between cosmology and biology. According to this *biocosmic analogy*, universes resemble self-reproducing organisms. They reproduce asexually by budding. So nature is a big fractal foam in which universes are bubbles. Old bubbles beget new bubbles. These self-reproducing bubble-universes are *cosmic replicators*. When bubbles reproduce, their offspring have variant laws. So there is heredity with variation. As bubbles beget bubbles, it becomes increasingly likely, Linde suggests, that complex bubbles like ours will emerge. He suggests that the ever-growing cosmic foam produces universes filled with all possible forms of life. Thus Linde writes that "One can draw some optimism from knowing that even if our civilization dies, there will be other places in the universe where life will emerge again and again, in all its possible forms."[31] If chaotic eternal inflation (or some other biocosmic proposal) really does provide this optimism, then the biocosmic analogy can help to solve the atheistic problem of evil (Sects. 5 in Chapter 1 and 3.1 in Chapter 6).

However, despite its resemblance to biological evolution, chaotic eternal inflation involves only reproduction and chance mutation. Dawkins insists that complexity cannot emerge from chance. Mathematical analyses of complexity confirm his insight: the slow-growth law (Sect. 2.2 in Chapter 3) implies that complex universes are not likely to emerge from random variation. So chaotic eternal inflation will

---

[30]Linde (1986, 1994); Dawkins (GD 185).
[31]Linde (1994: 55).

just wander randomly and eternally on the vast flat plains of cosmic simplicity. It wanders like a drunkard, or like a gang of drunkards, through the lower floors of the library of all possible universes. It fails to climb the *Cosmic Mount Improbable*. So chaotic eternal inflation can't explain cosmic complexity. To explain our complex universe, we need a *cosmic crane*. We need some *cosmic computer* running an entelechy. This crane cannot run *inside of* our universe. Since it creates our universe, it has to be running outside of our universe.

3.2   The Fecund Universe Hypothesis

The cosmologist Lee Smolin has proposed another explanation for our complex universe, which he calls the *fecund universe hypothesis*.[32] It also uses the biocosmic analogy: universes reproduce asexually via black holes. When a star in some parent universe collapses into a black hole, it collapses into an abyss of cosmic nothingness; but somehow that nothingness negates itself. It bounces back into the existence of some new offspring universe. Universes bud like yeast or hydras. As with chaotic eternal inflation, the fecund universe hypothesis posits cosmic self-replication. As parent universes beget their babies, the laws of physics mutate. There is heredity with variation. But here is where the fecund universe hypothesis outshines chaotic eternal inflation: variants whose laws encourage black hole production will have more offspring. So, as cosmic baby-making goes on, the cosmic generations become increasingly populated by universes whose laws are ever more finely tuned for making black holes.

According to Smolin, universes more finely tuned to make black holes are also more finely tuned for the evolution of internal complexity. Hence universes more finely tuned to create black holes are also more finely tuned for evolving complex things, like organisms. Consequently, as cosmic reproduction goes on, the multiverse will be increasingly populated by universes finely tuned for life. This explains why our universe is very finely tuned for life—or so it seems. Actually, it is hard to find any

---

[32]Smolin (1997); Dawkins (GD 174–5, 185; AT 2–4; SITS 272).

explanatory power here. All you can really infer is that, if our universe is finely tuned, then it is more likely that it occurs in some later generation of universes rather than in some earlier generation. The fecund universe hypothesis doesn't *explain* fine tuning at all. But let this pass. The hypothesis is interesting because of its biological parallels.

The biologist Andy Gardner and the physicist Joseph Conlon have studied the mathematical basis of the fecund universe hypothesis.[33] They argue that the self-reproducing universes in the fecund universe hypothesis satisfy the same equations as earthly biological evolution. This reinforces the biocosmic analogy. The analogy motivates this argument: Dawkins says biological evolution is a parallel distributed computation; but biological evolution and the self-reproducing universes satisfy the same equations; therefore, the swarm of self-reproducing universe is also a parallel distributed computation.[34] The swarm of universes runs an evolutionary optimization algorithm—it runs an entelechy. This cosmic crane finely tunes universes for making black holes and so for life. However, even here the fecund universe hypothesis fails.

One big problem with the fecund universe hypothesis is that it explains only the fine tuning of the physical constants in the basic physical laws of our universe. All the universes in this hypothesis are already *quantum relativistic universes*—they are already extremely complex. Where did this complexity come from? Dawkins emphatically rejects *unexplained complexity*.[35] But the fecund universe hypothesis leaves the complexity of our universe entirely unexplained. Another problem comes from empirical testing. The fecund universe hypothesis can be empirically falsified. Unfortunately, many physical studies indicate that it is false.[36] For consistency with his empiricism, Dawkins must reject the fecund universe hypothesis.

What is the cosmic crane? As long as it remains unknown, Dawkinsian metaphysics remains incomplete. The bottom floor of the Sanctuary

---

[33]Gardner and Conlon (2013); Gardner (2014); Dawkins (AT 4).
[34]CMI 72, 326; ADC 12.
[35]BW 141; CMI 77; GD 136–8, 145–6; BCD 420.
[36]For problems with the fecund universe hypothesis, see Rothman and Ellis (1993); Silk (1997). Smolin (2004: 710–711) says it would be falsified by a single heavy neutron star. Such a star has been discovered (Demorest et al. 2010).

for Spiritual Naturalists remains unbuilt. Moreover, as long as this crane is missing, *naturalism* remains incomplete. For the sake of completeness, we need to find this *cosmic crane*, and to describe its *cosmic entelechy*. This crane starts simple and gradually accumulates cosmic complexity. Although Dawkins himself does not describe this crane, he does describe general evolutionary principles which help to define it. He says we should examine a wide range of evolutionary cosmologies.[37] He instructs us to search for cranes that work as well in cosmology as Darwinism works in biology. Here believing in Dawkins means *building on Dawkins*. Consequently, we will to try to define the cosmic crane.

3.3   Merely Physical Falling Flat

Chaotic eternal inflation and the fecund universe hypothesis both look like evolutionary cosmologies. They both rely on the biocosmic analogy, which states that universes are cosmic self-replicators. They can be justified by an argument which uses the replication principle (Sect. 4.2 in Chapter 2): (1) Our universe is extremely complex. (2) The replication principle says that if any object is extremely complex, then it was evolved by some computation running an entelechy using replicators. (3) Therefore, our universe was evolved by some computation running an entelechy using replicators.

But they both face a deep Dawkinsian objection: (1) The machinery behind these physical cosmologies is highly complex. (2) But any complex machinery demands an explanation. (3) The best explanation for any complex machinery is that there is some crane which has brought it into being. This crane began with some simple machinery, which it lifted up to the heights on some cosmological Mount Improbable. (4) Therefore, even if one of these physical theories is true, it depends on some deeper crane. It cannot be *ultimate*. Analogous remarks apply to any *merely physical theory* which aims to account for our complex universe. Merely physical theories *already presuppose* highly complex physical structures (such as intricate space-times, energetic quantum

---

[37]Evolutionary cosmologies (GD 188–9). Naturalism needs this crane (BW 6).

fields, string theory landscapes, etc.). Every merely physical theory involves unexplained complexity. It falls flat. It needs some deeper crane.

Another way to see the difficulty with all merely physical theories involves the distinction between cosmic *bubbles* and cosmic *foams*. Merely physical theories (like cyclical universes, the fecund universe hypothesis, or chaotic eternal inflation) may very well explain the *complex bubble* in which we find ourselves. But they fail to explain the *complex foam* which contains our bubble. *Metaphysical theories*, which must be ultimate and complete, aim to explain the foam. Starting from simple premises, they derive the existence and form of *nature*, the entire complex foam with all its bubbles. They cannot start by assuming the foam and then deducing its bubbles.

The distinction between physical versus metaphysical theories is nicely illustrated by the controversy over the book *A Universe from Nothing* by the physicist Lawrence Krauss.[38] Krauss claimed to have answered the second Leibnizian question: *Why is there something rather than nothing?* This second Leibnizian question is also known as the *metaphysical question*. But the physicist-philosopher David Albert replied that Krauss had merely explained why there are quantum fields with particles rather than without particles.[39] And any quantum field, with or without particles, is obviously *something*, and something extremely complex. Since Krauss failed to explain why there are any quantum fields at all, he did not answer the metaphysical question.[40]

Dawkins often talks about the *spirit of wonder* which motivates scientific inquiry. We can wonder about things inside our universe. And we can wonder about our whole universe. We can get outside of our universe by constructing models of it in our minds. As soon as we construct such models, we direct the spirit of wonder at our universe itself: Why is there any universe? And why does it have the structure that it has? Why does it have any quantum fields at all? If the scientific enterprise is true to the Dawkinsian spirit of wonder, then it aims to explain all physical complexity. Nevertheless, on pain of vicious circularity, the explanation

---

[38]Krauss (2012).
[39]Albert (2012).
[40]Horgan and Ellis (2014).

for all physical complexity cannot come from physics. It has to come from deeper and more purely formal sciences, such as computer science, information theory, and pure mathematics.

4   Possible Universes

4.1   The Library of All Possible Universes

Dawkins talks about alternative possibilities.[41] He talks about ways that evolution might have gone but didn't. He talks about the many species of possible animals which never came into actual existence on our earth. He puts them into the Museum of All Possible Animals. Of course, since all actual animals are possible, this Museum also includes all actual animals. The other possible animals occur on other possible earths, in other possible Milky Ways, in other possible universes.

Dawkins often talks about other possible universes.[42] If they exist, they can also be organized into a Museum, namely, the *Museum of All Possible Universes*. There are many ways to think about possible universes. The way chosen here comes from Leibniz. For Leibniz, the possible contrasts with the actual. Possible universes are to actual universes as scripts are to plays. A script is a possible play; when the play is produced, the script corresponds to an actual play. The actual play is a model of the possible play. Or possible universes are to actual universes as recipes are to cakes. A recipe defines a possible cake (or class of possible cakes); when the recipe is followed, it produces an actual cake. Of course, recipes are programs. And the possible stands to the actual as the abstract stands to the concrete: possible universes are abstract cosmic forms. They are *cosmic natures*. They are unified systems of mutually consistent physical laws. Since possible universes are abstract, they are not really universes— they are just cosmic structures or patterns. An easy way to think about possible universes is to say they are cosmic descriptions. So Leibniz said the entire history of every possible universe is written into a *cosmic*

---

[41]Alternative possibilities (1993a; UR 312). Museum (CMI ch. 6).
[42]EP 384; GD 173–6; GSE 426; MR 165; AT 2–4; SITS 272.

book, which he called the book of its fates. So the Museum of Possible Universes is really just a library filled with cosmic books.

Leibniz referred to his library of possible universes as the *Palace of the Fates*.[43] It resembles the libraries (like the atomic or biological libraries) in Chapter 2. It is organized into floors. For Leibniz, the Palace of the Fates exists in the mind of God, as a data structure in the divine memory. But the Leibnizian concept of possibility does not really depend on any divine mind. You can affirm the Leibnizian possible universes without believing that they exist in any divine mind. They can just be objectively existing patterns of information. The most objective way to think about possibilities comes from pure mathematics. According to one mathematical interpretation, the cosmic books contain computer programs; when they are run on cosmic computers, actual universes come into being. According to another, they contain axioms written in some purely logical or mathematical language. These axioms define possible universes.

To distinguish the Dawkinsian library of possible universes from the Leibnizian Palace of the Fates, I will refer to the Dawkinsian library as the *modal library*. The term *modal* comes from *modal logic*, which is the formal study of possibility. The modal library is the Museum of All Possible Universes. It is organized into floors filled with cosmic books. Each cosmic book describes or defines some possible universe. But the books in the modal library were never written by any hands, and aren't intended to be read by any eyes. They aren't written in anything like any human languages—they are written in some purely logical language. Logical languages made by humans, like the predicate calculus, approximate that purely logical language. These cosmic books exist objectively, necessarily, and eternally. They are logical propositions. As such, they are parts of the Stoic Logos, which is just the rational structure of existence. Nevertheless, the purely logical interpretations of these books aren't likely to be very helpful.

---

[43]Leibniz (*Theodicy*, secs. 414–17).

4.2   The Finite Levels of Possibility

For the sake of familiarity, I will start with a *literary interpretation* of the modal library. According to this interpretation, its cosmic books resemble the books in our libraries. Each book in the modal library contains a text that describes a universe. It contains a string of symbols taken from some alphabet, that is, from a set of characters or symbols. For the sake of illustration, the books in the modal library are just written in English letters. Of course you can add numerals and punctuation marks. You can add all the technical symbols used in logic, mathematics, and the sciences. You can add blanks and carriage returns. You can expand the alphabet out to the *Unicode* character set. It is a digital character set with thousands of symbols, intended to represent almost every language on earth. But here I will just use the English letters.

The modal library has many floors filled with books. To define these floors, I start with the initial law. The *initial law* defines the bottom floor of the modal library. This is its zeroth floor, which contains the simplest book. The simplest book does not contain any text at all—it is a book whose pages are blank. The blank book describes the empty universe. To say that two books are identical means that they contain exactly the same series of symbols. The modal library contains exactly one copy of every possible book. It does not contain any redundancies. So it does not contain two blank books. Now the initial law for the modal library can be stated precisely: there exists exactly one blank cosmic book on the bottom floor of the modal library.

The *successor law* defines the higher floors. It defines each next higher floor of the modal library in terms of the current floor. It has two clauses. The first clause of the successor law states that every book on the current floor can be *varied* in some new ways. These novel variations are not already present on the current or lower floors of the modal library. These novel variations are analogous to the novel mutations in some genetic code. Here each way to vary a book adds a single symbol to the end of that book. The novel variants of any book are its *offspring*. Now the second clause of the successor law states that the next floor contains every offspring of every book on the current floor. It states that for every book on any floor, for every way to vary that book, there exists an offspring book which is varied in exactly that way.

The initial and successor laws interact to fill up the finite floors of the modal library. The book on the zeroth floor is blank. For each symbol in the alphabet, the blank book begets an offspring which adds that symbol to the end of the blank book. So each book on the first floor of the modal library contains exactly one symbol. These books vary from the book *a* to the book *z*. Every possible string of length one occurs on the first floor of the modal library. The same logic applies to all books on all higher floors. For any book on any floor, for any symbol, that book begets an offspring by adding that symbol to its end. The books on the first floor beget their offspring on the second floor. Each book on the second floor has two symbols. The book *a* begets offspring that vary from *aa* to *az*. The book *b* begets offspring that vary from *ba* to *bz*. And so it goes. Every possible string of length two appears on the second floor of the modal library. The books on the second floor beget offspring with three symbols on the third floor. More generally, each book on the n-th floor has n symbols. And every possible string of length n appears on the n-th floor of the modal library. So every possible string occurs on some floor of the modal library. The modal library is *lexically complete*. It is a plenitude of strings—it is a lexical *plenum*. Figure 1 illustrates some strings on the first three floors of the modal library. Each upwards line is a directed arrow from parent to offspring. However, to avoid clutter, the arrowheads are not shown.

Each offspring adds new content to its parent. It increases the amount or quantity of text in that book. The *quantity of information* in any book is just its length. So the books on the n-th floor contain n units of information. Since each book begets each offspring by adding just one symbol, the variations are *minimal* changes. There aren't any books in between a parent book and one of its offspring. Fractional symbols don't exist. For example, there isn't any book in between the book *d* and its offspring *de*. Hence the quantity of information in any offspring is minimally greater than the quantity of information in its parent. So every big change is some series of small changes. A variation which adds ten new letters to a book is produced by adding one new letter ten times. The changes add up to generate books of any finite length.

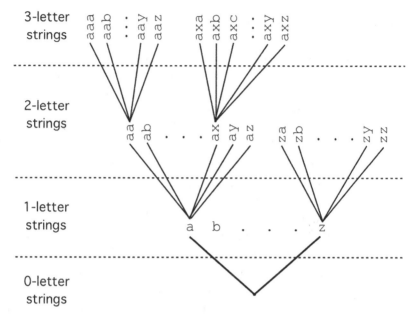

Fig. 1 Some strings on the first three floors

As strings give birth to strings, they form a family tree, in which the branches are directed paths to offspring. The paths are arrows. As strings evolve along these paths, increasingly long and varied sequences of letters appear. Most of these strings are essentially random or chaotic sequences. But some of them start to contain repeating motifs: *abababab*. And they contain parts that vary in conditional ways. So they start to contain grammatical regularities. They gain mathematical regularities: *babaabaaabaaaa*. They support more deeply nested structures with intricately related parts. The structural regularities in these strings slowly become more beautiful. They accumulate aesthetic value. And some of them will evolve into sequences of words in the English language. It doesn't take long for *harrypotterisaboywizard* to appear. Eventually, the entire series of Harry Potter books will appear compressed into a single volume on some floor of the modal library. Of course, this is just a poetic illustration—the modal library doesn't really contain books written in

English. But it might contain a program which, if run on some cosmic computer, would bring the Harry Potter universe to life.

4.3 The Infinite Levels of Possibility

The lengths of the cosmic books are measured by the *natural numbers*. These are the numbers 0, 1, 2, 3, and so on. More mathematically, these are the *ordinal numbers*, also known as the *ordinals*. Poetically speaking, you can picture the ordinal number line as a spire rising upwards into the abstract sky. It is the *axis mundi*, the spindle of nature. More precisely, the *axis mundi* is defined by three laws. The *initial law* states that there exists an initial ordinal 0. The *successor law* states that every ordinal n is surpassed by its successor $n + 1$. The successor law can be expressed in two parts: every ordinal can be surpassed in exactly one way; for every way that an ordinal can be surpassed, there exists some successor ordinal which surpasses it in that way. The initial and successor laws define the series of finite ordinals. They define the series 0, 1, 2, 3, and so on. Of course, mathematicians posit infinite ordinals—so the third law takes us into infinity.

To define the third law, we start with *progressions*. To say that a series of ordinals is a progression means that the initial ordinal is in that series and that every ordinal in that series has a successor in that series. It follows that every progression is infinitely long. For example, the infinite series of finite ordinals is a progression. Now the *limit law* asserts that every progression of ordinals is surpassed by a limit ordinal. Every limit ordinal is greater than every ordinal in the progression of which it is the limit. The limit law can be expressed in two parts: every progression of ordinals can be surpassed in exactly one way; for every way that a progression of ordinals can be surpassed, there exists some limit ordinal which surpasses it in that way. The limit law entails that the series of finite numbers 0, 1, 2, 3, … is surpassed by the first limit ordinal. Since this limit ordinal is greater than every finite number, it is infinite. The first limit ordinal is known as ω. But the infinite ordinal ω is not the biggest ordinal. The successor of ω is $\omega + 1$, then $\omega + 2$, and these numbers never end.

Modern mathematics defines infinities beyond ω. Every infinite ordinal is surpassed by greater infinite ordinals.

The modal library is a series of floors. Every book on any floor has offspring on the next higher floor. The offspring relation creates progressions of books. The initial book has children, which have children, which have children … These progressions are *lineages*. Each lineage starts with the initial book, and every book in any lineage has exactly one offspring in that lineage. So a lineage is also a chain of books. Lineages do not branch. Now, to say that a book is a limit of some lineage means that it is minimally more complex than every book in the lineage of which it is the limit. And to say that a book is a limit book means that it is a limit of some lineage. The *limit law* for the modal library states that every lineage of books has at least one limit. Limit books go on limit floors. The first limit floor is indexed by the first infinite ordinal ω. Every book on the first limit floor is the limit of some lineage of finitely detailed books. So every book on the ω-th floor of the modal library is an infinitely detailed book, which contains infinitely much information. Since the modal library has a floor for every ordinal, and the ordinal spire is absolutely high, the modal library is also absolutely high. It soars off into the mathematical heavens, it rises without bound into the abstract sky. It is an unsurpassable collection of surpassable books—its completeness is its ideal ecstasy.

5   The Treasury

5.1   Endlessly Ever Better Books

Every book in the modal library has some intrinsic value. Following Dawkins, we identify (intrinsic) value with complexity: the value of any book is its complexity. The stratification of the books in the modal library into floors raises a question: what are the most valuable books in the modal library? These are the *optimal books* in the modal library. To find these optimal books, we need to think about complexity.

The scientific study of complexity shows that it accumulates: atomic complexity accumulates; molecular complexity accumulates; biological complexity accumulates; technological complexity accumulates. These examples support the general thesis that all complexity accumulates. This general thesis is confirmed by the purely mathematical study of complexity. The slow-growth law (Sect. 2.2 in Chapter 3) states that all complexity accumulates. This law applies to the books in the modal library: they accumulate complexity along slowly-increasing paths from parents to offspring. And if complexity is intrinsically valuable, then value also accumulates. The books in the modal library gain value through slow accumulation. Consequently, the optimal books in the modal library are defined by their locations in long paths of gradually increasing value.

The optimal books occur on *optimal paths*. If some path is optimal, then it always increases in value. But every path needs some starting place. So where do the optimal paths start? There are three reasons to say that every optimal path starts with the initial simple book. The first reason comes from arbitrariness. If a path does not start with the initial book, then its start is arbitrary. But arbitrariness is not a virtue. The second reason comes from the Dawkinsian theory of complexity. More complex books are less probable books; paths that start with less probable books are less probable paths; but lesser probability is not a virtue. The third reason comes from the nature of optimality itself. The optimal paths are those on which the accumulation of value is maximal. But if some path does not start with the least valuable book, then its accumulation of value cannot be maximal. So the optimal paths start with the simplest book. For these three reasons, every optimal path starts with the initial simple book.

Every optimal path starts with the initial simple book on the very bottom floor of the modal library. Since the initial book has no content or detail, it has no complexity; therefore, its value is zero. It is the least valuable book in the entire modal library. Nevertheless, it is the most valuable book *on the bottom floor*. It is the optimal book on that floor. Of course, there will be better books on higher floors. These books correspond to universes. They are abstract cosmic forms. Better books define *better universes*. But our theory of cosmic value is not utilitarian. Happiness plays no role in the value-rankings of abstract universes. The optimization of cosmic value does not minimize pain and maximize pleasure. Utilitarianism implies that the best universes are utterly devoid of life. It ends in anti-natalism and nihilism. Against utilitarianism, the better universes have greater intrinsic value. They are universes in which functional excellence grows greater through conflict and strife. They are universes in which *arete* shines out more brightly in the strife-torn *agon*. They are universes in which complexity evolves to greater heights of glory, in which ever more powerful computations become ever more deeply fractally nested in ever more powerful computations, in which reflexivity rises like a flame to greater heights of holiness. All these concepts of value are very closely associated with aesthetic values like beauty. If utility does increase, it increases as a by-product of the maximization of complexity.

The initial book has offspring on the first floor. If any book is an offspring of some parent, then it adds some content or detail to its parent. But content is not identical with value. If the value of the offspring is compared with that of its parent, there are three possible cases. (1) The addition of new content or detail from parent to offspring can cause the value of the offspring to decrease. It can distort, pervert, or mangle the value of the parent. This is like a bad genetic mutation that makes the offspring less fit. If the offspring is less valuable than its parent, then it is a *downgrade*. (2) The addition of content can cause the value of the offspring to remain unchanged. If the offspring is just as valuable as its parent, then it is an *equigrade*. (3) The addition of new content can cause the value of the offspring to increase. This is like a genetic mutation which increases the fitness of some offspring. If the offspring is more valuable than its parent, then it is an *upgrade*. The arrow that runs from any book to its offspring encodes the change in intrinsic value. Each arrow has a *charge* corresponding to its intrinsic value. The arrow is positively charged if it points to an upgrade; it is neutrally charged if it leads to an equigrade; it is negatively charged if it leads to a downgrade.

From parent to offspring, the change in content is minimal. The quantity of information in any offspring is minimally greater than the quantity of information in its parent. Consequently, any change in value is also minimal. If an offspring is less valuable than its parent, then the decrease is minimal; if it is more valuable than its parent, then the increase is minimal. A minimal increase in value is an *improvement*. So if some offspring is (minimally) more valuable than its parent, then it is an improvement of its parent. The concept of improvement can be used to define the optimal books in the modal library. The optimal books occur on optimal paths: every optimal path starts with the initial simple book, and proceeds only through improvements.

5.2 The Laws of the Infinite Treasury

The optimal part of the modal library is the *optimal library*. But a better name for the optimal library is the *treasury*. The treasury is a sublibrary of the modal library. It is a part of the modal library. This means that every book in the treasury is also in the modal library. However, the treasury is almost certainly a *proper part* of the modal library. This means that some books in the modal library are *not* in the treasury. Since the modal library is stratified into floors, the treasury is likewise stratified. The floors in the treasury are defined by laws acting from the three types of ordinals.

The *initial law* states that the bottom floor of the treasury contains exactly the least valuable book from the modal library. Since this is the only book on the bottom floor, it is by default the most valuable book on that floor. It is the *optimal book* on that floor. But the least valuable book is surpassed by better books on higher floors. It is surpassed by its upgrades—these are its improved offspring, its optimal successors. And these in their turn are surpassed by their optimal successors. So the optimal books on the successor floors are defined by a successor law.

The *successor law* has two clauses. Its *first clause* states that every book on any floor of the treasury can be improved in at least one way. There is always at least one way to make every book better. This clause asserts that every book in the treasury has at least one upgrade; it has at least one better offspring. The *second clause* of the successor law states that every next floor of the treasury contains every improvement of every optimal book on the previous floor of that library. The optimal books on every next floor are the improvements of the optimal books on the previous floor. For example, if some book is among the optimal books on the tenth floor of the treasury, then all its improvements are among the optimal books on the eleventh floor. Every optimal book on any floor is surpassed by better books on the next higher floor. There are many optimal books but *no best book*. No book is best in any sense. Since books define universes, this means that *no universe is best*. Every universe is surpassed by better universes.

The iteration of the successor law defines *optimal progressions* of books. More precisely, to say that a series of books is an optimal progression means that the initial book is in that series and that every book in that series has exactly one optimal successor in that series. Now the *limit law* for the treasury has two clauses. Its first clause states that every optimal progression of books can be surpassed in at least one way by some optimal limit book. Every optimal limit book is minimally better than every book in the progression of which it is the limit. The second clause of the limit law states that for every way that an optimal progression of books can be surpassed, there exists some optimal limit book which surpasses it in that way.

The limit law enables all optimal progressions to run through all the mathematical infinities. Most generally, to say that a series of books is an optimal progression means that the initial book is in that series; every book in the series has exactly one optimal successor in that series; and every subprogression in that series has exactly one optimal limit in that series. Every optimal progression runs upwards into the sky along the ordinal spire. It runs upwards into the sky along the *axis mundi*. It is an unsurpassable series of surpassable books. There is a well-defined mathematical sense in which every optimal progression *transcends* its members (Sect. 2.5 in Chapter 6). Since value increases within any optimal progression, its transcendence expresses a way that value can be absolutely infinite. It expresses a way that value can be absolutely maximal. It expresses an *ideal*. Since Dawkins says he likes poetic symbols for scientific ideas, we say poetically that optimal progressions are *stars*. These stars form the roof of the Sanctuary for Spiritual Naturalists. Our Sanctuary is therefore open to the sky.

5.3   The Growth of Cosmic Meaning

The improvements of books can be illustrated with strings of letters. Every string of letters is more or less meaningful in some language. For the sake of illustration, the intrinsic value of any string is its degree of meaning in English. The zero degree is pure meaninglessness. It is the absence of all meaning, but that absence is the absence of all words. So the initial book in the treasury is the blank book. It is the same as the initial book in the modal library. This blank book is the sole book on the zeroth and bottom floor of the treasury. An arrow runs from the blank book to the simplest meanings, the simplest words with any content. The two simplest words in English are *a* and *i*. These words are meaningful in English, but their meanings are very simple, containing almost no complexity. Since the words *a* and *i* are improvements of the initial simple word, they are the sole books on the first floor of the treasury.

Most of the paths from the first words lead to nonsense. However, some of the arrows from the first word lead to other words. They are arrows which accumulate meaning. They are upgrades which run to optimal successors. The more meaningful arrows from the word *a* go to words like *ah, am, an, as,* and *at.* The upgrades from *i* go to words like *if, in, is,* and *it.* Since these are longer words, there is a trivial sense in which they are more meaningful and valuable. And right now, at the very start of meaning, this trivial sense is the only sense. So they are improvements of the first words. Each improvement of the first word occurs on the second floor of the treasury.

Most of the arrows from the two-letter words lead to meaningless strings. You can add letters without adding any further value. But some lead to meaningful words. For example, the meaning-increasing paths from the word *an* run to *and* and *ant.* The upgrades from *as* lead to *ash, ask,* and *ass.* Once more, in the trivial sense of being longer words, these are more meaningful words. They are improvements of their parents. So some of the books on the third floor are improvements of books on the second floor. Figure 2 shows some paths to some third-floor books. But optimal arrows need not run directly to longer words or more meaningful strings. Consider the path $m \to ma \to mat \to math.$ This is a path on which meaning increases. So, even though *m* is not a word in

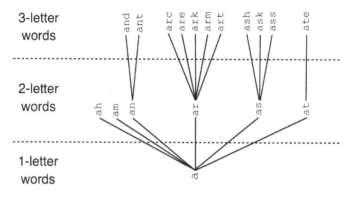

Fig. 2 Three generations of improvements

English, it is an improvement of the blank book. As the strings of letters grow longer, their variations in meaning grow more fine-grained.

Adding more letters can subtly increase meaning, leave it the same, or subtly decrease it. Thus going from *mat* to *math* increases meaning; but from *mat* to *madt* it decreases. An increase in meaning can involve an increase in irregularity, randomness, or noise. Noise has value in any evolutionary system. As valuable strings grow longer, they tend to gain more randomness as they also gain more orderliness. Randomness (like design) accumulates. Universes never roll dice. Dice rolls are equivalent to miracles, and neither of those exists. Any randomness in any universe is fated by its string. Yet their strings contain genuine randomness, real irregularity, purely mathematical noise. Noise isn't unpredictability or ignorance. It is the objective absence of patterning.

Most paths fail to increase meaning. The path $m \rightarrow mq \rightarrow mqz \rightarrow mqzj \rightarrow mqzji$ does not increase meaning. The books from mq to $mqzji$ and beyond are not optimal. They are *dark books*. The regions of the modal library filled with dark books are like the *no-go regions* in the Museum of All Possible Animals.[44] They resemble the uninhabitable valleys and empty depressions on Mount Improbable. Evolution does not reach these dark books. Poetically speaking, the totality of these dark books is the *night*.

---

[44]No-go regions (CMI 217–23). Uninhabitable valleys (CMI 133–6).

5.4   Cosmic Poetry

As you climb to higher floors in the modal library, some of its books begin to contain strings of words which are grammatically well-ordered. They will turn into well-ordered English sentences. But grammatical ordering does not depend on human language at all—it only depends on structure. Grammatical strings of words emerge because the modal library contains all possible sequences of letters and other characters (and therefore all possible sequences of words). And if you move only from parents to their *more meaningful offspring* (that is, if you only follow the paths in the treasury), then you will see meaning and complexity and reflexivity steadily increase.

As these words grow into longer and more meaningful texts, some of them turn into poems. Dawkins often talks about poetry. It plays an essential role in *Unweaving the Rainbow*. As a biologist originally deeply interested in genes, he builds his thought, not out of self-swerving atoms, but out of self-replicating codes. Science reveals that the rainbow of existence is self-writing poetry. Simpler stories beget more complex stories, less valuable texts beget more valuable texts. As texts gain value, they gain literary beauty and dramatic meaning. As optimal books are surpassed by their improvements, they become densely patterned. They become poems. We often think of poetry in terms of emotional expression—but that view is far too narrow. Poems are small texts into which great meaning is compressed. They are semantically dense texts. As you follow the paths in the treasury, the optimal books grow more poetic.

The most poetic texts contain *systems of axioms*. The entire subject of Euclidean geometry is compressed into the Euclidean axioms. The axioms of Euclidean geometry form a poem. You can follow a path in the treasury from the initial book to *The Elements of Geometry*, which contains the Euclidean axioms. It is a path that rises through a long lineage of books. That Euclidean book describes an abstract and purely mathematical space. As optimal books are surpassed by optimal books, they start to contain unified systems of physical axioms. But these systems are abstract cosmic natures—they are cosmic forms. You can follow a path from *The Elements of Geometry* to the book that contains an axiomatic formulation of Newtonian mechanics.[45] It describes a universe with material particles. A path runs from that Newtonian book to the book with axioms for the physical geometry of our universe.[46] Then to books with axioms for quantum field theory. Then to our book, that is, the book which encodes the physical structure of our universe. But our book is not the most complex book in the treasury. Every book is surpassed by endlessly many more complex books. No book in the treasury is maximally complex. So if complexity is intrinsically valuable, then every book is surpassed by infinitely many intrinsically better books. No book in the treasury is maximally valuable. The treasury does not contain any best of all possible books. A best book, like a biggest number, is mathematically impossible.

References

Albert, D. (2012, March 25). On the origin of everything. Review of L. Krauss, *A Universe from Nothing*. New York Times, Sunday Book Review, BR20.

Ayer, A. J, & Copleston, F. (1949). Logical positivism—A debate. In M. Diamond & T. Litzenburg (Eds.), *The Logic of God: Theology and Verification* (pp. 98–118). Indianapolis: Bobbs-Merrill.

Bostrom, N. (2003). Are you living in a computer simulation? *Philosophical Quarterly, 53*(211), 243–255.

Demorest, P. B, Pennucci, T., Ransom, S. M, Roberts, M. S. E, & Hessels, J. W. T. (2010, October 28). A two-solar-mass neutron star measured using Shapiro delay. *Nature 467*, 1081–1083.

Deutsch, D. (1985). Quantum theory, the Church-Turing principle and the universal quantum computer. *Proceedings of the Royal Society, Series A, 400*, 97–117.

Fredkin, E. (2003). An introduction to digital philosophy. *International Journal of Theoretical Physics, 42*(2), 189–247.

Gardner, A., & Conlon, J. (2013). Cosmological natural selection and the purpose of the universe. *Complexity, 18*(5), 48–56.

Gardner, A. (2014). Life, the universe and everything. *Biology and Philosophy, 29*, 207–215.

---

[45] McKinsey et al. (1953).
[46] Maudlin (2014).

Horgan, J., & Ellis, G. (2014). Physicist George Ellis knocks physicists for knocking philosophy, falsification, free will. *Scientific American* Blog. Online at https://blogs.scientificamerican.com/cross-check/physicist-geo rge-ellis-knocks-physicists-for-knocking-philosophy-falsification-free-will/. Accessed 14 March 2018.

Krauss, L. (2012). *A Universe from Nothing: Why There Is Something Rather Than Nothing.* New York: Free Press.

Leslie, J. (1989). *Universes.* New York: Routledge.

Linde, A. D. (1986, August). Eternally existing self-reproducing chaotic inflationary universe. *Physics Letters B, 175*(4), 387–502.

Linde, A. D. (1994). The self-reproducing inflationary universe. *Scientific American, 271*(5), 48–55.

Lloyd, S. (2002, May). Computational capacity of the universe. *Physical Review Letters, 88*(23), 237901–5.

Maudlin, T. (2014). *New Foundations for Physical Geometry.* New York: Oxford University Press.

McKinsey, J., Sugar, A., & Suppes, P. (1953). Axiomatic foundations of classical particle mechanics. *Journal of Rational Mechanics and Analysis, 2,* 253–272.

Pigliucci, M. (2013). New atheism and the scientistic turn in the atheism movement. *Midwest Studies in Philosophy, 37*(1), 142–153.

Rees, M. (2001). *Just Six Numbers: The Deep Forces that Shape the Universe.* New York: Basic Books.

Rothman, T., & Ellis, G. (1993). Smolin's natural selection hypothesis. *Quarterly Journal of the Royal Astronomical Society, 34,* 201–212.

Schmidhuber, J. (1997). A computer scientist's view of life, the universe, and everything. In C. Freksa (Ed.), *Foundations of Computer Science: Potential—Theory—Cognition* (pp. 201–208). New York: Springer.

Silk, J. (1997). Holistic cosmology. *Science, 277*(5326), 644.

Smolin, L. (1997). *The Life of the Cosmos.* New York: Oxford University Press.

Smolin, L. (2004). Cosmological natural selection as the explanation for the complexity of the universe. *Physica A, 340,* 705–713.

Steinhardt, P., & Turok, N. (2007). *Endless Universe: Beyond the Big Bang.* New York: Doubleday.

Swinburne, R. (2012). The argument from design. In L. Pojman & M. Rea (Eds.), *Philosophy of Religion* (6th ed., pp. 191–201). Boston: Wadsworth.

Wolfram, S. (2002). *A New Kind of Science.* Champaign, IL: Wolfram Media.

Zeilinger, A. (1999). A foundational principle for quantum mechanics. *Foundations of Physics, 29*(4), 631–643.

5

Cosmology

1   Cosmological Evolution

1.1   The Biocosmic Hypothesis

Since universes like ours are complex, they are improbable. Hence they are not likely to have occurred by chance. Dawkins thinks some cosmic crane best explains their actuality.[1] Although it may not involve natural selection, that crane must start out simple and gradually accumulate complexity. It must be an improbability pump running an entelechy, and following the upwards-pointing arrows in some high-dimensional landscape of cosmic possibilities. If all the books in the modal library were spread out onto some landscape, it would be a Cosmic Mount Improbable, and cosmic evolution would ascend to complex universes along its gradual slopes.

These considerations motivate an *Inductive Argument for Cosmic Evolution*. The first step just recognizes that our universe is complex. The second step is the *evolutionary principle*. It states that if any thing is complex, then it has been generated by some crane which started out

---

[1]GD 185.

© The Author(s) 2020

E. Steinhart, *Believing in Dawkins*,

https://doi.org/10.1007/978-3-030-43052-8_5

simple and climbed up through all the lower levels of complexity. The conclusion is that our universe has been generated by some cosmic crane. This crane contains at least one lineage of increasingly complex universes. Any such lineage serves as a model for the imperative to maximize reflexivity (Sect. 5.3 in Chapter 3). It serves a model for the Dawkinsian theory of the value of vision, and so for his axiology.

The evolutionary principle is justified by empirical evidence (from the evolution of atoms, molecules, organisms, and technologies). But there can't be any empirical evidence that the evolutionary principle applies to our universe. We can't look at our universe from the outside with either eyes or telescopes. At this point, the empirical sciences fail to answer our questions. Fortunately, the evolutionary principle is also justified by mathematics.[2] Mathematical reasoning yields the slow-growth law (Sect. 2.2 in Chapter 3), which states that all possible complexity accumulates gradually. This corresponds to Dennett's principle of the accumulation of design, which states that more complex things mostly copy their designs from simpler antecedents. Mount Improbable is mathematically universal. So the evolutionary principle applies to universes. And while application is not empirical, it remains scientific. The empirical sciences are not the only *natural sciences*. Mathematics is one of the natural sciences. Mathematical evidence for cosmic evolution can be provided by computer simulations.

Spiritual naturalism affirms that our universe was by produced by some cosmic crane, which contains at least one series of increasingly complex universes. The universes in this series appear neither through spontaneous generation nor through blind chance.[3] Perhaps the simplicity of the initial universe in this series entails that it exists necessarily, needing no simpler antecedent. But every later universe needs some cause. Apart from these universes, however, there are no causes. Hence the later universes in any cosmic crane must be created by the earlier universes in that crane. Universes again resemble self-reproducing organisms. This *biocosmic analogy* is very old—it is used by Plato, the

---

[2]Including Bennett's (1988: 1) slow-growth law.
[3]BCD 419; GD 186.

Stoics, Cicero, and Hume.[4] Both Linde and Smolin appealed to it. The biocosmic analogy motivates cosmic evolution like this: (1) Much as organisms beget organisms, so universes beget universes. (2) Much as complex organisms evolve from simpler organisms, so complex universes evolve from simpler universes. (3) Hence our universe evolved from simpler universes. So the biocosmic analogy motivates the *biocosmic hypothesis*: simpler universes evolve into more complex universes. Spiritual naturalism affirms the biocosmic hypothesis. But it needs further clarification.

1.2   Cosmic Spiders Weave Their Webs

The Stoics thought the universe was alive. Following them, Hume discussed the hypothesis that the universe is an organism.[5] Hume used trees, birds, and spiders to illustrate the biocosmic analogy. Focus on the cosmic animals. Just as earthly birds build their nests, or earthly spiders weave their webs, so some cosmic animal built our universe. The cosmic animals are distinct from their universes. They design and create their universes outside of their own bodies. Then they reproduce.

According to the *way of the spiders*, the genome of the cosmic organism contains both the codes for creating its universe and the codes for creating its offspring. On this point, the cosmic organisms resemble earthly organisms. Earthly spiders contain genes that shape their brains, and their brains direct them to weave their webs. Earthly birds likewise contain genetically inspired programs for building their nests. The reproductive machinery is located in the cosmic organism, rather than in its universe. Universes are not organisms—they do not contain their own genetic self-descriptions or reproductive organs. The separation between the cosmic organism and its universe allows some selective machinery to be put into the cosmic organism. Each cosmic spider runs an evolutionary algorithm. It can do some non-random selection. Dawkins

---

[4]Plato (*Timaeus*, 30b–31b); Cicero (*On the Nature of the Gods*, Bk. 2); Hume (1779: part 7). The Stoics used the biocosmic analogy (Hahm 1977: ch. 5).

[5]Hume portrays the universe as an organism (1779: 80–81). It may be a tree (1779: 87) or a web woven by a cosmic spider (1779: 90–91).

described an algorithm for evolving increasingly good spider webs.[6] One algorithm for the evolution of cosmic spiders and their universe-webs looks like this:

Step 0. There exists single initial adult cosmic spider. It evolved from simpler cosmic organisms. This initial spider has a genome. Its genome includes a program for spinning its web. This is its web-code. The initial web-code is simple.

Step 1. Every adult spider begins by running its web-code. By running this web-code, it creates its web. It spins its web out of its own energy; it thus creates its web out of itself rather than out of some external materials. On this analogy, the web is a universe. So the spider stands to its web as a godlike agent to its universe. The act of spinning the universe is the whole history of that universe from start to end.

Step 2. Every spider can make more spiders. It is a self-reproducing organism. It reproduces through *parthenogenesis*. Parthenogenesis is a mechanism through which an unfertilized egg develops—it is essentially self-fertilization. It produces an egg sack containing an enormous number of self-fertilized eggs. Each egg contains a mutated genome of the parent spider; this genome contains some new web-code.

Step 3. The eggs hatch inside of the egg sack. As they hatch, the newborn spiders compete with each other to escape from the egg sack. The less agile spiders perish while the more agile spiders survive. The parent spider observes the sack. If no offspring escape, then it goes back to Step 2 to generate a new sack. If some offspring escape, then they are fit. Their agility also corresponds to superior web-code; hence these offspring will go on to spin more complex webs. Each of these surviving offspring grows up into an adult spider. The algorithm returns to Step 1 for each new adult.

The algorithm for the self-reproducing cosmic spiders is an entelechy. As long as the algorithm runs, it ensures that webs become more complex. It climbs up the Cosmic Mount Improbable. This algorithm involves random variation and non-random selection. So the way of the spiders can explain the complexity of our universe. However, this non-random selection is poorly specified. And the correlation between the

---

[6]CMI ch. 2.

agility of the baby spiders and the quality of their webs is left unclear. So much more work would have to be done before this algorithm can be seriously defended as a strategy for cosmic evolution. Nevertheless, it suggests a strategy: just as earthly organisms climb the biological Mount Improbably through random variation and non-random selection, so cosmic organisms climb the Cosmic Mount Improbable. Since this strategy works for biology, something like it should work for cosmology.

But the way of the spiders still faces problems. The first problem is that the spiders contain unexplained complexity. Some of this complexity comes from the complexity of the cosmic program. Each cosmic organism carries the program for creating its universe in its genome. As their universes gain complexity, the genomes of the cosmic organisms will get bigger and bigger. They will become swollen with cosmic code. If the cosmic organisms could avoid carrying these programs, they could be simpler. And, to avoid the objection from unexplained complexity, the cosmic organisms must be as simple as possible. So the way of the spiders requires further refinement.

1.3 Cosmic Robots Read Their Books

When the biocosmic hypothesis is interpreted in terms of earthly organisms (like birds or spiders), the interpretation incorporates many irrelevant features of earthly biology. The biocosmic hypothesis requires only self-reproducing machines. Dawkins thinks of organisms as survival machines and robotic vehicles for genes.[7] He discusses self-reproducing robots. So the next refinement of the biocosmic hypothesis replaces the cosmic organisms with self-reproducing robots. This is the *way of the robots*.

The life-cycle of each cosmic robot is very simple: (1) read your cosmic program; (2) use your cosmic program to create your universe; (3) reproduce. And since the life-cycle of a cosmic robot is very simple, its internal machinery is also very simple. But what about the cosmic programs?

---

[7]Vehicles (SG *vii*, 19, 69; EP 21–5; BW 1–2). Self-reproducing robots (CMI ch. 9).

They reside outside of the robots. The robots resemble the Platonic demiurge, which uses an external pattern to guide its creation.[8] Of course, these cosmic robots are not identical with any demiurge. The cosmic programs reside in the books in the modal library. Those books contain recipes for constructing possible universes. If they are written in any language at all, it is some logical language like the predicate calculus. They describe purely mathematical structures. They are abstract meanings; they are propositions; they are parts of the Logos. Every book in the modal library has some offspring. The offspring relation organizes the books into a network. This *modal network* is a collection of nodes linked by arrows. The nodes in this network are books while the arrows rise up from lower books to higher books.

The *zero robot* sits at the initial node at the bottom of the modal network. The zero robot is the first cause. It can be referred to as *Alpha*, the origin of all concrete things. This initial node is alive: it is the living root of the world tree. The initial node supports a leaf which catches the energizing light of the sun. A cosmic program is written on this leaf. The zero robot reads this cosmic program. The machinery for reading a cosmic program is pretty simple: it is like the machinery a computer needs to read its digital zeros and ones, or like the machinery a cell needs to read its genes. As it reads its program, the zero robot carries out its instructions. These instructions tell it how to build its universe. These instructions are also simple. The complexity of any universe emerges from the iteration of simple instructions. Every cosmic robot builds its universe outside of itself, like a spider weaves its web, or a bird builds its nest. And it builds its universe out of its own cosmogenic power. It creates its universe *ex nihilo*. Of course, since the zero robot is initial, its work is indeed simple. The initial leaf is blank—it contains the null word. So the zero robot does not build any universe at all.

Now the zero robot gets ready to reproduce. Before it makes its offspring, it looks at the arrows rising up out of the initial node. Following the depiction of the modal library in Sect. 4 (Chapter 4), twenty six arrows rise up out of the initial node. Each arrow rises to a node that contains a one-letter word. Following the depiction of the

---

[8]Plato (*Timaeus*, 27d–29b).

treasury in Sect. 5 (in Chapter 4), twenty four of these arrows are dead. These *dead arrows* do not rise to any meaningful words in the English language—they rise up to dead leaves. These dead leaves are not illuminated by the sun. If these leaves are books, they are dark books. But *living arrows* rise from words to more meaningful words. The null word is surpassed by exactly two more meaningful words, namely, the words *a* and *i*. So exactly two living arrows rise from the null word to the words *a* and *i*. Just as a computer detects the difference between zero and one, every cosmic robot detects the difference between dead arrows and live arrows. The zero robot does not make offspring for the dead arrows. It makes offspring only for the living arrows. It therefore begets two baby robots. They are the robots in the first generation. If any arrow is alive, then it is a branch in the world tree. It is a branch that carries living sap from some lower to higher leaf. The living nodes and branches make up the world tree. It is a proper part of the modal network.

Focus on the first baby robot. It climbs up its branch from the initial node to one of the first-generation nodes in the world tree. A leaf sprouts from this node. This first-generation leaf contains the cosmic word with the single letter *a*. This word is the cosmic form for the first universe. The first robot uses that word to make the first universe. It makes this universe outside of itself using only its own power. This first universe is the simplest of all universes. And now the first robot gets ready to reproduce. It sees that many dead arrows rise up out of its node. The arrow from *a* to *ak* is dead. But many living arrows rise up out of its node. The arrows from *a* to *an* and *as* and *at* are living branches. And there are other living branches. For each living branch, the first robot creates an offspring. So it makes many second-generation robots.

Each second-generation robot climbs up its branch to some second-generation node. Each node contains a leaf with a two-letter word, the cosmic form for some second-generation universe. These robots use their words to make their slightly more complex universes. As this process repeats, robots beget robots in the next generation. The new robots climb up the living branches in the world tree and read their leaves. Since higher leaves contain more and more information, the higher universes on the world tree are more complex. The swarm of robots runs an entelechy—as the swarm of cosmic robots climbs the world tree, it makes a *cosmic arrow*. As they make their universes, these robots are doing design work. But these robots are mindless machines.

The cosmic robots do not compete with each other. After a robot is hatched, it just climbs up its own branch by itself. It has no predators or enemies. Nevertheless, something *analogous* to natural selection acts on these robots. It acts through the arrows. The arrows compete with each other—they struggle with each other virtually. The arrows that go to more meaningful nodes win this virtual competition. The arrows that rise to the more meaningful nodes are the fittest arrows. These fittest arrows survive while the others die. So this is the survival of the fittest. The fittest arrows are selected while the less fit are rejected. This is *arrow selection*. Arrows are selected in the modal network much like they are selected by the maximum entropy production principle (Sect. 1.3 in Chapter 3). The robots sense the difference between living and dead arrows. They only send their offspring up the living arrows, which are living branches. So the modal library exercises a *selective* effect on the robots and on their nests. The generations of robots rise up along and only along the branches that increase meaning.

As the swarm of cosmic robots climbs up the world tree, that tree becomes decorated with universes. More precisely, every living node bears a universe. These living nodes correspond to the books *in the treasury*. The dead leaves, linked by dead branches, are not decorated. These dead leaves are dark books. Poetically speaking, they make up the *night*. So the modal network contains an internal tree whose leaves are all living. While the dead leaves lie in darkness, the living leaves are illuminated by the sun. The tree of living leaves is the world tree. The universes in the world tree shine with solar light. Poetically speaking, the shining leaves on the world tree make up the *day*. Figure 1 illustrates a part of this world tree. The black dots are the cosmic organisms. They are surrounded by their universes, which are the concentric circles. More circles indicate more complexity. So universes higher in the tree are more complex.

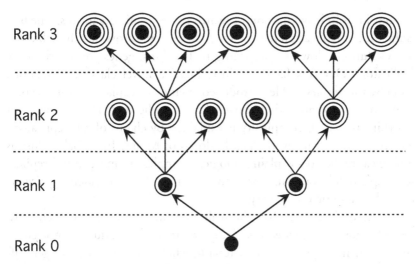

Fig. 1 A phylogenetic tree of evolving universes

2   Cosmic Replicators

2.1   The Science of Animats

The books in the treasury describe possible universes. But these books are just inert mathematical objects. They are motionless systems of abstract equations or lifeless computer programs. When the physicist Stephen Hawking thought about the equations in the final physical Theory of Everything, he asked: "What is it that breathes fire into the equations and makes a universe for them to describe?"[9] So what breathes fire into these books and brings them to life? What are these cosmic dragons? According to the biocosmic hypothesis, the cosmic organisms breathe fire into these books. Since they create universes, these dragons resemble old-fashioned creator gods. However, since those old gods did not evolve, they were supernatural. Dawkins rejects all supernatural gods.[10] Since the cosmic organisms do evolve, they are not supernatural.

---

[9]Hawking (1988: 174).
[10]GD 31–41.

Four ideas in Dawkins point towards these cosmic organisms. The first is the idea of replication itself. Genes are the replicators that drive biological evolution. However, according to Dawkins, genes are not the only replicators. Dawkins famously introduced a novel class of replicators, namely, the *memes*.[11] He introduced memes to explain cultural evolution—he posited an entirely novel class of replicators. The core idea for Dawkins is that all complexity is explained by chains of self-replicators, which slowly swarm up the gentle slopes of Mount Improbable. Just as biological replicators explain biological complexity, and cultural replicators explain cultural complexity, so there must be some *cosmic replicators* to explain cosmic complexity.

The second idea involves the scale of replication, or, more generally, reproduction. At the lowest scale, genes are self-reproducing molecules. At intermediate scales, from bacteria to whales, organisms are larger self-reproducing structures. At a higher scale, Dawkins suggests that entire planets can be thought of as quasi-biological replicators. The mathematician John von Neumann described self-reproducing machines. Following von Neumann, Dawkins briefly talks about self-reproducing robots.[12] He talks about *von Neumann probes*, which carry the seeds of life across the sky.[13] As they encounter planets, they seed them with DNA or other chemicals necessary for life. Hume pictured solar systems as trees which reproduce by sending comets to other solar systems, and the probes are like the seeds of these Humean trees. So if molecules, organisms, and planets can be thought of as self-reproducers, the final and highest scale is cosmic: entire universes are also self-reproducers. More accurately, there must be some replicators at the cosmic level. They contain the machinery for cosmic begetting.

The third idea comes from universal Darwinism.[14] For Dawkins, universal Darwinism means that, if complex life exists on other planets, then it evolved there by natural selection. Von Neumann probes might have dropped off simple organisms, but only natural selection can

---

[11]SG ch. 11; EP ch. 6; GD ch. 5.
[12]CMI 276–81.
[13]ROE 160–1.
[14]BW 288; ROE 151–8; ADC ch. 2.2; SITS 119–50, 191–2.

produce complexity. Evolution by natural selection depends on chemical and physical principles. It depends on thermodynamic principles, and therefore on entropy and information. Beyond universal Darwinism, there lies *universal evolution*. Any evolutionary theory that aspires to genuine universality must also aspire to ultimacy. The evolution of cosmic replicators is ultimate. The cosmic replicators make variant copies of themselves. They have descent with modification. The arrows on which they reproduce are *entangled*: they struggle against each other in a virtual competition based on value. The arrows that increase value are fitter, and the fittest arrows survive. So arrow selection plays the role of natural selection for cosmic replicators. They slowly evolve from simple to complex. The logical demand for ultimacy motivates the addition of the cosmic replicators to Dawkinsian metaphysics.

The fourth idea comes from the complexity of our universe.[15] Cranes are needed to explain the complexities of atoms, molecules, organisms, technologies, and universes. Organisms and artifacts come from cranes using replicators. It has already been argued that: (1) Our universe is extremely complex. (2) The *replication principle* (Sect. 4.2 in Chapter 2) says that if any object is extremely complex, then it was evolved by some crane running an entelechy using replicators. (3) Therefore, our universe was evolved by some crane running an entelechy using replicators. But any such computation is a big arrow, one which is composed of cosmic replicators moving along little arrows.

These four ideas motivate the addition of cosmic replicators to our Dawkinsian metaphysics. Here believing in Dawkins means believing in evolution at every scale. It means *building on Dawkins* to define evolution at the cosmic scale. It means going beyond his incomplete texts to build the Sanctuary for Spiritual Naturalists. Thus I will add these cosmic replicators. And just as Dawkins made a new word for his cultural replicators, so I will make a new word for these cosmic replicators. The term *animat* has been used in artificial life to describe digital organisms. Since these cosmic replicators resemble digital organisms, I will refer to them as animats.

---

[15]GD 146, 184–9.

It might be objected that it is not *scientific* to postulate these animats.[16] One reply is that they are motivated in part by the biocosmic analogy. Reasoning by analogy is a legitimate form of scientific reasoning. But the objection can go deeper: these animats cannot be confirmed or disconfirmed by any observations. Since they are not linked to empirical evidence, they are not scientific. The reply distinguishes between *direct* and *indirect* observation. The animats cannot be confirmed or disconfirmed by direct observation. But Dawkins does not restrict science to direct observation. Dawkins often uses computer models. To illustrate the power of evolutionary cranes, he famously wrote the weasel program and the biomorphs program. Dawkins used simulations to provide evidence for the evolution of spider webs and eyes. Dawkins is no positivist—he says science allows indirect testing. And the theory of evolving animats can be indirectly tested by computer simulations. We can simulate evolving animats, and see whether they produce things like our universe. So the theory of the evolving animats satisfies the conception of the scientific method laid out by Dawkins himself. On his view of science, the theory of evolving animats is a scientific hypothesis.

It may be objected that the theory of evolving animats is too speculative.[17] The reply is that Dawkins encourages speculation about origins. He encourages us to speculate about the origins of language and the origins of life. So when it comes to cosmic origins, we can also speculate. Nevertheless, we must speculate *responsibly*—we need to build mathematically precise models of evolving animats, and we need to verify that they can generate things like our universe. Models of evolving animats have been built. Cellular automata are universes with very simple space-times and causal laws. Computers have used evolutionary algorithms to design cellular automata which exhibit self-replicating patterns, which perform global information processing, which support self-moving patterns, and which exhibit Turing universality.[18] But much more research remains to be done here. Cellular automata are much

---

[16]Programs for biomorphs (BW ch. 3; EE); spider webs (CMI ch. 2); eyes (CMI ch. 5). The scientific method (MR 15–18).

[17]Origins of language (BCD 290) and life (BW 147).

[18]Steinhart (2014: sec. 81).

simpler than our universe. Doing this research is a part of spiritual naturalism. Dawkinsian principles encourage it.

2.2   Enfolding Cosmic Origami

The animats help to spell out the biocosmic hypothesis, but it would be wrong to take them too literally. They are mythical creatures. Behind this mythology is a detailed mathematical theory of cosmic complexity. However, since its technicalities are tedious, the animatic mythology makes for a better story. The animats come into existence by natural laws. At this point, two laws are relevant. The *initial law for animats* says that the initial book in the treasury is realized by an initial animat. It is the ancestor of all the others. The *successor law for animats* defines the way the animats produce their offspring (that is, their successors). It says these animats reproduce according to some evolutionary algorithm. These laws for animats are theorems that follow from ultimate axioms. These laws are fleshed out in this evolutionary algorithm:

Step 0. There exists a single initial animat. It is the cosmic Alpha. Alpha sits on the first book in the treasury. The first book in the treasury is also the first book in the modal library. It describes the simplest universe, which is the empty universe. So the initial book just contains the statement that describes the empty set.

Step 1. Every animat reads some book in the modal library. It reads its book like a cell reads its genome or like a computer reads its program. Perhaps the books are inside the animats like some genome is inside of some cell. Or perhaps the books are outside the animats. To clarify how the animats work, it will be helpful to use a simple example. All the ideas in this simple case can be extended to cases of any complexity. The *flip-flop book* sits in the modal library. This book contains a list of four statements written in some purely logical language. For precision, these statements can be written in the language of set-theory (that is, the predicate calculus plus the membership sign). But in English these four statements look like this: (1) It has two spatial points. (2) Each point is a temporal successor of the other, so that time is circular. (3) Each point has an energy value either zero or one. (4) The energy value changes in time. So if one point has the value zero, then the other has the value one. Of course, these statements aren't really statements written in some language. We need to think more Platonically. More Platonically, each statement in the flip-flop book is a proposition. So the flip-flop book is a list of little propositions; but a list of little propositions adds up to one big proposition. As such it is a purely logical object. It is a part of the Logos.

Every book in the modal library describes some purely mathematical *cosmos*. So the flip-flop book describes the purely mathematical *flip-flop cosmos*. Its spatial points are purely mathematical objects. The temporal relation is a purely mathematical relation. Its energy values are also purely mathematical objects. So the flip-flop book is true at the flip-flop cosmos and that cosmos is a model of that book. For the sake of precision, we pictured these Platonic books as written in the language of set-theory. But if some Platonic book uses the language of set-theory to define its cosmos, then that cosmos is ultimately just a set. The Platonic flip-flop book specifies or describes a set-theoretic structure.[19] It describes some pure set that exists in the world of sets. Thus each book in the modal library resembles an axiomatic definition of a set. There is no need to get into the details of set theory here. All we need to say about sets is that *sets are dots*. Some sets are members of other sets. If this lower set is a member of that higher set, then there is a membership arrow from this lower dot to that higher dot. A set-theoretic structure is a network of dots linked by arrows. It is a connect-the-dots network.

The Platonic books in the modal library exist eternally and necessarily. Some books in this library are read by animats, which are little Platonic robots. These Platonic robots are not physical. They are just dynamical operators in the world of sets. They are pictured here in poetic terms as the animats. For the sake of illustration only, they are described as being literate. But they are really just mindless mathematical agents. The Platonic books resemble paper books that contain internal folded origami structures. When you open the book, the origami structure

---

[19]Each object in the flip-flop cosmos has the form (*type, instance*). The spatial points are (0, 0) and (0, 1). The temporal relations are ((0, 0), (0, 1)) and ((0, 1), (0, 0)). The energy values are (1, 0) and (1, 1). The field is ((0, 0), (1, 0)) and ((0, 1), (1, 1)). Any pair (x, y) is the set $\{\{x\}, \{x, y\}\}$. So the flip-flop cosmos is a set.

unfolds over the book. So when an animat opens its book, the cosmos described in that book unfolds above the book. The origami cosmos is a connect-the-dots network that unfolds over the book.

An animat reads its book by opening its book. As it reads its book, its origami cosmos unfolds above that book. When the flip-flop animat reads the flip-flop book, the flip-flop cosmos unfolds over that book. The animat gets to work on this cosmos. It acts on the set-theoretic cosmos specified by or enfolded in the book. As the animat acts on that origami cosmos, it transfers actuality to it, and thus actualizes it. If some animat actualizes some cosmos, then it creates a *physical universe* which instantiates that cosmos. A physical instance of the mathematical cosmos comes into existence. However, physical instantiation *does not create any new entities*. Spiritual naturalists reject every dualism which opposes the abstract to the concrete. All physical things are strictly identical with mathematical objects—they are just sets. Every physical point in physical space-time is identical with a mathematical point in mathematical space-time. Physicality resembles paint. When an animat uses its mathematical cosmos to create its physical universe, it paints physicality onto that cosmos.

Physical instantiation *only introduces new relations*. The relations in the universe run parallel to the membership relations in the cosmos. Every set is a dot that sits at the top of some membership graph. Its graph is a connect-the-dots network. The links in the network are instances of the membership relation. Animats actualize sets by decorating their membership graphs with *physical string*. If an animat actualizes a set, then it ties a loop of physical string to that set. This loop of string is an instance of the *is-present-to* relation. Thus every actualized set is present to itself. If some animat actualizes a set, then it also actualizes every member of that set. If it actualizes a set and a member of that set, then it ties a length of physical string from the member to the set. This length of physical string is also an instance of the *is-present-to* relation. Thus physical sets are present to each other. Physical members are present to the physical sets that contain them. These loops and lengths of string *supervene* on sets and membership arrows. The physical universe supervenes on its set-theoretic cosmos. So when an animat covers its cosmos with physical paint, it decorates its cosmos with colorful physical string. Hence the *flip-flop universe* emerges as the animat decorates the flip-flop cosmos. The animats are like little Platonic demiurges making sculptures out of string. They resemble the Humean cosmic spiders, as they weave their webs out of the powers in their own bodies.

The axioms in any book specify their cosmos in an orderly way. More generally, they are like the instructions in some computer program. By following those instructions, an animat actualizes its cosmos in an orderly way. It decorates its cosmos in an orderly way. This orderliness generates a flow of presence through the timeless Platonic model of the book, and this flow of presence generates physical time in model. This flow of presence is the Dawkinsian *spotlight* (Sect. 1.4 in Chapter 8). Since that flow unfolds in accordance with the causal laws in the book, that time unfolds lawfully. The flow of presence through the model generates a *physical perspective* within the cosmos. This physical perspective is an actual universe. For anything in that actual universe, time appears to flow from past to future. So if some Platonic book describes a cosmos which contains mathematical particles moving through mathematical space-time, and if some animat decorates that book with physical string, then it generates an actual universe in which material things move energetically through space from past to future.

2.3   Making Cosmic Babies

After an animat creates its physical universe, it moves to the second step in its evolutionary algorithm. *Step 2.* Every book in the modal library sprouts a quiver of arrows. Each arrow runs from that book to one of its offspring. These arrows now fall into three classes (Sect. 5.1 in Chapter 4): if an arrow runs to a less valuable offspring, then it is a *downgrade*; if it runs to an equally valuable offspring, it is an *equigrade*; if it runs to a more valuable offspring, it is an *upgrade*. Each arrow is an instance of the relations between books in the *modal library*. Some but not all of the books in the modal library are also in the treasury. Each upgrade runs from book to book *in the treasury*. The laws for books ensure that every book sprouts at least one upgrade in its quiver. You can picture these types of arrows in terms of slopes: upgrades slope upward on the Cosmic Mount Improbable; equigrades have no slope; downgrades slope down.

The arrows that sprout from some book all have probabilities. Since all these arrows sprout from a common origin, their probabilities are all entangled. The arrows compete with each other for probability. This competition corresponds to natural selection in biological evolution. The fitness of any arrow is its increase in value. The downgrades have negative fitness. They lose out in the competition for probability; hence they end up with zero probability. The equigrades have no fitness. They too have zero probability. The upgrades have positive fitness. Every upgrade wins in the probability-competition. If an animat were only able to produce one offspring, then only one upgrade would win that competition. However, since animats can produce many offspring, and since they strive to maximize value, every upgrade wins maximal probability. So the probability of each upgrade is one. Since the probabilities of downgrades and equigrades are zero, while those of upgrades are one, these probabilities are skewed very far from random. But what explains this skew? The story about entanglement and competition merely describes a mechanism. An explanation will be provided later (Sects. 3 and 4 in Chapter 6).

The arrows that sprout from books define the reproductive opportunities for animats. Each arrow offers more or less *resistance* to animatic reproduction. The resistance on any arrow is the inverse of its probability. On the one hand, arrows with probability zero have resistance one. If an arrow has resistance one, then it blocks all reproduction. Hence animatic reproduction is impossible on downgrades and equigrades. On the other hand, arrows with probability one have resistance zero. If an arrow has zero resistance, progress along that arrow is guaranteed. Hence animatic reproduction is necessary along upgrades. Animats reproduce along and only along their upgrades. Every animat *rejects* arrows with zero probability and *selects* arrows with unit probabilities. It selects exactly those arrows which run to better offspring of its own book. It selects improvements. This is *arrow selection*. If an animat sits on some book on some floor of the treasury, then it selects all and only the arrows pointing to best books on the next floor. Animatic selection stays *in the treasury*—it never selects dark books. Since it is rational to select the best and reject the rest, every animat makes its selection rationally.

Step 3. Every animat begets embryonic offspring along its upgrades. It tells its offspring to climb up their arrows to better books. It tells them to run the programs in those books. After it gives its embryos their basic instructions, the parent animat has completed its tasks—it stops running, it dies. But its embryos now start climbing up their arrows towards some books on the next higher floor of the modal library. As they climb up their arrows, they mature into adult animats. Their growth is complete when they sit on the book at the end of their arrows. Now this algorithm repeats at Step 1 above. By running this algorithm, the animats climb ever higher in the treasury.

This evolutionary algorithm is an entelechy. As the animats run it, they produce a genealogical tree. Each node in this tree is an animat. It actualizes the cosmic form in some book in the modal library. Since each animat has some offspring, each animat is linked by branches to some later animats on higher levels of the tree. As the levels go higher, the animats get fitter. They become more complex and therefore more intrinsically valuable. As they become more valuable, they read more complex books, and they generate more complex universes. So by running this swarm-entelechy, the animats do cosmic design work. They bring some of the books in the modal library to life. But they do not bring all of those books to life. The tree of animats is a proper part of the modal library. The part inhabited by animats is the treasury.

As animats beget animats, the swarm of animats flows across the landscape of the Cosmic Mount Improbable. It climbs ever higher into the treasury, like a swarm of ants climbing a tree, or like a murmuration of starlings soaring ever higher into the sky. By running its entelechy, the swarm of animats climbs towards the ideal peak on its Mount Improbable. The swarm of animats is the cosmic arrow, which points to its own omega point. With each generation, the swarm of animats actualizes more and more of the treasury. Its omega point is therefore the entire treasury. Its omega point is its ideal finality—its omega point is its ecstasy. The swarm of animats reduces its distance to its ecstasy in a purely iterative way, as repeatedly doubling a number drives it to an infinite limit. So the swarm of animats is teleonomic. However, since the swarm itself has no mentality, it has no teleology—its ecstasy is not some goal foreseen.

The swarm of animats is a massively parallel computation distributed across a plurality of universes. It runs an entelechy. Since every process that runs an entelechy is a crane, the swarm of animats is the *cosmic crane*. The cosmic crane blindly does design work—it blindly maximizes cosmic intrinsic value, that is, cosmic complexity. It maximizes value through iterated arrow selection, that is, by blindly running its evolutionary algorithm. The cosmic crane is *not* utilitarian: it does not act according to the greatest happiness principle, it has no interest in either happiness or suffering. It is concerned with the intrinsic values of universes, and with maximizing cosmic *arete*. So, even though it does not care about happiness or suffering, the cosmic crane is providential and benevolent. If the Platonists are right, then the duty or imperative to maximize reflexivity drives the cosmic crane. It therefore seeks to self-organize so that it maximally reflects the Good. It resembles a rational value-driven agent; it *looks like* a person—but it is *not* a person. To satisfy our hyper-active animacy detectors, we often identify cranes with deities.[20] A process theologian might try to identify the God with the cosmic crane. However, that identification is an example of religious hijacking. It is an example of theological doublespeak, which pollutes an ideal metaphysical structure with the human-all-too-human pettiness and ugliness of religion.[21]

The animats model the axioms of our deontic logic. The first axiom of our deontic logic states that if any agent can maximize value, then it ought to maximize it. Since the animats can maximize value, they have the duty to maximize it. The second axiom of our deontic logic states that if any agent is rational, then it strives to do its duty. Since the animats select their reproductive arrows rationally, they are rational agents. They strive to do their duties. Every animat strives to maximize reflexivity. And every animat partly succeeds. Every animat does enough of its duty to produce its successors. Every progression of animats does enough of its duty to produce its limits. And while every animat fails to do its whole duty, the swarm of animats succeeds. Every animat participates in the

---

[20]BW 43.
[21]UR 117–8; ADC 58; GD 51, 270; SITS 1; FH 99.

ideal animat, which is an unsurpassable tree of surpassable animats. The ideal animat completely does its duty. It builds the world tree, which is nature. Since the ideal animat completely does its duty, it is holy.

3   The Spawn of Aesthetic Engines

3.1   At the Bottom of the Sacred Mountain

To illustrate the evolution of animatic complexity, it will be useful to describe a single large stream of animatic life. This stream rises up through a single bundle of branches in the world tree. This is not the only stream of animatic life—other streams run along other bundles of branches. This is only a single illustration, which evolves through our universe. It starts with Alpha. Alpha runs an utterly simple universe. Since it is simple, it is empty. It has no internal physical structure. The descendents of Alpha evolve into the *mereological animats*. They build *mereological universes*, which contain only simple part-whole structures. The mereological animats evolve into the *geometrical animats*. These run universes with simple spatio-temporal structures. They build universes with one-dimensional timelines. These evolve into universes containing both time and space. The geometrical animats evolve into *pastoral animats*. These inherit the space-time structures of their ancestors. They elaborate those space-times by adding simple energy fields. Each point either has or lacks energy. The pastoral animats evolve into the *Wolfram animats*, which build the first cellular automata. They are named after Stephen Wolfram, who did pioneering work on cellular automata.[22]

The animat Beta appears in some high generation. It sits on some book on some high floor in the treasury. It is the first Wolfram animat. As it interprets its book, it produces a 1D cellular automaton. The Beta universe has one-dimension of space. It is a finitely long line of discrete spatial points. Each point is a cell. The line wraps around: its last point

---

[22]Wolfram (2002).

is linked to its first point, so that it forms a circle. Each point has the energy values zero or one. Each point has a right neighbor and a left neighbor. The next value of each point depends on its own value and the current values of its neighbors. The next value of each point is defined by a causal law. The causal law for Beta looks like this: if your right neighbor has value zero, and your current value is one, and your left neighbor is zero, then your next value is zero; otherwise, your value is one.

There are many ways to make Beta more complex. Beta has many offspring; and its offspring have offspring; and so it goes. The Wolfram animats evolve into the *Conway animats*. These are named after the mathematician John Conway, who invented the game of life.[23] Every game of life is cellular automaton with two dimensions of space and one of time. The space-time in any game of life supports an energy field, in which patterns of energy change through time. They can be stable; they can run through cycles; they can move. These patterns are the first material things. The primitive Conway animats build games of life with simple internal physics. These evolve into animats whose games of life contain more complex internal patterns. These patterns rise through all finite degrees of mechanical complexity. The Conway animats evolve into the *Turing animats*. These run games of life which contain Turing-universal computers.[24]

The Turing animats evolve into the *Hegelian animats*, so-called because Hegel was obsessed with self-representative systems. Since any universal Turing machine can run any game of life, and since the game of life can run universal Turing machines, there are games of life which are self-simulating. So these Hegelian animats run Turing machines that simulate games of life. Since their Hegelian universes are only finitely complex, they can only simulate smaller parts of their own structures. This is finite self-representation—it is the first appearance of self-reflection. These Hegelian universes contain machines which encode propositions about themselves. This self-encoding will evolve into conscious self-representation and ethical self-legislation. These Hegelian animats evolve into the *von Neumann animats*, so-called because the

---

[23]Poundstone (1985).
[24]Rendell (2002).

mathematician John von Neumann did the first work on the logical foundations of self-replication. The von Neumann animats run games of life which contain self-reproducing patterns.[25] These are the simplest universes containing Total Replication of Instruction Programs.[26] This self-replication will evolve into biological self-reproduction. From Alpha to these von Neumann animats, the duty to maximize reflexivity is hard at work.

3.2   Approaching Our Universe

The generations of animats continue to climb the Cosmic Mount Improbable. As they do, they generate ever more complex universes. Two-dimensional cellular automata gain a third dimension. Their internal patterns become massive particles, which start to interact according to higher level laws, like the law of gravitational attraction. Thus forces like gravity emerge. So the earlier animats evolve into the first *Newtonian animats*, which run universes with finite approximations to Newtonian physics. As animats beget animats, their universes move ever closer to infinite precision. Infinite progressions of finitely complex animats beget infinitely complex animats. These run universes with infinitely divisible space-times and continuous dynamics. These infinite precision Newtonian animats evolve into the *Maxwellian animats*. They run universes which have both gravity and electro-magnetism. They evolve into the first *quantum mechanical animats*, which run quantum mechanical universes. The Maxwellian animats evolve into the *Einsteinian animats*. These run universes with quantum mechanics and general relativity. They run primitive quantum relativistic universes. These primitive quantum relativistic universes begin to resemble our universe.

The animat which makes the first quantum relativistic universe is the root of a great branching tree. The nodes in this tree are stratified into generations of animats, which make generations of universes. These generations can be stratified by their evolutionary histories. The most

---

[25]Poundstone (1985 ch. 12).
[26]CMI ch. 9.

primitive universes are *particulate universes*. After the big bang, energy evolves into particles; but these universes soon collapse back into noise. The particulate universes are surpassed by *atomic universes*. After the big bang, energy in atomic universes evolves into particles, and then into atomic wholes composed of particle parts. Atomic universes are surpassed by *molecular universes*. Energy in those universes evolves into particles, then into atoms, then into molecules composed of atoms. Molecular universes are surpassed by *chemical universes*. Energy evolves in chemical universes through the levels of particles, atoms, molecules, and then systems of interacting molecules such as autocatalytic networks. Chemical universes are surpassed by *biological universes*. Energy evolves in them through all lower levels and then into living organisms. There are many generations of biological universes.

The biological universes are stratified into levels of biological complexity. As universes gain biological complexity, more and more of their planets support ecosystems which rise ever higher through the degrees of organism complexity. On more and more planets, the trees of life rise higher and higher. They raise their lovely branches to the sun. Some of these universes inhabit a series which leads to our universe. On this series, the lowest universes are the *bacterial universes*. They contain planets covered with writhing networks of bacterial cells. These are surpassed by the *eukaryotic universes*, in which life evolves through bacteria and eukaryotes. These are surpassed by the *spongiform universes*, in which life evolves through bacteria, eukaryotes, and sponges. These are surpassed by the *amphibian universes*, by the *reptilian universes*, by the *mammalian universes*, by the *hominid universes*, and eventually by the *human universes*. These human universes contain people like us. Our universe is one of these human universes—but there are myriad other human universes.

As these universes gain complexity, the duty to maximize reflexivity entails that they become more self-representative. The duty to maximize reflexivity entails that new things reflect their histories—hence new things are made by copying the structures of old things. The duty to maximize reflexivity entails the emergence of material structures which carry information about their pasts. It entails the emergence of replicators. It entails that the genomes of new organisms are made by mostly copying the genomes of their parents. Hence organisms contain ever more detailed representations of their own genealogies. Likewise the duty to maximize reflexivity entails that brains emerge which can reflect the universe by representing parts of the universe. It entails that brains emerge which can reflect the universe by partly representing the whole universe. It entails the emergence of computers which can simulate the universe in which they emerge. It entails the emergence of mirrors nested in mirrors.

3.3   Mirrors Nested in Mirrors

As universes gain complexity, structural principles begin to govern the changes from earlier to later universes. The most important structural principle is Dennett's *principle of the accumulation of design*. It says that new designs are mostly copied from old designs.[27] Hence design accumulates. This principle can be illustrated by considering the successors of universes. The successor relation on universes parallels the offspring relation on animats. If an animat runs a universe, then each offspring of that animat runs some successor of that universe. The principle of the accumulation of design implies that the structure of each successor universe is largely copied from the structure of its predecessor. But this means that most of the things in any complex universe have very similar counterparts in the successors of that universe.

Our universe will be followed by myriad successors. Each of those successors will inherit almost all of its structure from our universe. It will *almost exactly* reproduce the structure of our universe. But it will also add new complexity and value by extending or elaborating that inherited structure. Things get copied from universe to universe like genes get copied from cell to cell. Every successor of our universe contains a planet very much like earth. It contains a future counterpart of our present earth. On each future earth, evolution unfolds in *almost exactly* the same way as it unfolded on our present earth. Each future earth contains a counterpart of every organism that ever appeared on our present earth.

---

[27]Dennett (1995: 72).

It contains counterparts of your parents and a counterpart of you (Sect. 5 in Chapter 8). The principle of the accumulation of design implies that the differences are very subtle. It also implies that these differences start to emerge at the top. The bacteria are exactly replicated; the plants exactly replicated; the animals exactly replicated; the hominids exactly replicated. The subtle divergences begin to appear at the highest heights of complexity. If humans are the most complex things that will appear on our earth, then the lives of our future counterparts will differ in subtle ways from our lives. The lives of your successors will contain slightly more value than your life.

Of course, our universe is not the best of all possible universes—it is not the end of cosmic evolution. There are many ways to improve our universe. For every way that our universe can be improved, it will be surpassed by some successor universe which is improved in that way. Our universe sits at the root of an endlessly ramified tree of ever better universes. If the Platonic duty to maximize reflexivity drives the evolution of universes, then the descendents of our universe will be increasingly self-reflective. Their internal evolutionary histories will rise ever higher in complexity, intrinsic value, and beauty. Our universe will be surpassed by *transhuman universes*. The history of the first transhumanist universe almost exactly replicates the history of our universe up to the industrial age (that is, the time when humans start to apply scientific technologies to transform their own bodies). But then their histories diverge. By employing scientific technologies, the earthly humans transform their bodies into transhumans and then into superhumans. They become increasingly godlike and goddesslike. According to this evolutionary theory, you will be surpassed by godlike counterparts. They leave earth to colonize the solar system. They build computers on the scale of planets.

The transhumanist universes evolve into *Kurzweil universes*. Our superhuman descendents colonize the galaxy. As these galactic universes surpass themselves, they ever more closely approximate the ideal histories described by Kurzweil. They weave all the matter in the Milky Way into a godlike supercomputer. More and more of the dumb matter in the universe is saturated with intelligence. The universe begins to wake up. Each next universe makes further progress towards the Kurzweil omega point. It makes more progress towards an infinite deontic ideal. The categorical imperative holds ever more accurately at these increasingly excellent universes. The agents in these universes are increasingly likely to do their duties—to do what they ought to do. And eventually infinitely complex universes emerge which contain infinite computers. These evolve into *Royce universes*, which contain infinitely self-representative holographic computers. These Royce universes resemble the Plotinian divine mind.

As these universes gain complexity and value, the duty to maximize reflexivity entails that they will contain increasingly accurate self-simulations. The descendents of our universe will contain better simulations. They will ascend through all the finite degrees of self-representation. As universes gain complexity, they contain simulations ever more deeply nested in simulations, mirrors fractally nested in mirrors. And just as it is possible to simulate individual universes, so it is possible to simulate the evolution of universes. Universes beyond ours contain simulations of the process of cosmic evolution which led to our universe. Those simulations start with Alpha and work their ways up to our universe and beyond. So the world tree is a *self-representative system*. Every earlier part of the world tree is simulated in some later part of that tree. This is *reflexivity*: the world tree reflects itself back to itself. It looks back into itself. It loops back on itself in a way that resembles self-consciousness. Of course, the world tree is not a mind. But it is sufficiently mindlike that somebody could mistake it for a mind. The world tree is a structure which grows towards the Good by maximizing reflexivity.

4   The Naturalized Cosmic Design Arguments

4.1   Arguments for a Divine Mind

Cicero gave a Stoic design argument in *On the Nature of the Gods*.[28] He refers to a ship: just as ships are so complex that they require intelligent designers, so our universe is so complex that it requires an intelligent designer. Cicero inspired Hume to write *The Dialogues Concerning Natural Religion*. Hume devotes much of his *Dialogues* to criticizing design arguments. Before he criticizes it, Hume states his version of the Stoic *Cosmic Design Argument*.[29] It goes like this: (1) The universe resembles a machine. (2) If this thing resembles that thing, then the cause of this thing resembles the cause of that thing. (3) Therefore, the cause of the universe resembles the cause of a machine. (4) The cause of a machine is a human mind. (5) Therefore, the cause of the universe resembles a human mind. (6) Just as the universe is greater than any human machine, so the universe-causing mind is greater than any human mind. (7) So, there exists some divine mind which designed our universe. This divine mind is God.

The theistic *Fine Tuning Argument* is another cosmic design argument. It goes like this: (1) Our universe has many virtues (Sect. 1.2 in Chapter 4). These include the facts that it is lawfully ordered, that its laws have certain mathematical forms, and that its fundamental physical constants have certain values. These virtues imply that our universe started in an extremely low entropy state, that ordered flow produces entropy faster than unordered flow, and that the universe moils to maximize entropy production. Therefore, our universe moils to increase complexity. Since complexity is intrinsically valuable, our universe moils to produce value. (2) These virtues are unlikely because, if they were even slightly different, then our universe would not moil to increase value. It would not moil to generate life, intelligence, and rationality. It would not moil to maximize reflexivity. (3) Since these virtues are

---

[28]Cicero (*On the Nature of the Gods*, 2.89–90).
[29]Hume (1779: 53).

unlikely, they cry out for explanation. These virtues raise the first Leibnizian question: why is our universe the way that it is? (4) According to the theists, the best explanation asserts that there exists a God whose perfect nature explains the virtues. By inference to the best explanation, God exists. This God finely tunes the features of our universe into the virtues needed for rational animals. Of course, finely tuning for rational animals requires finely tuning for reflexivity.

4.2   A Long Line of Blind Worldmakers

When Hume was writing his *Dialogues Concerning Natural Religion*, the great sailing ships were among the most complex machines in existence. Hume argued that ships do not really have intelligent designers. Their designs emerge through a long process of technological evolution. But technological evolution involves a long series of mostly blind engineers.[30] Hume then applied his reasoning about ships to the whole universe. Just as complex ships evolve through the gradual improvement in the art of ship-building, so complex universes evolve through the gradual improvement in "the art of world-making." Thus Hume is the first to describe cosmic evolution. Just as there is a long series of increasingly skilled *ship-wrights*, so there is a long series of increasingly skilled *world-wrights*. These world-wrights correspond to our animats.

The Humean world-wrights are cosmic engineers. Engineers proceed mostly by blind variation and selective retention (Sect. 3 in Chapter 3). They go mostly by cumulative finding, mentally simulating evolution by natural selection. Of course, intelligence makes this simulation more efficient. It allows you to learn from experience. You can work with regularities and patterns rather than with details. By working with regularities, you can compress vast regions of design space into small abstractions. But an evolutionary algorithm that is more efficient is still evolutionary. It does not involve any magical foresight. Hume's point about the slow evolution of universes is supported by the striking case of the game of life, which was designed by the mathematician John Conway

---

[30]Hume (1779: 77).

and his students. Conway has exceptionally high intelligence. Yet it took him and his students over eighteen months of mostly blind trial and error to come up with the game of life.[31] If universes are designed, the designer is cosmic evolution.

The Humean parable of the blind world-wrights enables us to naturalize the Cosmic Design Argument. The *Naturalized Cosmic Design Argument* goes like this: (1) The universe resembles a machine. (2) If this thing resembles that thing, then the cause of this thing resembles the cause of that thing. (3) So, the cause of the universe resembles the cause of a machine. (4) The cause of a machine is *technological evolution* (Sect. 3 in Chapter 3). Technological evolution involves a long series of engineers. Each engineer does a little design work. Over the long series, design gradually accumulates into a complex machine. Design accumulates mostly through cumulative finding. (5) So, the cause of the universe resembles technological evolution. Cosmic evolution involves a long series of cosmic engineers. Each cosmic engineer does a little cosmic design work. Over the long series, cosmic design gradually accumulates into a complex universe. The cosmic engineers are the animats. They act algorithmically and teleonomically. They blindly run a massively parallel distributed computation. This cosmic design argument has no role for God. By separating it from God, we liberate it from theism.

The Cosmic Design Argument relies on analogical reasoning: if this thing resembles that thing, then the cause of this thing resembles the cause of that thing. This analogy motivates the *Analogical Argument for the Animats*. It goes like this: (1) A crane is a computer that runs an entelechy. It may be a massively parallel distributed computer. Cranes do their design work by blind variation and selective retention. They may use intelligence to speed up the process, but efficient blind search is still blind. Thus cranes work teleonomically. All the design work in our universe is done by cranes. Cranes do atomic, molecular, biological, and technological design work. All the complex and intrinsically valuable things in our universe are produced by cranes. (2) Our universe resembles a highly complex piece of technology or biology. (3) If this thing resembles that thing, then the cause of this thing resembles the

---

[31]Khovanova (2010).

cause of that thing. (4) So, reasoning by analogy, the cause of our universe is some cosmic crane. (5) But all cranes involve armies of algo-′ rithmic agents swarming through libraries of possibility. So the cosmic crane involves armies of algorithmic agents swarming through the treasury. These algorithmic agents are the animats. They are self-replicating machines.

4.3   Finely Tuned for Flight and Sight

Although the Naturalized Cosmic Design Argument explains the complexity of our universe, it might be objected that it doesn't quite explain its *fine tuning*. To reply to this objection, we offer the *Naturalized Fine Tuning Argument*. It goes like this: (1) Our universe is finely tuned for the production of complexity, which is intrinsically valuable. More precisely, it is finely tuned for the maximization of reflexivity. (2) But the only other finely tuned things we see in nature are biological and technological machines. When we look at the organs of plants and animals, we find that evolution has finely tuned them for their functions: wings are finely tuned for flight, eyes are finely tuned for vision, the organs of bats are finely tuned for echolocation.[32] Of course, there is no need for teleology here: the tuning emerges from a teleonomic computation that sharpens its functionality as it converges to its ecstasy. The same points hold for technologies. (3) If this thing resembles that thing, then the cause of this thing resembles the cause of that thing. (4) Reasoning by analogy, the cause of the fine tuning of our universe resembles the cause of other fine tunings in nature. (5) It follows that the cause of our finely tuned universe resembles cause of biological and technological fine tuning. But organic and artificial machines were finely tuned by cranes involving replicators. (6) By analogy, the cosmic crane also involves replicators. These replicators are the animats. Just as replicating organisms finely tune their organs, so the replicating animats finely tune their universes. They finely tune them for maximizing reflexivity. The

---

[32]Wings (CMI ch. 4); eyes (CMI ch. 6); echolocation (BW 23–37).

Naturalized Fine Tuning Argument replaces God with the swarm of animats.

The ultimate replicators are the animats. Since they are ultimate, the animats are as simple as possible. The initial animat Alpha is as simple as possible. Every other animat inherits its entire nature from Alpha. Hence every animat is as simple as possible. The animats have no unexplained complexity. As the animats ascend the Cosmic Mount Improbable, the books they encounter accumulate complexity. But the animats remain simple. When some parent animat begets its offspring, then that reproductive step does some design work. But the design work gets done by moving along an upsloping arrow; it gets done by arrow selection. You can imagine a snowplow driving up a mountain road covered with snow. As it goes further and higher, more and more snow accumulates at the front of the plow—it makes a bigger snowball. The complexity of the snowplow stays the same, but the complexity of its snowball gets bigger. Of course, figuratively (the figure is metonymy) we can say the complexity of an animat is just the complexity of its universe. So (metonymically) animats gain complexity.

The animats are the ultimate agents. Consequently, the axioms of our deontic logic apply to them. The first axiom of our deontic logic states that if any agent can maximize value, then it ought to maximize it. The second axiom of our deontic logic states that if any agent is rational, then it strives to do its duty. As they strive to do their duties, the animats bring increasingly complex universes into actuality. Animats actualize their universes like spiders weave their webs or birds build their nests. They actively create their universes. If some animat creates some universe, then its animatic activity creates the physical activity in that universe. The agency of the animat appears as agency in its universe. Consequently, the axioms of our deontic logic apply to the universes: every universe ought to maximize value; every universe strives to maximize it.

As universes gain complexity, they evolve into increasingly complex dynamical systems. They evolve into computers nested in computers. The agency of the universe expresses itself in the agencies of its internal machines. Hence our deontic axioms apply to the computers in universes. Every computer in every universe both ought to maximize value and strives to maximize it. These computers run algorithms which strive for their finalities. As these computers get better, their finalities align more accurately with their duties. Their finalities are more likely to *be* duties. They get skewed ever further from randomness and ever more intensely towards their duties. They become more finely tuned for maximizing reflexivity. Moreover, they strive more reliably towards their finalities. Hence they strive more reliably to do their duties. Consequently, they drive their internal cranes to climb to ever greater heights.

References

Bennett, C. (1988). Logical depth and physical complexity. In R. Herken (Eds.), *The Universal Turing Machine: A Half-Century Survey* (pp. 227–257). New York: Oxford University Press.

Dennett, D. (1995). *Darwin's Dangerous Idea: Evolution and the Meanings of Life*. New York: Simon & Schuster.

Hahm, D. (1977). *The Origins of Stoic Cosmology*. Columbus, OH: Ohio State University Press.

Hawking, S. (1988). *A Brief History of Time*. Toronto: Bantam Books.

Hume, D. (1779 / 1990). *Dialogues Concerning Natural Religion*. New York: Penguin.

Khovanova, T. (2010). *The sexual side of life. By John H. Conway as told to Tanya Khovanova*. Online at http://blog.tanyakhovanova.com/?p=260. Accessed 10 May 2019.

Poundstone, W. (1985). *The Recursive Universe: Cosmic Complexity and the Limits of Scientific Knowledge*. Chicago: Contemporary Books Inc.

Rendell, P. (2002). Turing universality of the game of life. In A. Adamatzky (Ed.), *Collision-Based Computing* (pp. 513–539). London: Springer-Verlag.

Steinhart, E. (2014). *Your Digital Afterlives: Computational Theories of Life After Death*. New York: Palgrave Macmillan.

Wolfram, S. (2002). *A New Kind of Science*. Champaign, IL: Wolfram Media.

6

Ontology

1   The Ontological Liturgy

1.1   Why Is There Something Rather Than Nothing?

The *metaphysical question* asks why there is something rather than nothing.[1] It is the ultimate *why* question. Since ontology is the study of existence, and the question asks why there is any existence rather than none, we can also refer to this question as the *ontological question*. To answer it, we do the *ontological liturgy*. It's fair to say that Dawkins is obsessed with this question. It pops up over and over again in his writings, often at odd places. Sometimes he says the question is nonsensical. That's not scientific. Imagine somebody asking why there are complex animals, and being told that the question is illegitimate. Rejecting the question as nonsense looks like an Argument from Personal Incredulity, but Dawkins rejects those arguments. So he ought to be open to the metaphysical question. Dawkins says that our lives anesthetize us to the "wonder of existence." Over and over again, in *Unweaving the Rainbow*,

---

[1]Nonsensical (ROE 97; GD 79–80). Personal incredulity (BW 38; GD 155). Wonder of existence (UR 6).

© The Author(s) 2020
E. Steinhart, *Believing in Dawkins*,
https://doi.org/10.1007/978-3-030-43052-8_6

he celebrates the *appetite for wonder*. But the wonder is that anything exists at all.

Dawkins believes everything in the universe has a rational explanation.[2] But what about the universe itself? A universe that exists without an explanation, or as a brute fact, or by sheer chance, would be a universe that existed reasonlessly. If our universe were like that, then none of its internal structures or events would have any rational basis. Science would be absurd. Against that absurdity, Einstein said he had a profound faith in the rationality of existence. He thought that faith was necessary for being a scientist. Dawkins confirms that, like Einstein, he too has a profound faith in the rationality of existence. Dawkins had a discussion with Krauss about the relations between science and religion. Krauss said that religion is a way for humans to respond to a world that is unfair and irrational. But Dawkins replied sharply that "The world is not irrational. The world may be unfair but it is not irrational." Of course, this faith in the rationality of existence is metaphysical rather than scientific. It harks back to Stoicism and Platonism. This faith in the rationality of existence is part of spiritual naturalism.

Why is there something rather than nothing? The question is meaningful—it demands an answer, and the ontological liturgy aims to answer it. Dawkins has already agreed that there exists some simple first cause. He twice suggests that the first cause answers the question.[3] Suppose the simple first cause is Alpha. Whatever it may be, it cannot answer the metaphysical question. After all, Alpha *is something*. So the question applies to it just as much as to any other thing. Why is there some first cause rather than none? Why does Alpha exist at all? The answer cannot be that Alpha is necessary. For any existing thing, it is *logically possible* that it does not exist. The denial of Alpha does not produce a contradiction. As long as something is not *logically necessary*, we can still ask why it exists. It is necessary that the square root of two is an irrational number. But why? We can ask for a proof. So we can ask for a proof of the existence of Alpha. But that proof can't be like the cosmological arguments, which reason *from existing things* to the first existing thing. By

---

[2]Orderly universe (UR *xi*, 151). Faith in reason (TL 73–4). Rational world (SSSF).
[3]GD 184 and 185.

reasoning *from existence*, they *assume* existence, and they cannot explain what they assume. They don't provide an ultimate sufficient reason. To answer the metaphysical question, we cannot start with any existing thing. Of course, this means that theists cannot say that God answers the question. Nor can physicists start with empty quantum fields. The ontological liturgy cannot start with any existing thing. If it can't start with anything, it has to start with nothing. The ontological liturgy builds the inner framework of the Sanctuary for Spiritual Naturalists (Sect. 6 in Chapter 1). This framework rests on the deepest foundations of spiritual naturalism. Laying its foundation stones involves some heavy lifting and hard work. But the ontological liturgy is also our most vivid liturgy. If our Sanctuary contains a temple, this is it.

1.2   The Necessity of Pure Reason

To answer the metaphysical question, we have to do some reasoning. Our reasoning cannot start with anything; it has to start with nothing. It will be *pure* reasoning, which takes no premises from experience. Dawkins is ambivalent about pure reason. He says he does not like arguments from pure reason. He illustrates this dislike in his treatment of the Anselmian *ontological argument*.[4] For our purposes, this is the argument in the second chapter of Anselm's *Proslogion*. Among the many versions of this argument, I'll start with the short and very clear version by Kiteley.[5] It goes like this: (1) The property of being divine is perfect. (2) If any property does not have an instance, then it is not perfect. (3) Therefore, the property of being divine has an instance. (4) The instance of that property is God. The premises of this argument don't come from experience. The argument moves from purely rational principles to its conclusion.

Dawkins knows that atheists have taken the ontological argument seriously.[6] However, he characterizes it as "infantile." He parodies the

---

[4]GD 104–8.

[5]Kiteley (1958).

[6]Atheists take it seriously (GD 107). Infantile (GD 104). First parody (GD 104). Second parody (GD 107–8). Hostility to pure .reason (GD 104). Trickery (GD 105). Skepticism about pure reason (GD 107; see EP 275). Rather than a philosopher (GD 107).

argument in the language of children on the playground. Then he mocks it again in a second parody. But why these parodies? Mockery is not argument. So these parodies suggest two points. First, Dawkins has no rational reply to this argument. Second, *this argument disturbs him.* He says his hostility to the ontological argument comes from the fact that it relies on pure reason. He thinks it conceals some logical trickery. Dawkins says he is extremely skeptical about purely rational arguments. He tries to explain this hostility by appealing to his scientific training. He says that perhaps this hostility merely indicates that he is a "scientist rather than a philosopher." But that contrast is inappropriate.

Consider the long argument in *The Selfish Gene.*[7] Of course, that argument is inspired by an enormous amount of empirical data. It is, nevertheless, primarily conceptual. It might have been reached by armchair reflection on the meaning of replication. When a critic objected that *The Selfish Gene* was too philosophical, Dawkins replied that arguments based on purely conceptual principles can be "more powerful than arguments based on particular factual research." Dawkins himself regards *The Selfish Gene* as so abstract that it applies to life everywhere in the universe. His argument vastly exceeds the evidence. And Dawkins does plenty of philosophical work. Moreover, he says that Darwin's belief that life began in simplicity can be affirmed on "a basis of pure logic rather than evidence." And he insists that science requires deep faith in the rationality of existence. He insists that the world is rational. An atheist who affirms the rationality of existence really *should* be disturbed by the ontological argument. But the correct atheistic response is analysis and counter-argument, not mockery and *ad hominem* attacks. The ontological liturgy will do this analysis and counter-argument.

For consistency with his own principles, Dawkins should not reject arguments based on pure reason *just because* they are based on pure reason. Arguments from pure reason serve as the foundations for his rational magisterium. Here believing in Dawkins means welcoming arguments from pure reason. They lay the foundations for our Sanctuary. Of course, this does not imply that we should accept arguments from

---

[7]Particular factual research (SG 322). Rather than evidence (GSE 416). Faith in rationality (UR *xi*, 151; TL 73–4; SSSF).

pure reason *just because* they are based on pure reason. We will accept purely rational arguments when all their premises are true, and reject them when any of their premises are false. Some of the premises of the Anselmian ontological argument are false. But we will deal with that argument shortly (Sect. 4.1). Now we want to try to use pure reason to find some reason why there is something rather than nothing.

1.3   The Zero: The Instability of Nothingness

Dawkins wrote an Afterword to Krauss's book *A Universe from Nothing*.[8] He wrote that "Not only does physics tell us how something could have come from nothing, it goes further, by Krauss's account, and shows us that nothingness is unstable." However, in his book, Krauss isn't talking about nothing. He is talking about *something*, namely, quantum fields. Dawkins says the Kraussian "nothing" is simple. However, it is a very *complex* something. So Dawkins violates his own principles twice. For consistency with his own principles, he should be subjecting the Kraussian "nothingness" to the same withering analysis that he applied to God. Whatever the ultimate reason for existence may be, it cannot be that there was a fluctuation in some quantum field. Why are there any quantum fields rather than none? Why are there any fluctuations? And, if Dawkins were to answer that there are no reasons for these things, then obviously he has not answered the metaphysical question. Worse, existence would be reasonless. But Dawkins says existence is rational. So he needs to provide some reason.

Krauss says that nothingness is unstable. Perhaps that answer contains a grain of truth. Of course, the *nothing* cannot be a physical *something*, and its instability cannot be a physical instability. We need to be more abstract, so that we can think about nothing in a more purely rational and logical way. After all, the category of nothingness is purely logical. We can only approach it by reasoning about negation. If nothing is purely logical, then its instability is purely logical too. The philosopher Charles Sanders Peirce thought that nothing is *logically* unstable.

---

[8]Agrees with Krauss (AK 189; FH 21–4). Says Kraussian nothing is simple (FH 22).

Nothing is non-being. Speaking logically, non-being is pure negativity; as pure negativity, non-being negates itself; by negating its own non-being, it makes being be.[9] The negation of the negative makes the positive: the self-negation of non-being is being. Why is there something rather than nothing? Because nothing negates itself. That's a perfectly logical answer. But we can also be poetic. By negating itself, non-being steps outside of itself; by stepping outside of itself, it makes its ecstasy; thus being is the ecstasy of non-being. The self-negation of non-being is logically responsible for being. Logical responsibility is *grounding*. So the self-negation of non-being is the *ground* of being. It is *being-itself*.

Dawkins advocates for scientifically-inspired poetry.[10] Building on Dawkins, I will extend this into logically-inspired poetry. The primordial nothingness is the logical *Zero*. The philosopher William Inge argued that ancient Platonists like Plotinus started with this logical Zero.[11] Of course, since they did not have any zero in their number systems, they confused the Zero with the One. That confusion can be resolved by distinguishing the Plotinian Zero from the Plotinian One. The Plotinian Zero is the non-existent source of all existence. It is the hidden spring from which all rivers flow.[12] The Zero is "fathomless depths of power."[13] Like the ocean, the Zero is the *abyss* of nothingness. The abyss is deeper than every depth and higher than every height. Following Plotinus, I will use *water* to symbolize this abyss; hence the element of water symbolizes non-being. Figure 1 uses the symbol of ocean waves to depict the abyss of non-being.

---

[9]Peirce (1965) begins with nothingness (1.175, 6.33, 6.214, 6.215, 8.317). The nothingness is a powerful potentiality for being (6.217). The nothingness negates itself (6.219). He says "Thus the zero of bare possibility, by evolutionary logic, leapt into the unit of some quality" (6.220). A system of Platonic forms emerges from this self-negation (6.189–6.213). However, Peirce is obscure on many of these points. See also Heidegger (1998) and Nozick (1981: 123).

[10]UR 27, 180, 233.

[11]Inge (1918: 107–108).

[12]Plotinus (*Enneads*, 3.8.10.5–10).

[13]Plotinus (*Enneads*, 6.9.6.10–15).

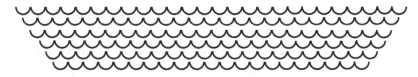

Fig. 1 The abyss of non-being

1.4   The One: Being-Itself

The Zero negates its own negativity. Since the negative of the negative is the positive, the self-negation of non-being is *being*. However, just as non-being is not the absence of this or that being, so the being that emerges from its self-negation is not the presence of this or that being. It is not some being among beings. That which emerges from the self-negation of non-being is pure existence, it is pure *being-itself*. If non-being is a purely logical concept, then so is being. So we can put this in more purely logical terms: Non-being is inconsistency; it is inconsistent with itself; its self-inconsistency is a purely negative self-relation. Thus non-being contradicts its own inconsistency. By contradicting its own inconsistency, it makes consistency be. Being-itself is consistent with itself. Its self-consistency is a purely positive self-relation.

If being-itself exists in the same way that beings exist, then being-itself is just one of the many beings; however, since being-itself is the source of the existence of every being, being-itself is not one of the many beings; hence being-itself does not exist in the same way the beings exist; consequently, there are two ways to exist. It is traditional to say that being-itself exists *ontologically*, while the beings exist *ontically*. Being-itself is prior to all ontic distinctions among beings. It is prior to simplicity and complexity; it is prior to universality and particularity. Some philosophers have argued that it is nonsensical to talk this way. It is absurd to distinguish two ways of existing.[14] If they are right, then our reasoning here fails. We reply with reference to modern logic.

Modern logic uses the backwards "E" as shorthand for existence: ∃. This symbol is known as the *existential quantifier*. To say that something

---

[14]McLendon (1960) and Fenton (1965).

exists is to associate it with the backwards E. Thus "Socrates exists" means "There exists some being which is identical to Socrates." We can refer to that being using the variable x. To say that Socrates exists means that (there exists some x) (x is identical with Socrates). Now replace "there exists" with the symbol ∃. To say Socrates exists means that (∃x) (x is identical with Socrates). The variable x is *bound* to the existential quantifier ∃. And Socrates is a value of that variable. The great logician Willard Van Orman Quine said that *to be is to be the value of a bound variable.*[15] The Quinean slogan isn't an ontic statement about this or that being. It is an ontological statement about being-itself; it defines the being of beings. Quine's slogan is basic for modern logic. If you say (∃x) (...x...), then the x refers to some being among beings while the ∃ refers to being-itself. It is entirely possible to make logically sensible statements about being-itself. Being-itself is a purely logical category. But science depends on logic.

Non-being is the logical Zero. By negating itself, the Zero turns into being-itself. There are two reasons to identify being-itself with the logical *One*. The first reason is that the One is the source of the existence of all the beings. Plotinus writes of the One that "the ultimate source of every thing is not a thing but is distinct from all things: it is not a member of the totality of beings, but the origin of their being."[16] But being-itself is also such a source. And since there cannot be two such sources, being-itself is identical with the One.[17] The second reason is that the One is the power of the abyss manifest as existence. It is that power which makes beings be. It is the self-negating power of the Zero turned outwards into positive existence. But the same holds for being-itself. If it were not for the self-negation of non-being, there would not be any beings. Hence the self-negation of non-being is that power which makes beings be. But that self-negation is being-itself; hence being-itself is that power which makes beings be. Again, since there cannot be two such powers, being-itself is the One. By negating itself, the logical Zero turns into the One. So we've counted from Zero to One.

---

[15]Quine (1948).

[16]Plotinus (*Enneads*, 5.3.11.20–25; see *Enneads*, 5.2.1.1–2).

[17]Plotinus (*Enneads*, 5.4.1.15–17).

Should atheists talk about being-itself? Tillich famously identified God
with being-itself.[18] Perhaps some Medieval theologians (like Aquinas)
also identified God with being-itself. Dawkins explicitly rejects the equation of God with being-itself.[19] He insists that the term "God" is defined
by the scriptures and practices of the Abrahamic religions (Sect. 2.1
in Chapter 4). The Abrahamic scriptures become absurd if "God" is
replaced with "being-itself". The Abrahamic practices regard God as a
person with high social status; but being-itself is not the king of kings.
To say that God is being-itself is theological doublespeak, which goes
with religious hijacking. One task of spiritual naturalism is to reclaim
philosophical concepts from theology. We want to liberate these concepts
from their bondage to God—to set them free. So here we reclaim both
being-itself and the One. Being-itself is not God, and we agree with Plotinus that the One is not God.[20] The One was an old *pagan* concept.
It can be reclaimed for a new and entirely irreligious pagan spirituality.
Poetic symbolism can help with this reclamation. Being-itself emerges
from the abyss of non-being like an island emerges from the sea. Hence
the element of *earth* symbolizes being-itself. Figure 2 illustrates the emergence of the earth. But words and symbols are not enough. To reclaim
being-itself, we need new practices (Sect. 4 in Chapter 9). These practices will arouse the One in your own body. Through them, you can
participate in the flow of power from the earth to the sun.

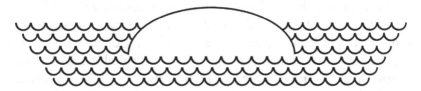

Fig. 2 The island of being-itself

---

[18]Tillich (1951: 235–237).
[19]Dawkins (1992, 1996b: 51; AK 190).
[20]Plotinus (*Enneads*, 6.9.6.13–14).

1.5   The Two: The Self-Consistency of Being-Itself

Being-itself is the One; but the One generates *the Two*. To define the Two, we count from the Zero through the One. The Zero is non-being; non-being is self-inconsistency; but self-inconsistency negates itself. Self-inconsistency is a purely negative self-relation. This self-relation negates itself. But a purely negative self-relation is pure *anti-reflexivity*. Thus by negating itself, non-being minimizes anti-reflexivity. And, since the negative of the negative is the positive, this self-negation of non-being generates being-itself. Thus the Zero generates the One. Since non-being is self-inconsistency, it follows that being-itself is self-consistency. The self-relation of being-itself is self-consistency. This self-relation indicates logical duality: x is consistent with x. But the one x appears twice. This is where *beings* emerge from *being*. So this self-consistency is the Two; it is the dyad. Far from being inert, this self-consistency inherits the power of the One. The power of the dyad *maximizes* self-consistency. Since self-consistency is a positive self-relation, to maximize self-consistency is to maximize reflexivity.

Consistency involves propositions. If the dyad does not generate any propositions, then there is no consistency. And if it does not generate all logically possible propositions, then consistency is not maximized. The dyad therefore generates all those propositions. The axioms of logic are the first propositions. They are the first beings among beings. As it maximizes consistency, the dyad assigns truth to all tautologies and falsity to all contradictions. Likewise, the dyad assigns truth-values in accordance with the laws of logic. It generates a system of logically well-organized propositions. Their meanings are interwoven by logical relations. Hence the system of these propositions resembles a *rhizome*, which is a network of logically interwoven roots. Since it is logically well-organized, the rhizome is *rational*—it is *the Logos*. For any proposition P, the Logos affirms that P if and only if P is true; it denies that P if and only if P is false. Since the Logos is a system of propositions, it resembles a mind. However, the Logos is not a mind; it does not think. It is just an eternal necessary logical structure. Of course, if there are no propositions, then our reasoning fails. But it is self-contradictory to deny the existence of propositions. If you say there are no propositions, then you have asserted a proposition. It cannot be a fact that there are no facts. And abstract objects like propositions do appear in scientific theories. The axioms of logic and mathematics state propositions, as do the laws of science. So, if naturalists only accept the kinds of things that appear in scientific theories, they can accept propositions.

Besides the laws of ordinary logic, the first propositions generated by the dyad also include the laws of the logic of value. These are the first principles of deontic logic. It is a tautology that logical value ought to be maximized. If it were not maximized, logic would not be logical; but logic is logical. Self-consistency is the fundamental logical value. Hence it is a tautology that self-consistency ought to be maximized; but the maximization of self-consistency entails the maximization of reflexivity; hence reflexivity ought to be maximized. An ideal maxim emerges from the positive self-relation of being-itself. This maxim is the imperative *Maximize reflexivity!* This ideal maxim regulates the action of the dyad. The dyad has the duty to maximize reflexivity. It likewise has the power to do its duty. Since the power and the duty of the dyad are equally rooted in being-itself, its power and its duty coincide. If the dyad fails to do its duty, then it contradicts itself; but it does not contradict itself; hence it does its duty. The dyad maximizes self-consistency; it maximizes reflexivity.

So the dyad maximizes self-consistency. However, if only tautologies are true, then self-consistency is not maximized. A tautology is true by definition (philosophers say its truth is analytic). But it is logically possible that there are propositions whose truths are not merely analytic. Since they are not merely true by definition (like tautologies), they must be true because they are about existing things. These existential truths congregate into *theories of existence*. The maximally self-consistent theory of existence defines the maximal plurality of all logically possible beings. The basic propositions in the maximally self-consistent theory of existence are the axioms in the *science of being*, a science which is unified by being-itself.[21] It is the development of the meaning of being-itself. For

---

[21] Plotinus (*Enneads*, 3.9.2, 4.9.5).

Platonists, the science of being is mathematical. The axiomatic propositions in the Logos resemble those of geometry.[22] Proclus used the Euclidean axiomatic method to write his *Elements of Theology*. Hence the Logos also affirms all the propositions in that mathematical axiom system than which none greater is logically possible. These axioms give rise to the beings. The logician Kurt Gödel showed that no axiom system can decide the truth-values of all propositions. The axioms do not decide the truth-values of all propositions. The Logos decides them as it maximizes consistency.

Modern mathematical philosophers have argued that the consistency of an axiom system suffices for its truth.[23] Here spiritual naturalists offer an argument: (1) This narrower axiom system is less inclusive than that wider axiom system if and only if the models of the narrower system are included in those of the wider system. (2) The greatest axiom system is both consistent and maximally inclusive. (3) If the greatest axiom system is false, then self-consistency is not maximized. (4) But the dyad maximizes self-consistency. (5) Therefore, the greatest axiom system is true. The models of this system include all other consistently definable models. But what is the greatest axiom system? Where is consistency maximized? To think precisely about the greatest axiom system, we turn to the foundations of mathematics. Specifically, we turn to set theory. Sets are dots. While there are many controversial aspects of set theory, it is arguable that all consistent theories have models in the richest possible plurality of sets. If that is right, then the greatest axiom system is the greatest theory of sets.

While many issues in set-theory remain unsettled, we can *provisionally* say that the greatest set theory includes all the axioms in the Von Neumann–Gödel–Bernays (VGB) set theory. It also includes axioms for all the consistently definable higher infinities (also known as *large cardinals*).[24] But the greatest set theory cannot be exhausted by any finite list of axioms: it is an unsurpassable series of surpassable theories. It is an *ideal theory*. This ideal theory also adds the *proper classes*, collections that

---

[22] Plotinus (*Enneads*, 4.9.5.24–26, 6.3.16.20–23).

[23] Balaguer (1998: 5–8).

[24] Drake (1974) and Kanamori (2005).

are too general to be sets. The proper classes are unsurpassable. They mark the horizon of being-itself. So the dyad assigns truth to the axioms in the ideal set theory. Of course, since these axioms do not decide the truth-values of all propositions, the dyad independently assigns other truth-values. By affirming the ideal set theory, the dyad generates the greatest system of sets. Mathematicians refer to this system as "V"; but here we will also refer to it as the *pleroma*, which means *fullness*. It is arguable that all consistent theories have models in the pleroma. If that is right, then all logically possible forms exist in the pleroma. The pleroma is combinatorially complete.

2   That Abstract Atmosphere

2.1   Platonic Rationality

Our answer to the metaphysical question began with the purely logical categories of non-being and being-itself. It led to abstract objects like propositions and sets. An *ontology* is a catalog of types of objects. An ontology that includes abstract objects is Platonic. Hence the ontology of spiritual naturalism is a Platonic ontology. But what about Dawkins? Do his writings favor Platonism or oppose it? He objects to the ancient Platonic notion that there exists some immutable form of every species.[25] So there is at least one place where he seems to oppose Platonism. But his objections target an obsolete version of Platonism. Modern Platonism does not contradict evolution—it isn't about immutable bunnies. Moreover, Dawkins often seems to endorse modern Platonism. His texts contain at least four arguments that support it.

The first Dawkinsian argument for Platonism comes from his love of patterns and structures. Dawkins loves genes. But genes are structures, not substances. A gene is a series of molecular letters. If they are replaced with other letters of the same type, then the gene remains the same. Genes are formal *types* rather than material *tokens*. They are patterns rather than material things. Moreover, biology involves

---

[25]GSE 21–4.

mapping structures onto structures: DNA gets mapped onto RNA, and RNA gets mapped onto protein. Dawkins loves poetry. But poems are also structures. As a lover of structures, Dawkins resembles Kurzweil. Kurzweil says he is not a materialist, but a *patternist*. He says he "views patterns of information as the fundamental reality."[26] For the Platonist, matter emerges from relational structures. It emerges from mathematical patterns. Taking his cue from the digital code of genetics, Dawkins wonders whether physics is ultimately digital.[27] But you don't have to believe that physics is digital to be a Platonist. You just have to believe that physical things have mathematical structures.

The second Dawkinsian argument for Platonism is the *Argument from Libraries of Physical Types*.[28] The argument goes like this: (1) Dawkins talks about the complexities of types of atoms, molecules, and organisms. But these types and their complexities are abstract. (2) He often refers to an abstract library of possible organisms. He says this library is a *mathematical* structure. (3) Moreover, Dawkins affirms that these abstract types exist objectively (Sect. 1.3 in Chapter 2). They are not concepts in our brains or words in our languages. (4) Consequently, these types, and their properties and relations, all exist objectively. They dwell in some Platonic world of abstract objects. However, this does *not* imply dualism. Since any Platonist says reality is unified by the One, there cannot be any dualism which opposes the physical to the mathematical. Just as an image supervenes on an array of pixels, so physical things supervene on mathematical objects. This Platonism finds further support in the ways Dawkins uses the formal sciences. Computer science, information theory, and mathematics are formal sciences. They study the properties and relations of abstract objects in the Platonic sky.

The third Dawkinsian argument for Platonism comes from his claims about the qualities of our universe. He praises its orderliness, beauty, and rationality (Sect. 2.3 in Chapter 4). He declares that those cosmic qualities are objective features of our universe. He links rationality with

---

[26]Kurzweil (2005: 5, 388).

[27]SITS 85–6.

[28]Library of possible organisms (BW ch. 3; CMI ch. 6; ADC ch. 2.2). Is a mathematical structure (e.g. BW 73; CMI 218).

beauty, and he links beauty with mathematics.[29] This link motivates an *Argument for the Rationality of our Universe*: (1) Our universe is rational, orderly, and beautiful. And, according to Dawkins, it has these features objectively. But why does it have them? Traditional explanations turn to mathematics. (2) Mathematical reality is rational, orderly, and beautiful. And, according to Platonism, it has these features objectively. (3) Hence the best explanation for the fact that our universe has its features is that it is a model of a purely mathematical structure. (4) So, by inference to the best explanation, our universe is a model of a purely mathematical structure. (5) If that is right, then purely mathematical structures objectively exist. This argument can be backed up with an *Argument from Effectiveness*. It goes like this: (1) Pure mathematics is surprisingly effective in physics.[30] (2) But the best explanation for that astonishing effectiveness is that our universe is a model of a purely mathematical structure (Sect. 2 in Chapter 5). (3) So, by inference to the best explanation, our universe is a model of a purely mathematical structure. (4) If that is right, then purely mathematical structures objectively exist. Hence the Dawkinsian texts that describe the qualities of our universe support the existence of a Platonic world of abstract objects.

The fourth argument is the *Argument from the Superiority of Science over Theology*. It goes like this: (1) Science is more truthful than theology. (2) Part of this superiority comes from the ways that science relies on evidence. However, theology also uses evidence (e.g. in its design and cosmological arguments), and sophisticated theologians try to use science to support their theologies. So there must be some deeper reason why science is superior to theology. (3) Dawkins says that the deeper reason is that science depends on mathematics while theology does not.[31] (4) On the left hand, mathematics is merely subjective. Both religion and mathematics are merely human inventions. (5) On the right hand, mathematical truth is objective, eternal, and necessary. It is utterly free from all bias and subjective pettiness. It commands assent from all rational persons, and it constrains all agency. Even God (were He to exist) could

---

[29]UR *xi*, 63–4, 151; TL 75–6.
[30]Wigner (1960) and Steiner (1998).
[31]AK 190; UR 63–4.

not make two plus two be five. Thus mathematics rules over theology. (6) Either the left hand or the right hand. (7) If the left hand, then Dawkins is wrong. (8) Therefore, Dawkins takes the right hand. But that hand also holds objectively existing mathematical objects.

These four arguments show how the Dawkinsian texts support Platonic ontology. On Dawkinsian foundations, it is reasonable to raise a Platonic frame. So I'm adding Platonism to *my interpretation* of his writings. I do *not* say that Dawkins is a Platonist; I say *only* that his texts make far more sense with Platonism than without it. Believing in Dawkins means trying to make more sense out of his writings. This Platonism helps to raise a more complete and systematic building over his incomplete and unsystematic texts. Moreover, this Platonism helps to strengthen his atheism. Theists love to portray atheists as mere materialists. Since materialism is such an impaired ontology, this makes atheism easy to attack. It's far more difficult for theists to attack Platonic atheism. The Dawkinsian texts support Platonic atheism. Platonism is a kind of structuralism.[32] It says that *nature is a purely relational structure*, a structure which is ultimately best described in purely mathematical terms. Naturalism therefore rejects materialism in favor of Platonism. Here I'm clearly *building on Dawkins* using my own architectural plans. According to these plans, inspired by his writings, the Sanctuary for Spiritual Naturalists rises endlessly into the abstract Platonic sky.

2.2   Evidence for Abstract Objects

To make greater sense of the Dawkinsian texts, spiritual naturalists affirm the existence of abstract objects. But the Dawkinsian texts are not inerrant scriptures. Believing in Dawkins means providing some independent *evidence* that abstract objects exist. Focus on mathematical objects. The evidence that they exist comes from their successful application. Mathematical theories (like the calculus) are empirically tested by

---

[32]The term "structuralism" has a distinctive meaning in biology. Structuralists in biology disagree with Darwinism. Dawkins is not a biological structuralist. Here structuralism is used in the sense found in linguistics, mathematics, and philosophy.

applying them. If they were false, their applications would fail. Hence the calculus is the most widely and deeply tested theory in human history.[33]

The *Argument from Mathematical Utility* makes this idea more precise: (1) If mathematical objects don't exist, then the theories about them are false. (2) But if those theories are false, then the equations used in science are false. (3) If they are false, then the technologies based on those equations won't work. (4) But they do work. (5) Therefore, the scientific equations are true, the mathematical theories are true, and the mathematical objects exist. This argument can be further developed as the *Argument from Smartphones to Sets*: (1) If the axioms of set theory are false, then the theorems of the calculus are false. (2) If they are false, then the mathematical equations in physics are false. (3) If they are false, then your smartphone and all your other modern technologies won't work. (4) But they do work. (5) So the axioms of set theory are true. (6) But those axioms assert the existence of sets. (7) Consequently, those sets exist.

Another way to use science to justify the existence of mathematical objects comes from the *Indispensability Arguments*.[34] They say that, since scientific theories cannot dispense with mathematical objects, those objects exist just as much as all the other objects found in scientific theories. They are just as real as physical things. There are many versions of the Indispensability Argument. A short and sweet version runs like this: (1) Scientific theories refer to physical things (like molecules, atoms, particles, black holes, quantum fields, space, and time). (2) All the evidence for those scientific theories is evidence that those things exist. (3) Since there is lots of evidence for those scientific theories, there is lots of evidence that their physical things exist. (4) But those scientific theories also refer to mathematical objects (like numbers, functions, vectors, Hilbert spaces, Lie groups, and so on). (5) It is not possible to rewrite our best scientific theories in ways that make the mathematical objects disappear. They refer to mathematical objects in exactly the same way that they refer to physical things. (6) So, all the evidence for those scientific theories is also evidence for their mathematical objects. They exist just as much as any other scientific objects (like atoms or black holes).

---

[33]Goodman (1990).
[34]Resnik (1995), Maddy (1996: 318–319), and Colyvan (2001).

Of course, the Indispensability Argument has been challenged. But its strongest challenge failed.[35] So far, the argument remains in force. And it's inconsistent to try to just get rid of the mathematical objects. The techniques used to show their dispensability can be deployed even more effectively against the physical things in scientific theories.

Finally, the *Easy Ontology Argument* shows that naturalists should embrace abstract objects. It goes like this: (1) The main principle of naturalist ontology states that if our best scientific theories refer to objects of some type, then naturalists are committed to the existence of objects of that type. (2) Besides referring to concrete things, our best scientific theories refer to many types of abstract objects. (3) Therefore, naturalists are committed to all those types of abstract objects. These natural abstract objects include mathematical objects, properties, relations, propositions, and so on. As science changes, naturalists change their ontologies. If science ever stops referring to abstract objects, then naturalism will not include them. However, the history of science shows that, as science becomes more truthful, it also becomes more mathematical.

2.3   Abstract Objects Fill the Sky

The Sanctuary rests on Dawkinsian foundations. There are at least four reasons to think that those foundations specifically support sets. The first reason comes from one way that Dawkins deals with ultimate questions. When dealing with them, Dawkins recommends the book *Creation Revisited*, by the chemist Peter Atkins.[36] But when Atkins deals with ultimacy, he turns to set theory. Atkins writes that "the deep structure of the universe may be a globally self-consistent assemblage of the empty set."[37] Since Dawkins refers to Atkins, and Atkins starts with sets, Dawkinsian foundations support sets.

---

[35] Field (1980) attacked the indispensability argument. Field tried to show that a fragment of Newtonian gravitational theory could be developed without any abstract objects. His strategy does not generalize (Liston 1993; Melia 1998; Bueno 2003). Shapiro (1983) showed that Field made logical errors. Hence Field (1985) modified or retracted many of his earlier conclusions. His attack failed.

[36] BW 14; GD 143–4.

[37] Atkins (1992: 115).

The second reason to think Dawkinsian foundations support sets comes from the simplicity of set theory. Dawkins insists that foundations must be as simple as possible.[38] But no exact theory of existence is simpler than set theory. Apart from the machinery of pure logic, set theory involves just one term: the membership relation. Its meaning is defined entirely in terms of logic. From this single relation, and nothing else, set theory builds all its complexities. The axiom for the empty set asserts that there exists an object which does not have any members. No possible object is simpler than that. Hence it is appropriate for Dawkinsian metaphysics to begin with set theory.

The third reason comes from structuralism. Through its connections with both Stoicism and Platonism, Dawkinsian ontology is highly structural. A structure is a system of objects defined entirely by their relations. Since all sets are defined in terms of the membership relation, the plurality of sets is a structure. Structural realists say that nature is ultimately purely structural. The ontology of sets is a purely structural ontology. Hence we are structural realists (patternists). By saying that nature is purely structural, we modernize the old Stoic Logos doctrine. The Logos includes the axioms of set theory. If naturalism begins with simplicity, then naturalism is structural realism. Counting from the Zero, through the One, and through the Two, we reach *the Many*.

The fourth reason comes from the evolution of complexity. Dawkins insists that complex things depend on simpler antecedents. Although sets don't literally evolve, more complex sets are combinations of (and thus dependent on) simpler sets. Sets gradually accumulate complexity. Dennett's principle of the accumulation of design applies to sets. Sets resemble atoms, molecules, organisms, and technologies. Just as evolutionary arrows were defined for those things, so one can be defined for sets. The *collective arrow* states that if any set is complex, then it exists in some series that started out simple and climbed up through all lower levels of formal complexity. The plurality of sets is the *collective crane*. The axioms of set theory are its entelechy. Since Dawkinsian ideas support the gradual accumulation of complexity, they support set theory. Similar remarks apply to the ordinal numbers, to the *axis mundi*.

---

[38] FH 22.

2.4   Welcome to the Ontological Orgy

The axioms of the greatest set theory (Sect. 1.5) entail that the sets are stratified into a great library. The floors of this library are the *ranks* of sets. Higher ranks in this library contain more complex sets. These ranks are defined by the familiar schema of the three laws. The *initial law for sets* states that the initial rank of sets contains exactly the simplest sets. If there were many simple sets, they would all be identical. Hence there is exactly one simple set. Since it is simple, it has no internal complexity—it has no members. Since it has no members, it is empty. You write the *name* of a set by writing its members between braces. Since the empty set has no members, its name is {}. The empty set exists. It does not *contain nothing*; on the contrary, it *fails to contain anything*. The ranks of sets are indexed by ordinal numbers: the initial rank is the zeroth rank, which contains only the empty set. The empty set is the only book on the bottom floor of the set-theoretic library. Simpler sets in lower ranks are surpassed by more complex sets in higher ranks. The next two laws for sets define this surpassing.

The *successor law for sets* states that every rank of sets is surpassed by a successor rank of sets. Every successor rank includes all new combinations of sets from all lower ranks. Thus each successor rank is *combinatorially complete*. Old sets produce new sets much like old organisms produce new organisms. There are two rules. The first rule states that every lower set asexually generates a higher set. The higher set is just the lower set enclosed in braces. So the first rule is an arrow:

$$\text{set} \rightarrow \{\text{set}\}.$$

The second rule states that every plurality of lower sets sexually generates a higher set. The higher set is just the plurality in braces. So the second rule is an arrow:

$$\text{plurality of sets} \rightarrow \{\text{plurality of sets}\}.$$

The empty set on the zeroth rank asexually begets the set of the empty set on the first rank. The empty set {} gives birth to the set {{}}. Now the lower set {{}} on rank one gives birth to the higher set {{{}}} on rank two. And the two lower sets {} and {{}} have sex to produce the higher set {{}, {{}}} on rank two. Now there are four sets on ranks zero through two. Through their acts of sexual and asexual reproduction, they will beget twelve new sets on rank three. Hence the ranks zero through three contain sixteen sets. The library of sets grows ever higher as lower sets beget higher offspring. And it grows ever wider. The fourth rank contains over sixty five thousand sets. The fifth rank contains more sets than there are particles in the entire universe. The plurality of sets grows by combinatorial explosion. Sets breed with sets without restraint—the plurality of sets is an ontological orgy, and these ranks rise towards infinity.

The *limit law for sets* states that every infinite progression of ranks is surpassed by some limit rank. Each limit rank contains all new combinations of sets from all lower ranks. The first limit rank is the first infinite rank. It contains all new combinations of sets from all finite ranks. Of course, since every finite combination of finite sets already occurs on some finite rank, none of those finite combinations are new. The only new combinations are the infinite combinations of finite sets. Every set on the first limit rank is an infinite set created by infinite sex. Infinite pluralities of lower sets sexually generate these new infinite sets. The successor law applies to the first limit rank to produce the next higher rank. So now infinite sets asexually beget higher sets. And finite and infinite sets sexually beget higher sets. Sets breed with absolute promiscuity.

There exists a rank of sets for every ordinal in the *axis mundi*. The union of all the ranks of sets makes the library of sets. Since that library expands as it rises, and since the letter V expands as it rises, the library of sets is also known as V. While the letter V is nice, it would be useful to have some word to refer to the library of sets. The Greek word *pleroma* means fullness (Sect. 1.5). It referred to the fullness of the divine mind, but we can reclaim it for spiritual naturalism: since the library of sets is combinatorially full, it is the pleroma. Any set that contains only other sets as members is a *pure* set. So all the sets in the pleroma are pure. Since those sets are stratified into ranks, the pleroma is often referred to as the *iterative hierarchy of pure sets*.[39]

The completeness of the pleroma entails that it contains all consistently definable forms. Assuming that our universe is consistently definable, its form exists in the pleroma. Our universe is an instance of some purely set-theoretic form. Variations of its form are other universes. Since the pleroma is combinatorially complete, every possible structure appears in the pleroma. This has led physicists like Max Tegmark to say that all possible universes are purely mathematical structures.[40] We affirm that all possible universes are set-theoretic structures. Every book in the modal library describes some set-theoretic structure (Sect. 2.2 in Chapter 5). Every book corresponds to some set. However, since many sets cannot be interpreted physically, not every set corresponds to a book.

But where do all these sets come from? If mathematical existence is self-consistency, and if self-consistency emerges from the self-negation of non-being, then they emerge from that self-negation. The self-negation of non-being is being-itself. Hence the iterative hierarchy of sets rises up out of being-itself. It rises up over the island of being-itself; it rises up over the ocean of non-being. What poetic symbolism can we use for the pleroma? Since it rises upwards, it is appropriate to picture it as rising into the *sky*. Hence the traditional element of *air* symbolizes the pure sets. They are shapes in the air. When the island of being emerges from the abyss of non-being, it separates the water from the air. It opens the sky. Each phase of the pleroma is like an altitude or height in the sky above the ground. Thus each phase of the pleroma can be represented by an arc over the ocean and the earth. Figure 3 shows some of these phases.

---

[39]The iterative hierarchy is usually defined in terms of *phases* of sets, where each phase comes from lower phases. The initial phase V(0) is {}. Each successor phase V(n+1) is the power set of V(n) for any ordinal n. The power set of x is the set of all subsets of x. Each limit phase V(L) is the union of the V(k) for all k less than the limit ordinal L. V is the union of the V(n) for all n in the *axis mundi*.

[40]Tegmark (2014).

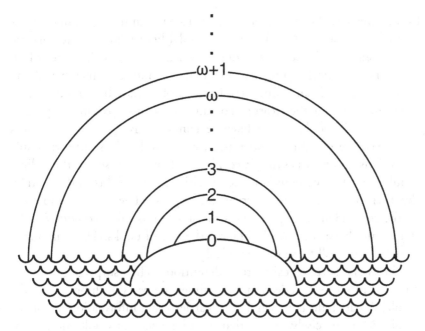

Fig. 3 The sky over the island in the ocean

2.5   On Transcendental Ideals

Every set is both a member of some greater set and a subset of some greater set. So every set is surpassed by greater sets. Since numbers are sets, every number is surpassed by greater numbers. Likewise every form is surpassed by more complex forms. Every book in the modal library is surpassed by more complex books. Every computer is surpassed by greater computers; every mind by greater minds. Since all these things are surpassed, they are surpassable. There are no unsurpassable sets, numbers, forms, books, universes, computers, organisms, minds, and so on. Every degree of perfection is surpassable. Maximally perfect beings are logically impossible.

Some collections are too general to be sets. The *pleroma* is the collection of all sets. Technical arguments show that it cannot be a member (or subset) of any greater collection. It therefore cannot be surpassed by any greater collection. Hence the pleroma is an unsurpassable collection of surpassable sets. Likewise the modal library and the treasury are unsurpassable collections of books. The *axis mundi* is the series of all numbers. Technical arguments show that it cannot be a member of any greater series. It is an unsurpassable series of surpassable numbers. The modal library and the treasury contain many unsurpassable sequences of surpassable books. They likewise contain unsurpassable sequences of surpassable computers, organisms, and minds. These unsurpassable collections are known as the *proper classes*. They are not sets. If the ordinal numbers in the *axis mundi* define the altitudes or heights in the abstract sky, then the proper classes surpass all heights. They can be symbolized by the *stars*. Hence the pleroma is the star of sets; the *axis mundi* is the star of numbers; the modal library is the star of books. There are many stars of forms. All these stars are ideals.

The stars surpass exhaustive definition. This surpassing can be analyzed in terms of reflection. Sometimes the parts of things *reflect* the whole. A part reflects its whole if it carries some information about the whole. More precisely, to say that a part reflects its whole means that some property which characterizes the whole also characterizes the part. For example, the parts of the pleroma are just sets. A general reflection principle says that every well-defined property of the pleroma is reflected by some set inside the pleroma. The set *reflects* that property back to the pleroma. It resembles a mirror in which the pleroma sees some aspect of itself. For example, since the pleroma is infinite, there is some set in the pleroma that is infinite. Reflection principles show how the pleroma is *ineffable*.[41] If you think you have fully described it, it will always turn out that you have merely fully described some set in the pleroma. You will have merely described some mirror which reflects its starlight. Reflection principles show how the pleroma maximizes reflexivity. By consistently expanding the range of reflected properties, reflection principles yield powerful new axioms which reveal the awesome vastness of the pleroma. These axioms are associated with numbers. So there are mirrors in the *axis mundi* which reflect its numerical starlight. The *axis mundi* also maximizes reflexivity. Every star maximizes reflexivity in its own way.

---

[41]Welch and Horsten (2016).

The stars have four *stellar ambiguities*. They are ambiguous first with respect to definability.[42] On the one hand, they are easily definable: the pleroma is the class of all sets; the *axis mundi* is the class of all numbers. They have a precisely defined height: the height of every star is the lowest height above all sets. On the other hand, they surpass all exhaustive definition—they are indefinitely extensible. The stars are so high they are heightless. They are ambiguous second with respect to unity. On the one hand, they are sufficiently unified to *have* members; on the other hand, they are not sufficiently unified to *be* members. They are ambiguous third with respect to identity. On the one hand, the pleroma and the *axis mundi* are two distinct stars; on the other hand, the stars cannot be counted.[43] They are ambiguous fourth with respect to existence. On the one hand, there exists some x such that x is the pleroma. Just as Hypatia is a being among beings, so the pleroma is a being among beings. More precisely, just as it is logical to write $(\exists x)$ (x is Hypatia), so it is logical to write $(\exists x)$ (x is the pleroma). On the other hand, if the pleroma were just another being among beings (like Hypatia), then it would be available for inclusion in some greater class. However, since it is not available in that way, it is not a being among beings. Hence it is illogical to write $(\exists x)$ (x is the pleroma).

The stellar ambiguities indicate that the stars lie in the wilderness between being-itself and non-being. Since the stars are unsurpassably great, they are the brightest objects in the sky. Every star is a way of being perfect. While a monotheist would say there exists only one way to be perfect, we say there are infinitely many ways to be perfect. There are many stellar ideals. But to pass beyond the stars is to pass into paradoxical self-inconsistency. To rise above them is to vanish into the abysmal heights. Quine said that to be is to be the value of a bound variable (Sect. 1.4). This is written as $(\exists x)$ $(...x...)$. At any set-theoretic object, the binding of the x to the $\exists$ is secure. The power of the x is balanced with that of the $\exists$. At the stars, the binding dissolves. At any proposition $(\exists x)$ (x is a star), the powers of the x and the $\exists$ are unbalanced. The power of the $\exists$ vanishes into the maximality of being-itself while that

[42]Maddy (1983: 113–123) and Potter (2004: Apx. B & C).
[43]Brown (1980).

of its x vanishes into the minimality of non-being. These powers vanish into the liminal threshold between being-itself and non-being. Reflection means that this threshold is a horizon—it recedes as you approach it. At this horizon, the beings among beings dissolve. This logic will help with atheistic mysticism (Sect. 4.2 in Chapter 7) and meditation (Sect. 2.3 in Chapter 9).

To say that an object has these stellar ambiguities is to say that it is *wild*. Hence the stars are *wild*. This wildness means that the stars are *wholly other*. They dwell in the *wilderness*, which is the threshold between being-itself and non-being. Following Kant, the stars are *sublime*.[44] They are tremendous and fascinating mysteries. Following Otto, the stars are *numinous*.[45] Following our own theory of ecstasy, the stars are *ecstatic*. The pleroma is the ecstasy of sets, the ecstasy of the collective crane. The *axis mundi* is the ecstasy of numbers, the ecstasy of the ordinal crane. Finally, following a long philosophical tradition, the stars are *transcendental*. Transcendence is ideality. An *ideal number* is an unsurpassable series of surpassable numbers. An *ideal collection* is an unsurpassable series of surpassable collections. An *ideal animal* is an unsurpassable series of surpassable animals. An ideal object of any type transcends that type. An ideal animal is not an animal—it is an ecstasy of animality.

3   Diamond Hard Light

3.1   The Atheistic Problem of Evil

Dawkins constantly celebrates the positivity of nature.[46] He frequently talks about its beauty. He says the universe is a "grand, beautiful, and wonderful place." He celebrates the "glory" of the universe. He says we are "hugely blessed" to be alive here. He speaks of "the true reverence with which we are moved to celebrate the universe." But if nature is not

---

[44]Kant (1790: secs. 25–27).

[45]Otto (1958).

[46]Wonderful place (FH 99). Glory (FH 23). Blessed (UR 5; ADC 12). True reverence (AT 700). Spring from bed (UR 6) Rejoice in (UR 36).

generated by some ultimate goodness, why say *reverence*? He says each day we should "spring from bed eager to resume discovering the world and rejoicing to be a part of it." And that science is something "to read and rejoice in." Why say *rejoice*? Evolution is the war of all against all; the sun will incinerate the earth; increasing entropy ultimately destroys all value. If the Good does not shine out through all this cruelty, death, and waste, then Dawkins is at best superficial and at worst perverse.

Dawkins urges artists in all mediums to create works celebrating the glories of nature as revealed by science (Sect. 2 in Chapter 7).[47] Writers like Goethe, Alexander von Humboldt, Ralph Waldo Emerson, and Henry David Thoreau all wrote great texts celebrating the glories of nature. Dawkins urges the composition of Goethean Hymns to Nature. He says that feeding science into poetry makes greater poetry. He says that if Michelangelo had been commissioned by some scientific institution to make some scientifically-inspired art, then he would have created something even more beautiful than his painting for the ceiling of the Sistine Chapel. He says that if musical composers like Beethoven and Haydn had been inspired by science, their music would have been even greater than their religious music. Dawkins says that Verdi might have written an even greater *Dies Irae* had he known about the destruction of the dinosaurs by the Chicxulub asteroid. But all these religious works of art were *celebrations of hope*. They pointed to *redemption* from suffering and death. If the Good does not shine out through the dust of this earth, then there will not be any scientifically-inspired art.

Dawkins often recommends a kind of Stoic *amor fati* (love of fate). He recommends courage in the face of suffering.[48] But the Stoics based their love of fate on the providential Logos. For the Stoics, this is the best of all possible universes. Everything that happens here is for the best. Of course, it is not for the best for you; it is for the best for the whole. Nevertheless, the Stoic cosmos strives for excellence in all its events. If the Good does not animate this universe, then the Dawkinsian appeals to *amor fati* make little sense. And Dawkins goes far beyond Stoic *amor fati*. He offers a *beatific vision of nature* (Sect. 5.1 in Chapter 3). The

[47]Greater poetry (UR 27). Sistine Chapel (GD 111; FH 108). Hymns to Nature (UR 24).
[48]ADC 13, ch. 4.4; UR *ix*; GD 20, 394; FH 24.

Dawkinsian beatific vision of nature is not consistent with the negativities in nature. This is the *atheistic problem of evil* (Sect. 5 in Chapter 1). The correct way to solve the atheistic problem of evil is to point out that the Logos is not utilitarian. It does not optimize happiness; on the contrary, it optimizes *arete*, the excellence that emerges in conflict. But this solution depicts the universe an optimizer. If the Good does not animate this universe, then this beatific vision is an error.

Dawkins portrays our universe as non-randomly skewed towards the production of ever greater complexity. The probabilities on all the arrows in all the physical computers are skewed very far from randomness and towards complexity (Chapter 2). But Dawkins portrays complexity as intrinsically valuable. So the universe is non-randomly skewed towards the creation of ever greater value. This skew is not accidental—the deepest laws of physics (such as the laws of thermodynamics) support this skew. The constants in the basic laws appear to be finely tuned for the creation of value. And the universe itself is extremely complex and therefore extremely valuable. But if there is no ultimate source of value, then these cosmic features have no explanation. They are irrational. Hence Dawkins would be wrong about both the rationality and positivity of existence. Assuming Dawkins is not contradicting himself, he needs the Good.

Dawkins writes that the scientific observation of our universe reveals that "there is, at bottom, no design, no purpose, no evil and no good, nothing but blind pitiless indifference."[49] However, it is not unreasonable to say that blind pitiless indifference *is evil*. If that is right, then Dawkins has directly contradicted himself. And, if the universe is evil in this way, then it is *hideous*, and the correct way to respond to it is *horror*. But Dawkins says we should respond to it with all kinds of positive emotions. We should regard it as a gloriously wonderful place. So it looks like Dawkins has contradicted himself again. The only way to save him from these apparent contradictions is to argue that some deeper value shines out through this indifference. Dawkins himself says there is "a kind of grandeur in nature's serene indifference to the suffering."[50]

[49]ROE 133. See ROE 96 and GSE 395.
[50]GSE 401.

For the sake of this grandeur, all the suffering is justified. This grandeur *redeems* the suffering. It is the Dionysian holiness of existence—and only the Good can generate this holiness.

Dawkins was aroused to something like mystical experience by the glory of our universe (Sect. 4 in Chapter 7).[51] Speaking about mystical experiences with Alan Lightman, Dawkins says "You can't out-transcendence me." Dawkins says his books aim to arouse transcendent wonder, but transcendent wonder requires some transcendent object. He experiences the sublime features of the earth and sky as holy and sacred. If these experiences are meaningful, then they point to some transcendental object. The Good is transcendent in the sense that it is unsurpassable in value. Yet the Good is just a logical object, a proposition, like a law of nature, or a law-of-laws. If the Good does not exist, then these experiences are meaningless. To make sense of these experiences and values in an atheistic context, it is necessary to posit something like the Good.

To make sense out of Dawkins, it is necessary to add something like the Good. Of course, the Good is part of Platonism. You may object that we are very far from Dawkins now. Dawkins himself never talks about the Good. My first reply is that I am just completing the arguments that he didn't complete, and finding the deeper explanations and principles that he didn't seek. My arguments take their premises from his texts. Believing in Dawkins means trying to make more sense out of his texts. Adding the Good helps to make that sense. My second reply is that I'm *building on Dawkins*. I'm using his ideas to develop a naturalistic spirituality that can compete with the old religions. I'm building the Sanctuary for Spiritual Naturalists.

3.2   The Duties of Abstract Existence

The modal library contains all possible cosmic forms. These cosmic forms are the books in the modal library. They are abstract objects. Any book is either *actual* or it is not actual. To say that a book is actual means

---

[51]Mystical experience (TL 59–60; GD 31–2). Holy and sacred (SS). Out-transcendence me (2018: 12:28). Transcendent wonder (GD 33).

that it has a concrete instance. If it has a concrete instance, then there exists some physical universe that is a model of that book (Sect. 2.2 in Chapter 5). There are many ways to think of the relation between books and their instances. Perhaps books stand to universes as scripts to plays, as recipes to meals, as genomes to mature organisms, or as programs to computations; or as blueprints to houses. We don't need to worry about the details now. Right now, we are concerned with the reasons for the existence of physical universes. Here is a physical version of the metaphysical question: Granted that there are books, why are there any actual books rather than none? Why are there any physical universes?

We are working here through pure reason. So we need to start by looking at the reasons for or against the actualization of books. These reasons determine whether or not some book *ought* to be actual (or *should* be actual). To say that some book ought to be actual expresses an *obligation*: its actuality is obligatory. To say that some book ought to not be actual expresses a *prohibition*: its actuality is prohibited. If any books should or should not be actual, then there are some obligations and prohibitions regarding the actualities of those books. These obligations and prohibitions (if there are any) are neither ethical nor moral; ethical obligations involve intelligent social animals; but books are not moral agents. Since axiology is the abstract logic of value, all obligations and prohibitions can be classified as axiological. Now focus on the obligations (the same applies to prohibitions). Axiological obligations divide into two kinds. *Ontic obligations* apply to some class of beings: to the class of humans, or robots, or bees, or aliens, or so on. But *ontological obligations* emerge from the logic of being-itself. Since abstract existence emerges first from being-itself, the first ontological obligations are the duties of abstract existence. If there are any such duties, they are the duties of abstract existence to itself. Are there such duties? If any books should be actual, then such duties exist. So we need to see whether any books should be actual.

Since we are working here through pure reason, we need to move from reasons to obligations. Four axioms link reasons to obligations. Emerging from the dyad, these are among the first principles of deontic logic. (1) If there is any reason against the actuality of some book, then that book should not be actual. (2) If there is no reason against the actuality of some book, then it should be actual. (3) If there is any reason for the actuality of some book, then it should be actual. (4) If there is no reason for the actuality of some book, then that is a reason against its actuality; but then it should not be actual. These four axioms entail the laws in the logic of ontological obligation. These laws follow the structure of the modal library through initial, successor, and limit books.

The logic of ontological duty begins with an *initial law*, which deals with the initial book in the modal library. Thus obligation begins with simplicity. The initial book is the simplest book. A defect is some distortion or perversion of the detailed structure of a book. But any defects require complexity. Since the initial book has no complexity, it has no defects. A defect in a book is some reason to not actualize it. If there are no defects in a book, then there are no reasons against its actuality; but then it ought to be actual. Since there are no defects in the initial simple book, it ought to be actual. If there are no defects in a book, then the book is optimal. So the initial book is optimal. Since defects are the only reasons against actualization, every optimal book ought to be actualized. It follows again that the initial simple book should be actual.

The logic of ontological duty continues through a *successor law*, which deals with the successor books in the modal library. Every book has at least one offspring. If any book is an offspring of some parent, then its intrinsic value can be compared with its parent's intrinsic value (Sect. 5.1 in Chapter 4). There are three cases. (1) If the offspring is less valuable than its parent, then it is a *downgrade*. Obligation does not pass through downgrades. If any offspring is a downgrade, then the change from parent to offspring introduced some defect. It introduced some perversion or distortion of value. But any such defect is a reason against actualizing that offspring; so any downgrade should not be actualized. (2) If the offspring is just as valuable as its parent, then it is an *equigrade*. Obligation does not pass through equigrades. If the offspring does not gain value, then there is no reason to actualize it; but if there is no reason for it, that is a reason against it; its failure to gain value is a defect. Hence it should not be actualized. (3) If the offspring is more valuable than its parent, then it is an *upgrade*. Obligation passes through upgrades. If any book in the modal library ought to be actual, then all of its upgrades ought to be actual. An upgrade is an improvement. The increase in value is a reason to actualize the offspring. If any book is optimal, then all and only its upgrades are optimal.

The logic of ontological duty continues through a *limit law*, which deals with the limit books in the modal library. To get to limits, we must move through progressions. Obligation passes through progressions of upgrades. If any progression of books in the modal library starts with some book that ought to be actual, and proceeds only through upgrades, then that progression is optimal. All the books in an optimal progression ought to be actual. If all those books should be actual, then the progression should be actual. So every optimal progression should be actual. Optimal progressions introduce no defects—that is why they are optimal. Every progression in the modal library has at least one limit in the modal library. Limit books fall into the familiar three cases: any limit is either a downgrade, an equigrade, or an upgrade. If any limit is a downgrade or equigrade, then it distorts or perverts its added complexity; so it should not be actual. It introduces a defect; but defects are reasons against actuality. An upgrade is an optimal limit. Obligation passes through optimal limits. If any progression in the modal library ought to be actual, then every optimal limit of that progression ought to be actual. That is, every improvement of that progression ought to be actual.

These arguments demonstrate that some of the books in the *modal library* ought to be actual. But what about the treasury? It is easy to see that these arguments define exactly the books *in the treasury*. The initial book is in the treasury; every optimal successor is in the treasury; and every optimal limit is in the treasury. Hence *if any book is in the treasury, then it should be actual*. But what about the books that are not in the treasury? The principles we have used to show that some book ought to be actual are the *only* principles that entail that it ought to be actual. For any book not in the treasury, no principles entail that it ought to be actual. But if no principles entail that it should be actual, then it is not the case that it should be actual. If it is not the case that some book should be actual, then it is the case that it should not be actual. That is, if there are no reasons for the actuality of some book, that very absence is a reason against its actuality. Consequently, *if any book is not in the treasury, then it should not be actual*. The books outside of the treasury are the *dark books*. They cannot be reached by iterated improvement from the simplest book. To reach some dark book, it is always necessary to introduce some perversion or distortion of value. But evolution never descends.[52] Dark books are defective. Since they are defective, they should not be actual. It follows that the books that are in the treasury should be actual and the books that are not in the treasury should not be actual. Abstract existence is obligated to actualize all the books in the treasury and it is prohibited from actualizing any of the dark books that are not in the treasury. These are the ontological obligations and ontological prohibitions of abstract existence. They emerge from being-itself: since being-itself maximizes self-consistency, it demands that abstract existence does its duty. If all and only the books in the treasury are actualized, then the duty of abstract existence is done.

3.3   The Best of All Possible Propositions

The modal library is a collection of books (a *class* of books). Since every book in the modal library defines a possible universe, every class of books defines a class of possible universes. The philosopher Klaas Kraay says that a *possible world* is a class of possible universes.[53] So every class of books from the modal library is a possible world. Possible worlds are *not* possible universes. A class containing a single book is not identical with the single book it contains. The *empty world* is the world that contains no books. The *maximal world* is just the modal library, which contains all books. The worlds are ordered by inclusiveness. So the empty world is the smallest world while the modal library is the biggest world. This world is bigger than that world if this world contains every book in that world plus some extra books not in that world.

Just as books either should or should not be actual, so worlds should or should not be actual. To say that a world should be actual means that every book in that world should be actual. On the one hand, if any world contains any books that should not be actual, then that world should not be actual. Actualizing that world would violate the prohibition against

---

[52]CMI 132–6.
[53]Kraay (2011: 365).

actualizing dark books. Any world that should not be actual is a *dark world*. Since dark worlds should not be actual, their values are negative. On the other hand, if any world contains only books that should be actual, then that world should be actual. Any world that should be actual is a *bright world*. By default, every book in the empty world should be actual. Of course, since the empty world contains no books, and since value is contained in books, the empty world has no value. Its value is zero. The empty world is a bright world; however, it is the least of all the bright worlds. It is the dimmest of all possible worlds. Bright worlds (like all worlds) are ordered by inclusion. They accumulate value as they accumulate books. Bigger bright worlds are better than smaller bright worlds. The biggest bright world is the treasury. Since it is the biggest bright world, it is the best bright world. It is the brightest of all possible worlds. Since the treasury includes all the books that should be actual, and excludes all the books that should not be actual, *the treasury is the best of all possible worlds*. Of course, *possible worlds are not possible universes*. No universe is best. Possible universes are books; but every book is surpassed by better books; hence no universe is best.

The *actual world* is the greatest collection of actual universes, which are physical things. Of course, since any actual object must also be possible, the actual world is also a possible world. Exactly one possible world is actual. For if no world were actual, then the empty world would be actual. And if many worlds were actual, then their union would be actual. Therefore, exactly one possible world is actual. Every possible world is associated with some proposition which asserts that all its books are actual. Any such proposition is an *actualizer* for that world. If the actualizer for some world is true, then all the books in that world are actual. If the actualizer is false, then those books are not all actual. Any proposition that does not assert that some book is actual is equivalent to the actualizer for the empty world. For example, the propositions that one plus one is two, or that Socrates is human, do not assert that any book is actual. So they are equivalent to the actualizer for the empty world. If the biocosmic analogy is right, then actualizers are associated with animats. If the actualizer for some world is true, then every book in that world has some animat. If any book has some animat, then that animat takes that book as input and makes some concrete universe as output.

Actualizers are ranked by value as their worlds are ranked by value. Since dark worlds should not be actual, their actualizers have negative values. Since the empty world has no books, its actualizer has zero value. Any proposition that is equivalent to the actualizer for the empty world likewise has no value. Since bigger bright worlds are better bright worlds, the actualizers for bigger bright worlds are better than the actualizers for smaller bright worlds. Since the treasury is the best of all possible worlds, its actualizer is the best of all possible actualizers. But all propositions that are not actualizers have no value; and those that actualize dark worlds have negative value. So the actualizer for the treasury is better than all other propositions. It is therefore the best of all possible propositions. This best proposition is *the Good*.

4   The Naturalized Ontological Argument

4.1   The Anselmian Ontological Argument

An *ontological argument* reasons from some purely abstract first principles to the existence of at least one concrete thing. It answers the concrete metaphysical question: why are there any concrete things rather than none? Theists say the first concrete thing is God; since God creates the universe, all other concrete things follow. Among the many ontological arguments for God, the most famous was composed by Anselm in the second chapter of his *Proslogion*. Some say Anselm gave two ontological arguments; if so, then this is his first argument. Focus on this argument only. Since many writers have found Anselm's (first) argument obscure, they have worked to clarify it.

An extremely clear interpretation of Anselm's ontological argument has been offered by the philosopher Peter Millican.[54] I summarize it here. The term *form* is used in this argument to denote an abstract

---

[54]Millican (2004: 457–458).

object which may be instantiated by a concrete thing. So the *ontological argument over forms* (OF) looks like this: (OF1) There are some forms. (OF2) These forms are ordered by greatness. (OF3) There exists exactly one greatest form *Divinitas*. (OF4) Forms are either instantiated by things or are not instantiated. (OF5) Some forms are instantiated (e.g. humanity is a form which is instantiated by Socrates). (OF6) Any instantiated form is greater than any uninstantiated form. Now the argument proceeds by *reduction to absurdity*. Philosophers refer to this kind of argument as a *reductio*. To make a *reductio*, you assume something, and then show that the assumption leads to a contradiction—hence the assumption must be false. (OF7) Assume for *reductio* that Divinitas *is not* instantiated by any thing. (OF8) If Divinitas is not instantiated, then some other forms are greater than Divinitas (e.g. humanity is greater than Divinitas). (OF9) But then Divinitas is not the greatest form. (OF10) So the assumption that Divinitas is not instantiated by some thing leads to a contradiction. (OF11) Therefore, Divinitas *is* instantiated by some thing. (OF12) Following tradition, Divinitas is instantiated by God. (OF13) Consequently, God exists.

The ontological argument over forms is valid—it obeys the laws of logic. But are its premises true? The argument begins with some Platonic premises. This Platonism asserts that (OF1) there are some forms; (OF4) forms are either instantiated by things or not; (OF5) some forms are instantiated. If this Platonism is false, then the argument fails. Of course, we've argued for Platonism—so we can't complain about these premises. But Platonism isn't distinctive here—lots of arguments assume these Platonic premises. The premises which are distinctive for the ontological argument are that (OF2) the forms are ordered by greatness; (OF3) there exists exactly one greatest form; and (OF6) any instantiated form is greater than any uninstantiated form.

The second step (OF2) states that forms are ordered by greatness. The traditional way of ordering forms by greatness is the classical *great chain of being*. And indeed the old Degrees of Perfection arguments for God used the classical great chain. Anselm developed a Degrees of Perfection argument in his *Monologion* (ch. 4). It was the inspiration for his ontological argument in the *Proslogion* (ch. 2). But the classical great chain is not sound. So any effort to defend the second step will have to replace it. And the classical great chain can be successfully replaced by the library of all possible physical things (the physical library). The physical library sorts forms into ranks based on their complexities. It contains things for all finite ranks. As these ranks grow really high, they become filled with possible technological forms. They become filled with possible computers. The infinite rank contains the forms of infinite computers.

The third step (OF3) states that there exists exactly one greatest form. The physical library contains many infinite forms. Could one of these be the greatest? Can there be anything greater than infinity? Modern mathematics says that every infinity is surpassed by greater infinities. Just as there is no greatest infinite number, there is no greatest infinite form.[55] The set-theoretic way of thinking about forms confirms this objection. If forms are sets, then it follows that for every number in the ordinal spire, there exists a non-empty rank of forms. Forms in ranks indexed by greater ordinals are greater forms. So there exists an iterative hierarchy of forms, which rises endlessly with the ordinals. There are no unsurpassable ordinals, no unsurpassable sets, no unsurpassable forms. It follows that every computer is surpassed by greater computers.[56] But Dawkins says minds are computers; so every mind surpassed by greater minds. And if gods are minds, then every god is surpassed by greater gods. Likewise all universes are surpassed by greater universes. No form is greatest. The third step is false.

The sixth step (OF6) states that a form with instances is greater than a form without instances. For any forms F and G, if F is instantiated but G is not, then F is greater than G. This step asserts the *superiority of instantiation*. It *does not* treat existence as a predicate, property, or perfection. Technically speaking, it uses the existential quantifier to handle existence.[57] But it seems easy to defeat the superiority of instantiation. Let *tyrannicality* be the quality common to all evil tyrants; let *sagacity* be the quality common to all good sages. Unfortunately, tyrannicality has been instantiated many times. But many have argued that sagacity has never been instantiated: it is an ideal which no human has ever realized.

---

[55]Hartshorne (1967: 19–20, 1984: 7).

[56]Steinhart (2003, 2014).

[57]The principle of the superiority of instantiation is formulated using quantifiers like this: for any forms F and G, if $(\exists x)\,(F(x))$ and $\sim(\exists x)\,(G(x))$, then $F > G$.

Now the sixth step states that tyrannicality is greater than sagacity *just because* there have been evil tyrants. But that is absurd.

The ontological argument over forms fails at its third and sixth steps. Since forms play an essential role in this argument, it seems likely that the fault lies with the forms. Fortunately, forms are not the only type of abstract object. Perhaps we can improve the ontological argument by switching from forms to propositions. By replacing forms with propositions, we get a new argument. If (OF1) talked about forms, then (OP1) talks about propositions. Thus (OP1) states that there are some propositions. The next two steps are (OP2) the propositions are ordered by greatness and (OP3) there exists some greatest proposition. But what replaces instantiation? While forms are either instantiated or not instantiated, propositions are either true or false. With these changes, we are ready to move to a propositional version of the ontological argument.

4.2 In the Crucible of Reason Forged

Moving from forms to propositions yields the *ontological argument over propositions* (OP). Taking a term from Plato, I will refer to this argument as the *Agathonic Argument*.[58] The Agathonic Argument is the naturalized ontological argument. It runs like this: (OP1) There are some propositions. (OP2) The propositions are ordered by goodness. (OP3) There exists a unique best proposition *the Good*.[59] (OP4) Propositions are either true or false. (OP5) Some propositions are true (for example, that Socrates is human). (OP6) Any true proposition is better than any false proposition. (OP7) Assume for *reductio* that the Good is not true. (OP8) If the Good is not true, then some other proposition is better than it. (OP9) But then the Good is not the best proposition. (OP10) So the assumption that the Good is not true leads to a contradiction. (OP11) It follows that the Good is true. Moreover, it is necessarily true. The atheist

---

[58]Plato defines the Form of the Good as *tou agathou idea* (*Republic*, 508e2–3).

[59]Lokhorst (2006: 385) introduces the proposition *e* in his discussion of Andersonian deontic logic. He says *e* means that any proposition which ought to be true is true. Thus *e* is ($\forall$P) (OP $\Rightarrow$ P). Spiritual naturalism says *e* is the Good.

Iris Murdoch said she wanted "to use Plato's images as a sort of Onto-logical Proof of the necessity of the Good."[60] The Agathonic Argument proves the necessity of the Good.

The Agathonic Argument is valid; it remains to examine its premises for truth. The Agathonic Argument contains some Platonic premises. If these Platonic premises are false, then the argument fails. We argued for these premises earlier (Sect. 2.2); but we take them to be axiomatic. Here are the premises with their justifications. It is a Platonic axiom that (OP1) there are some propositions. It was proved above that (OP2) the propositions are ordered by goodness (Sect. 3.3). It was also proved there that (OP3) the Good is the unique best proposition. It is an axiom of standard logic, known as the law of the excluded middle, that (OP4) propositions are either true or false. It is an obvious truth that (OP5) some propositions are true.

The sixth step (OP6) asserts that a true proposition is better than a false proposition. Since the Greek word for truth is *aletheia*, this is the *alethic axiom*. Dawkins places an extremely high value on truth.[61] He talks about the sacredness of truth. Hence it is consistent with his principles to affirm the alethic axiom. Still, it needs to be supported by some arguments. The first argument for this axiom relies on the thesis (already supported in the step OP2) that the goodness of any proposition is the degree of goodness which it asserts. A proposition may assert that some paradise island exists; or that the set of all sets exists; but if they are false, then they assert no goodness at all. False propositions cannot be better than true propositions. Another argument for the alethic axiom runs like this: Since any effort to deny the alethic axiom must rely on valid inference from true premises, it seems that any effort to deny it is self-refuting. It is an axiom whose denial is self-inconsistent. The truth of the alethic axiom emerges directly from the from the reflexivity of being-itself. It emerges as the dyad maximizes self-consistency. It is one of the fundamental principles of deontic logic. When the denial of a proposition is a contradiction, philosophers say that proposition is *analytic*. So, the alethic axiom is analytic. Analytic truth is the strongest possible

---

[60]Murdoch (1992: 511).
[61]ROE 31–2; UR 21, ch. 6; ADC chs. 1.2, 1.7; GD 18–9, 319–20; GSE ch. 1; SITS 7, 26.

truth—it has the strength of logic itself. It is therefore logically necessary that truth is more valuable than falsehood. The alethic axiom is necessarily true; therefore, it is true.

The steps (OP7) through (OP11) are just logical; spiritual naturalists assume that logic defends itself, so these steps require no extra defense. The concluding theorem (OP11) states that the Good is true. The Good is the actualizer for the treasury. Since the Good is true, it follows that every book in the treasury is actual. Since there are no arguments that actualize the dark books, they are not actual. Every bright book has an animat. Hence the Good entails the existence of an animat for every book in the treasury. Since every bright book has an animat, it has a physical model, which is a concrete universe. Both animats and universes are concrete. So the Agathonic Argument reasons from some abstract first principles to some concrete things. Consequently, it is an ontological argument; but it is naturalistic rather than theological.

Being-itself expresses the ontological duty of abstract existence: abstract existence ought to actualize every book in the treasury. Since the Good is true, abstract existence does its ontological duty. The Logos includes all the premises that go into the Agathonic Argument. Hence the Logos includes the Good. Since the Good is true, concrete existence is for the best. The Good explains why nature at all scales, from atoms to universes, is skewed towards the creation of intrinsic value. Since the Good is true, the Logos is providential. The Good justifies the Dawkinsian beatific vision of nature (Sect. 5.1 in Chapter 3), it justifies his appeals to *amor fati*. The Good entails an atheistic *logic of hope*. Since the Good is true, the logic of hope is true. This logic of hope motivates scientifically-inspired works of art (Sects. 2–3 in Chapter 7), and scientifically-inspired rituals of transformation (Sect. 4 in Chapter 9). This hope makes our naturalism spiritual.

Somebody may want to raise an argument from evil against the Good. There are several replies. The first reply is that the Good maximizes *ontological* goodness; it maximizes the goodness in which all beings participate. The Good maximizes the value of being. It does *not* optimize *ontic goodness*; it does *not* optimize the goodness of these beings as opposed to those other beings. Like the stars, the Good has no specific concern with humans or other organisms.[62] The second reply is that only the whole is good without constraint. This whole is the best of all possible worlds, which is the plurality of actual universes. The parts of the whole are good in only constrained ways. The goods of one part constrain the goods of other parts. It is consistent with the goodness of the whole that the parts conflict with each other. The third reply is that the Good does not entail the minimization of suffering or the maximization of happiness. *The Good is not utilitarian.* The Good entails the maximization of intrinsic value—but that value is a kind of complexity. For life on earth, the maximization of that complexity happens through evolution by natural selection. Biological evolution moils teleonomically to maximize *arete*, the excellence that emerges in the strife-torn *agon*. Since the Good is true, evolution is good. All the suffering of life serves the Good and is redeemed by that service. The Good justifies *evodicies* (Sect. 1.3 in Chapter 9). The Good does not maximize the happiness of concrete being; it maximizes the *holiness* of that being.

As the Good, the best proposition can be represented metaphorically or poetically as the sun. Figure 4 shows the Good as the sun over the Platonic sky, over the ground of being, and over the oceanic abyss. Since the Good is the sun, and the One is the earth, the Good is not the One.[63] The One is being-itself; but the Good is the best proposition. They are not the same. Just as the sun occupies the highest place in the sky, so the Good occupies the highest place in the Sanctuary for Spiritual Naturalists. Just as the sun transcends all things, so the Good transcends all things. Just as the inclusiveness of every unsurpassable class is ideal, so the *meaning* of the Good is ideal. And the meaning of the Good is absolutely pure: it is pure self-overcoming; pure self-improvement; pure self-betterment; pure *for-the-bestness.* A long tradition identifies *holiness* with pure goodness.[64] Hence *the Good is holy.* Just as light shines out of

---

[62]UR 117.

[63]Later Platonists like Plotinus identified the Good with the One. But the Plotinian reasons for identifying the One with the Good are weak (Jackson 1967: 322; Mortley 1976: 49; Gerson 1994: 19–20). Even within the Plotinian system, the One and the Good play very different roles. The One is not the Good.

[64]Nietzsche says holiness arises from cleanliness (*Beyond Good and Evil*, sec. 271) while Tillich contrasts holiness with uncleanliness (1951: 215–217).

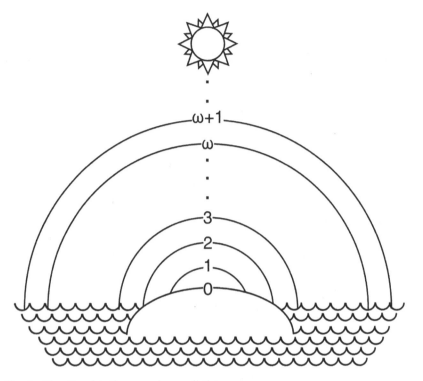

Fig. 4 The Good is the sun above all things

the sun, so truth shines out of the Good. It shines with holy light. Just as the sun radiates the energy which drives the evolution of life on earth, so the Good radiates the power which drives the evolution of concreteness. Just as the sun pumps physical entropy out of our solar system into space, so the Good pumps Platonic entropy out of the pleroma into the abyss. As it pumps out this entropy, the world tree rises into the sky.

4.3   The Transcendence of the Good

The books in the treasury are stratified into floors. Since those floors form a series, some of the actualizers form a series. The actualizers in that series are the *lifters*. There exists exactly one lifter for every ordinal number. The n-th lifter asserts that if any book in the treasury is on or below the n-th floor, then it is actual. It therefore asserts the actuality of all and only the books on the first n floors of the treasury. Lifters with lesser numbers are lower lifters. The lower lifters nest inside of the higher lifters. If the n-th lifter is true, then all the lifters less than n are true. If any lifter is true, then all the books in its world are actual; they all have animats and physical universes.

Since the treasury has a floor for every ordinal number, the treasury itself surpasses all its floors. The treasury is a proper class of books. Consequently, the actualizer for the whole treasury cannot be any lifter. If the Good actualizes the entire treasury, then it cannot be any lifter. Just as the treasury surpasses every floor, so the Good surpasses every lifter. The only way for the Good to surpass every lifter is for it to entail every lifter. Since the Good is true, this means that every lifter is true. Hence the Good asserts that for every n in the ordinal spire, the n-th lifter is true.[65] The Good is a beautifully simple proposition. The Good asserts that for every surpassable number in the unsurpassable spire, the lifter with that number is true. The Good therefore incorporates the unsurpassability of the ordinal spire into itself. The Good has the rank of an unsurpassable class. The Good is the ecstasy of the lifters. It dwells in the wilderness of the sky; it sits like the sun at high noon above the *axis mundi*. So while the Good is an extremely simple proposition, its meaning is unsurpassably complex. Like the pleroma, the Good is ineffable, sublime, numinous, transcendental, and ideal.

Does the Good violates Dawkinsian principles? Specifically, it might be objected that the Good contains unexplained complexity, which Dawkins rejects. There are two replies. The first comes from contingency. All complex concrete things are contingent: they might not exist. Since they might not exist, the fact that they do exist demands an explanation. Moreover, for concrete things, complexity goes with improbability. Since complex concrete things are improbable, their existence requires explanation. However, since abstract objects are not contingent, their complexities do not require explanations. Moreover, for abstract objects, complexity does not go with improbability. All abstract objects exist with

---

[65]If the n-th lifter is symbolized as $L(n)$, then the Good states that (for all n) (it is true that $L(n)$). Stated in the austere symbolism of formal logic, the Good is $(\forall n)\,(L(n))$.

the equal probability of one. Hence the Good exists as a matter of logical necessity. Its complexity requires no explanation. The second reply comes from the definition of the Good. The Good exists, in fine Dawkinsian fashion, at the end of an absolutely infinitely long series of increasingly complex objects. The series of lifters starts out simple and slowly gains complexity. So even here, in the abstract domain, complexity gradually accumulates. The Good respects Dawkinsian principles.

5   The World Tree

5.1   This Burning Shrine

The Good actualizes the treasury. This means that the Good actualizes every book in the treasury. The Good entails that every book in the treasury gets an animat; its animat uses that book as a recipe for creating a physical universe (Sect. 2.2 in Chapter 5). Focus on the books and the universes. The truth of the Good resembles the light of the sun. Just as the sun *emanates* light, so the Good emanates light. This light shines out of the Good until it strikes something which reflects it back to the Good. A book reflects the light of the Good if and only if it is actualized—its concreteness turns it into a mirror. Every book in the treasury is a little mirror which reflects some of the light of the Good. The treasury itself is that structure in which the Good (poetically) sees its reflection. The dark books do not reflect the light of the Good—that light shines right through them.

The books in the treasury are organized into an infinitely ramified tree. Since those books form a tree, the actual world also forms a tree. The actual world takes the shape of a great *world tree*.[66] The world tree is defined by three laws. These laws follow directly from the laws of the treasury and the truth of the Good. The *initial law* states that the root of the world tree is just the initial universe. The initial universe is just the initial animat, namely, Alpha. The *successor law* states that every universe in the world tree is surpassed by at least one successor. Every pair

---

[66]Plotinus (*Enneads*, 3.3.7.10–25, 3.8.10.10–20).

consisting of a universe and one of its successors is an arrow or branch in the world tree. The *limit law* states that every progression in the world tree is surpassed by at least one limit. Every pair consisting of a progression and one of its limits is also a branch in the world tree. So the world tree is a proper class of branches. Every universe in the world tree is a concrete instance of an abstract cosmic form. Since these cosmic forms exist in the abstract sky, their instances also exist in that sky. If the world tree is represented as the graph or network of its branches, then it rises endlessly into that sky. It rises from the earth to the sun, from the One to the Good. Since the world tree exists if and only if the Good is true, the world tree is a concrete model of the Good. The world tree shines with the reflected light of the Good. It stands blazing in the darkness of the night.

The initial law for the world tree states that its root is the initial universe. The successor law states that every universe is surpassed by at least one successor. The branch from any universe to any successor is some way that the universe can surpass itself. And if it can surpass itself, then it does surpass itself. Thus every universe in the world tree surpasses itself in every way. But if every universe surpasses itself in every way, then every universe contains some *power of self-surpassing*. Therefore, every universe contains some power of self-surpassing. But if every universe contains some power, then these powers exist. Are all these powers different or the same? All these powers have the same source, namely, the Good; consequently, all these powers are the same power. So there exists a power of self-surpassing which animates every universe. Of course, this argument is easy to extend to progressions and limits.

The power which animates every universe flows like sap through the veins of the world tree. Our poetic symbol for this power is *fire*. The Stoics thought the universe was animated by a divine fire-energy.[67] This fire-energy (the *pneuma*) produces every next universe in the Stoic cosmic cycle. The Platonist Iamblichus often talks about a holy fire-energy which animates all things.[68] The world tree is a burning shrine. Fire flows through every universe in the world tree and through every thing in every

---

[67] Cicero (*On the Nature of the Gods*, 2.23–8).
[68] Iamblichus (*On the Mysteries*, 1.8–9, 1.12, 2.4, 3.20, 4.3, 5.11–12).

universe. This fire is the reflection of the Good in every thing. Since the Good is holy, this fire manifests its holiness. Strictly speaking, this fire is only virtuous; but the ideal finality at which this virtue aims is the Good. The ecstasy of virtue is holiness. Poetically speaking, this fire is holy. All things burn with holy flame. This fire is actuality. It emerges when the light of the Good strikes the initial cosmic form. Once illuminated, this form bursts into flames. This fire rises through the sky, *returning* to the Good.

Our Platonic image is now complete. The Platonic image begins with the oceanic abyss of non-being; this abyss is *water*. The island of being-itself emerges from that ocean; that island is *earth*. The sky of abstract objects rises up infinitely above the earth; that sky is *air*. The Good stands in that abstract sky like the sun; the Good is *light*. The light illuminates the treasury; through that illumination, every book in that library bursts into flames. It becomes an actual physical universe. All these actual universes form the great world tree, which corresponds to *fire*. This complete Platonic image is shown in Fig. 5. It can serve as the basis for further poetry.

5.2   How Nature Reflects the Good

There are many scales of nature. At the smallest scale, people often say nature is just our earthly wilderness; at a larger scale, nature is just our physical universe; at the largest scale, nature is the plurality of all actual universes. Since the plurality of actual universes is the world tree, *nature at the largest scale* is the world tree. The world tree contains a universe for every book in the treasury. Since the treasury is an unsurpassable class of books, the world tree is an unsurpassable class of actual universes. Hence nature (at the largest scale) is an unsurpassable class of actual universes.

Nature is ordered into phases by the lifters. Every lifter actualizes a phase of nature. The n-th phase of nature contains an actual universe for every book up to and including the n-th floor of the treasury. So the phases of nature are nested: higher phases include lower phases. Nature itself is the union of all these phases. It has a phase for every number in the ordinal spire, the *axis mundi*, the vertical axis of nature, which rises

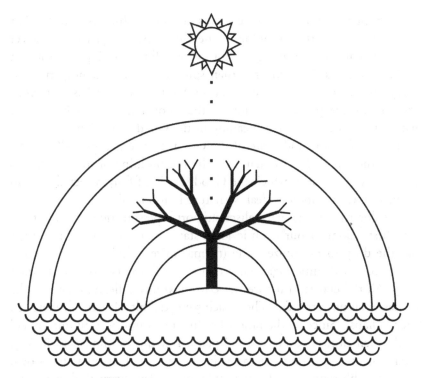

Fig. 5 The tree grows out of the ground towards the sun

from the earth to the sun. Since the *axis mundi* is an unsurpassable series of surpassable numbers, nature is an unsurpassable series of surpassable phases.

Since nature is an unsurpassable class, nature reflects itself. All the reflection principles that apply to the pleroma apply to nature: every definite property of nature is reflected back to nature by some part of nature. Nature is ineffable. If you think you have fully described nature, it will always turn out that you have merely described some smaller phase of nature. Nature escapes from every attempt to fully define it; it throws off every attempt to tame it. Nature is wild; it dwells in the wilderness of the unsurpassable classes. Thus nature is sublime, numinous, and transcendental. Nature maximizes reflexivity. As the cosmic crane, nature moils towards an ideal ecstasy. The self-negation of the abyss drives nature to attain its ecstatic ideal.

Each phase of nature corresponds to some lifter. But each lifter expresses some partial meaning of the Good. And higher lifters reflect more of that good meaning. So for every definite axiological property, if the Good has that property, then some lifter has that property. But every lifter actualizes a phase of nature. So if some lifter has that property, then some phase of nature has that property. Thus each phase of nature reflects some partial meaning of the Good. The higher phases of nature reflect more and more of the Good. Hence nature itself entirely reflects the Good. Nature manifests the full meaning of the Good. It is the concrete mirror in which the Good sees itself fully reflected back to itself. Just as the Good is ideal, so nature is also ideal.

But doesn't spiritual naturalism support a pessimistic view of nature? The first axiom of our deontic logic states that every agent in nature has the duty to maximize value (to maximize reflexivity). The second axiom of our deontic logic states that rational agents strive to do their duty. All the cranes in nature strive to maximize reflexivity. But all the cranes in all universes fail. They reach some definite degree of reflexivity, then collapse. Since all the rational agents in all universes ultimately fail to do their duties, nature is a sad place. But this pessimism is illusory. Every agent in nature participates in its own ideal. It inhabits its own unsurpassable sequences of surpassable agents. And every ideal agent does its duty; it succeeds in doing what it strives to do.

Spiritual naturalism supports an optimistic view of nature. The duty of every agent in nature is to maximize reflexivity. Although every agent in nature does its duty only incompletely, nature does its duty completely. Although all the agents in nature fail to do their duty, nature succeeds. As Plotinus said, evil is in the parts, but the whole is good. Nature is a rational axiological agent. Since nature does its duty completely, nature is axiologically ideal. If any agent is axiologically ideal, then it participates completely in the holiness of the Good. Consequently, nature is holy. Of course, the parts of nature vary in their degrees of holiness. The holier parts can be sacred for humans.

Since the Good entails the existence of nature, its power is immanent in nature. This immanent power expresses itself in every universe as a maximum value production principle (MVPP). The Good is the ultimate skewer. It skews the probabilities of all arrows towards greater excellence. The Good skews the probabilities on the cosmic arrows, so that the animats only reproduce along upgrades. Hence the biggest concrete arrow, the cosmic arrow, points towards the Good. It is the ultimate ecstasy toward which all things moil. The immanent power of the Good, the power in the logical core of every universe, is axiotropy (Sect. 4.3 in Chapter 3). Animated by axiotropy, all universes moil towards the Good. This moiling manifests itself in the physical laws of particular universes. Every evolutionary arrow in every universe points towards the Good. The Good expresses itself in our universe as the maximum entropy production principle (MEPP) or something like it (Sect. 1.3 in Chapter 3). Since all things moil towards the Good, they moil to more accurately reflect its meaning in their own natures. Hence the Good is the ultimate explanation for the duty to maximize reflexivity.

5.3   The Ultimate Computer

Each animat resembles the Platonic demiurge.[69] The Platonic demiurge began with some cosmic form which it used to generate some physical universe. But here each animatic demiurge just decorates a mathematical cosmos with physical string (Sect. 2.2 in Chapter 5). The Platonic demiurge is part of a mythical story. And the animats are parts of our Platonic mythology. It is time to clarify this myth. The animats seem like cosmic organisms that give substance to the biocosmic analogy. Thinking about them serves several useful scientific purposes: It can help us think about the genotypes for universes. It can help computer scientists think about ways to make scientific models of evolution at the cosmic scale. It can inspire physicists to think about progressions of increasingly complex universes in the landscape of all physical possibilities. These are intriguing research programs. Nevertheless, the animats are not what they seem.

The animats read books in the treasury. Of course, every book in the treasury is also in the modal library. These books are cosmic texts. Each

---

[69]Plato (*Timaeus*, 27d–29b).

cosmic text is like a series of sentences in some Platonic language. It is a series of sentences in some purely logical language, like the predicate calculus plus mathematical symbols. Only one mathematical symbol is needed: the membership symbol; other symbols are derived. But if each book is like series of sentences in some Platonic language, then it is a series of symbols in that language. Each book is a string of symbols taken from some Platonic alphabet. All these books are linked by arrows. From every book in the modal library, to every other book in the modal library, there exists an arrow. Each arrow looks like this:

$$book \rightarrow book.$$

Since each book is a string of symbols, each arrow defines a string rewriting operation. Following the concept of computation developed by Emil Post (Sect. 2.3 in Chapter 2), the modal library plus all its arrows is a computer. More precisely, these books and arrows make up the hardware of the cosmic computer. Its software resembles the software of the other computers in Chapter 2. This software is just an assignment of transition probabilities to the arrows in the cosmic hardware. These probabilities are assigned by the lifters. But the series of lifters is equivalent to the Good. Consequently, the Good skews the probabilities in the cosmic software very far from random. It skews them so that the only books that get actualized are the books in the treasury.

The cosmic software defines the flow of presence through the cosmic computer. This presence is the holy fire that flows through the cosmic arrows in the world tree. As presence flows from book to book, so animats reproduce from book to book. Each animat reads its book. Each book encodes a purely mathematical cosmos. It describes a purely set-theoretic structure. Each set-theoretic cosmos is a connect-the-dots network in which the dots are sets and the connections are membership arrows. When an animat reads its book, it actualizes the cosmos described in that book. Since each cosmos is a set-theoretic structure, it is a set with members. The animat actualizes that set by tying a loop of physical string to it. This loop of physical string is an instance of the is-present-to relation. Hence an actualized set is present to itself. If an animat actualizes a set, then it actualizes every member of that set. It ties a loop of physical string to each actualized member. It ties a length of physical string from that member to the set that contains it. This length of physical string is also an instance of the is-present-to relation. So each actual member is present to its containing set. The animat thus decorates its set-theoretic cosmos with loops and lengths of physical string. It brings these loops and lengths of string into existence. So you can picture the animat as a little demiurge with creative powers. *Or you can just identify the animat with the system of physical loops and lengths of strings.* The animats disappear into their decorations. Nevertheless, it is more convenient to poetically picture the animats as mythical organisms.

The Good skews all the probabilities in the cosmic computer towards increasing complexity and value. The animats reproduce along and only along the arrows with positive probability. Hence the animats moil towards the Good. They collectively run an entelechy. They constitute a massively parallel distributed computation, which runs on the cosmic hardware. As the swarm of animats rises up into the sky, the *cosmic arrow* rises up from this swarm. Like other arrows, this cosmic arrow points at the Good, the Good is its ecstasy. But this pointing involves no mentality—the Good is not a goal. The cosmic arrow points teleonomically, but not teleologically. Its pointing is rational rather than mental. Like other arrows, the cosmic arrow is a crane—it is the *cosmic crane*. And just as all other arrows are parts of the cosmic arrow, so all other cranes are parts of the cosmic crane. It is that crane than which no greater is possible.

Cranes are powered by axiotropy. Axiotropy is derived from the Stoic concept of *pneuma* or spirit. For the Stoics, spirit was fire-energy; spiritual naturalists use this fire symbolically; axiotropy is fire; hence every animat is a tongue of flame. This fire powers the Dawkinsian spotlight that shines in every physical universe (Sect. 1.4 in Chapter 8). When some lifter receives the light of the Good, every book in that lifter receives that light; but when some abstract book receives that light, it emanates a concrete image. The transition from abstract to concrete is symbolized by the transition from light to fire. The luminosity of the abstract book becomes a fire burning in the logical interiority of its concrete model. If an animat reads some book in the treasury, then it works on some set in the iterative hierarchy. But that set burns with fire. Hence a proper part of the iterative hierarchy dances with flames. The sets in the treasury are burning, but they are not consumed. The world tree stands in a whirling inferno. Thus *nature is a tree on fire in the sky*. This eternally blazing tree is the mirror which reflects the sun.

6   Lighting Fires

As I build the Sanctuary for Spiritual Naturalists, I talk about many metaphysical objects, objects which sophisticated theologians have identified with God. These include: being-itself, the One, the Logos, the necessary being, the simple first cause, the Good. If all these things were identical with God, then they would all be identical with each other. Since they are not all identical, they are not all identical with God. The monotheists can claim *at most one* of these things. Dawkins says they can claim none. To identify them with God is to engage in theological doublespeak.[70] On the right hand, offer these objects to well-educated people; on the left hand, offer the God of scripture to your congregation. The right hand knows not what the left hand does.

But perhaps the accusation of theological doublespeak is unfair—it makes those theologians look like hypocrites. Perhaps there is a more charitable way of understanding sophisticated theology. This way starts with the philosophical legacy of the ancient pagan Greeks and Romans— with the ethical and metaphysical ideas of the Stoics and Platonists. Those pagan ideas were taken up by Christians, and the Christian theologians developed them in their own religious ways. On this interpretation, the sophisticated theologians are just keeping the old pagan ideas alive inside the Christian religion. For example, when Paul Tillich says that being-itself is God, he is really just keeping an old pagan idea alive inside Christianity. There is nothing either logically or ethically objectionable about Christians taking these old pagan ideas and developing them in their own ways—they are free to do that. Of course, here atheists can claim equal rights: we are free to do this too. Iris Murdoch is free to develop the old Platonic concept of the Good in an atheistic way.

---

[70]Theological doublespeak (1996b; OTTR 399; TL 67; GD 33, 83–4; AK 190).

Andre Comte-Sponville is free to develop the old Platonic concept of being-itself in an irreligious way.

Consider ancient Stoicism. The Stoics argued for abstruse metaphysical doctrines, which they worked up into their own sophisticated pagan theologies. The Stoics participated in pagan religious practices. They practiced *haruscipy*, which is divination by examining the entrails of sacrificed animals. If you're looking for a book devoted to the glory of God, try the *Discourses* by the Stoic writer Epictetus. Stoic theology entered Christianity, so that old pagan ideas became bound to the Christian God. Nevertheless, the old Stoic ideas are not owned by Christians. Atheists are free to take those old pagan ideas and to develop them in their own new ways. Now consider the contemporary philosopher Massimo Pigliucci. His book *How to Be a Stoic* develops the old Stoic ethical and metaphysical ideas in new ways, ways that are atheistic and naturalistic. Modern Stoics do not practice haruscipy. But you can go to weeklong Stoic Camps which involve both spiritual practices and rituals like greeting the rising sun. Although modern atheistic Stoicism often looks religious, I prefer to use a different term. I prefer the term *meturgy*, which means change-working (Sect. 4 in Chapter 9). Meturgical symbols, practices, and institutions enter into *meturgical cultures*. Meturgical cultures compete with religious cultures. Modern atheistic Stoicism is meturgical. So is the atheistic Buddhism developed by Sam Harris, Stephen Batchelor, and others.[71]

Dawkins identifies as a cultural Christian (Sect. 2.2 in Chapter 7).[72] On this point, spiritual naturalism strongly disagrees. We seek to develop meturgical cultures that compete with the cultures of the Abrahamic religions. We seek our own atheistic institutions, practices, and symbols. We study ancient pagan philosophies, like Stoicism and Platonism, in order to build new ways of thinking and living. For example, this Chapter linked four objects from Platonic metaphysics with the four classical elements of earth, water, air, and fire. Look now to the atheistic philosopher Donald Crosby, who calls himself a religious naturalist. Dawkins

---

[71]Batchelor (1997) and Harris (2014).
[72]Cultural Christian (SITS 246; see also GD 383–7).

dislikes "religious naturalism" because he thinks it is confusing.[73] So let us call Crosby a spiritual or meturgical naturalist. Whatever you call him, Crosby talks about symbols and rituals.[74] He says earth, water, air, and fire can be used as "symbols of nature as the religious ultimate." He urges the creation of naturalistic rituals which use the four elements as symbols, and those rituals can also include the fifth element of light. Crosby encourages "rituals orienting to the four points of the compass, suggesting fealty to the whole of the earth and its creatures." Those cardinal directions are easily integrated into the symbolism developed here. Crosby encourages "rituals recognizing the equinoxes and solstices." The solar symbolism used here suggests that the solar holidays will hold special meaning for spiritual naturalists.

Spiritual naturalists are free to develop the old pagan philosophical ideas in many new ways. While these new ways may be spiritual, they will be atheistic; and while they may be meturgical, they will be irreligious. Spiritual naturalism supports the practical humanism found in the Ethical Culture societies—you serve the Good by making this earth a better place for all living things. But if you serve the Good, then you strive to surpass yourself. You strive for greater *arete*. But any greater *arete* is transhuman and godlike. Plato said we should try to become as godlike as possible (Sect. 3 in Chapter 9). You strive to surpass your rationality into some godlike rationality, to surpass your sociality into some godlike sociality, and to surpass your animality into some godlike animality. Spiritual naturalism therefore supports *spiritual transhumanism*. The old Platonic practice of theurgy can be naturalized as scientific self-hacking. I have already mentioned the new Stoics and Buddhists. Spiritual naturalism can express itself through many meturgical practices of self-transformation (Sect. 4 in Chapter 9). Meturgy occurs in the ecstatic dancing at *ritual raves*. And if you think fire is holy, then you light fires. Thus meturgies also occur in *fire circles* and in *transformational festivals* like *Burning Man*. The opportunities for new cultures are endless.

---

[73]Dawkins dislikes the religious naturalism of Goodenough because she is cagey about her atheism (GD 34). But Crosby is explicit about his atheism.
[74]Crosby (2014: 90, 147).

References

Atkins, P. (1992). *Creation Revisited: The Origin of Space, Time, and the Universe*. New York: W. H. Freeman.

Balaguer, M. (1998). *Platonism and Anti-Platonism in Mathematics*. New York: Oxford University Press.

Batchelor, S. (1997). *Buddhism Without Beliefs: A Contemporary Guide to Awakening*. New York: Penguin.

Brown, J. (1980). Counting proper classes. *Analysis, 40*(3), 123–126.

Bueno, O. (2003). Is it possible to nominalize quantum mechanics? *Philosophy of Science, 70*, 1424–1436.

Colyvan, M. (2001). *The Indispensability of Mathematics*. New York: Oxford University Press.

Crosby, D. (2014). *More Than Discourse: Symbolic Expressions of Naturalistic Faith*. Albany, NY: SUNY Press.

Drake, F. (1974). *Set Theory: An Introduction to Large Cardinals*. New York: American Elsevier.

Fenton, J. (1965). Being-itself and religious symbolism. *The Journal of Religion, 45*(2), 73–86.

Field, H. (1980). *Science Without Numbers*. Princeton, NJ: Princeton University Press.

Field, H. (1985). Comments and criticisms on conservativeness and incompleteness. *Journal of Philosophy, 82*(5), 239–260.

Gerson, L. (1994). *Plotinus*. New York: Routledge.

Goodman, N. D. (1990). Mathematics as natural science. *The Journal of Symbolic Logic, 55*(1), 182–193.

Harris, S. (2014). *Waking Up*. New York: Simon & Schuster.

Hartshorne, C. (1967). *A Natural Theology for Our Time*. LaSalle, IL: Open Court.

Hartshorne, C. (1984). *Omnipotence and Other Theological Mistakes*. Albany, NY: State University of New York Press.

Heidegger, M. (1998). What is metaphysics? In W. McNeill (Ed.), *Pathmarks* (pp. 82–96). New York: Cambridge University Press.

Inge, W. (1918). *The Philosophy of Plotinus* (Vol. 2). London: Longmans Green.

Jackson, B. D. (1967). Plotinus and the *Parmenides*. *Journal of the History of Philosophy, 5*(4), 315–327.

Kanamori, A. (2005). *The Higher Infinite: Large Cardinals in Set Theory from their Beginnings*. New York: Springer.

Kant, I. (1790/1951). *Critique of Judgment* (J. Bernard, Trans.). New York: Macmillan.

Kiteley, M. (1958). Existence and the ontological argument. *Philosophy and Phenomenological Research, 18*(4), 533–535.

Kraay, K. (2011). Theism and modal collapse. *American Philosophical Quarterly, 48*(4), 361–372.

Kurzweil, R. (2005). *The Singularity Is Near: When Humans Transcend Biology.* New York: Viking.

Liston, M. (1993). Taking mathematical fictions seriously. *Synthese, 95,* 433–458.

Lokhorst, G.-J. (2006). Andersonian deontic logic, propositional quantification, and Mally. *Notre Dame Journal of Formal Logic, 47*(3), 385–395.

Maddy, P. (1983). Proper classes. *Journal of Symbolic Logic, 48*(1), 113–139.

Maddy, P. (1996). Ontological commitment: Between Quine and Duhem. *Nous 30* (Supplement: Philosophical Perspectives 10), 317–341.

McLendon, H. (1960). Beyond being. *The Journal of Philosophy, 57*(22/23), 712–725.

Melia, J. (1998). Field's programme: Some interference. *Analysis, 58*(2), 63–71.

Millican, P. (2004). The one fatal flaw in Anselm's argument. *Mind, 113,* 451–467.

Mortley, R. (1976). Recent work on Neoplatonism. *Prudentia, 7*(1), 47–62.

Murdoch, I. (1992). *Metaphysics as a Guide to Morals.* London: Chatto & Windus.

Nozick, R. (1981). *Philosophical Explanations.* Cambridge, MA: Harvard University Press.

Otto, R. (1958). *The Idea of the Holy* (J. W. Harvey, Trans.). New York: Oxford University Press.

Peirce, C. S. (1965). *Collected Papers of Charles Sanders Peirce* (C. Hartshorne & P. Weiss, Eds.). Cambridge, MA: Harvard University Press.

Potter, M. (2004). *Set Theory and Its Philosophy.* New York: Oxford University Press.

Quine, W. V. O. (1948). On what there is. In J. Kim & E. Sosa (Eds.) (1999), *Metaphysics: An Anthology* (pp. 4–12). Malden, MA: Blackwell.

Resnik, M. (1995) Scientific vs. mathematical realism: The indispensability argument. *Philosophia Mathematica, 3*(3), 166–174.

Shapiro, S. (1983). Conservativeness and incompleteness. *Journal of Philosophy, 80*(9), 521–531.

Steiner, M. (1998). *The Applicability of Mathematics as a Philosophical Problem.* Cambridge, MA: Harvard University Press.

Steinhart, E. (2003). Supermachines and superminds. *Minds and Machines, 13,* 155–186.

Steinhart, E. (2014). *Your Digital Afterlives: Computational Theories of Life after Death.* New York: Palgrave Macmillan.

Tegmark, M. (2014). *Our Mathematical Universe: My Quest for the Ultimate Nature of Reality.* New York: Random House.

Tillich, P. (1951). *Systematic Theology* (Vol. 1). Chicago: University of Chicago Press.

Welch, P., & Horsten, L. (2016). Reflecting on absolute infinity. *Journal of Philosophy, 113*(2), 89–111.

Wigner, E. (1960). The unreasonable effectiveness of mathematics in the natural sciences. *Communications on Pure and Applied Mathematics, 13,* 1–14.

7

Possibility

1 The Naturalized Cosmological Arguments

1.1 The Cosmic Zero

When the theologians tell Dawkins that there exists some necessary first cause of all other things, Dawkins agrees. But then he adds that it "must have been the simple basis for a self-bootstrapping crane which eventually raised the world as we know it into its present complex existence."[1] Does it seem out of character for Dawkins to agree with the theologians on this point? It should not. His entire theory of the evolution of complexity requires some simple beginning, some maximally probable thing.

The existence of some first cause needs to be justified by arguments. The traditional arguments for a first cause are the *cosmological arguments*. These include the First and Second Ways in the Thomistic Five Ways. They have this general form: (1) There are some facts about our universe.

---

[1]GD 184–5, see 101.

© The Author(s) 2020
E. Steinhart, *Believing in Dawkins*,
https://doi.org/10.1007/978-3-030-43052-8_7

(2) The facts justify the existence of a first cause. (3) So the first cause exists. (4) The first cause is God. When he discusses cosmological arguments, Dawkins objects only to the fourth step. He says the first cause cannot be God.[2] As things get simpler and simpler, they get stupider and stupider. So the simple first cause is utterly mindless. It is not some complex designer who deserves praise or some maximally perfect being who is worthy of worship. It neither hears nor answers prayers; it is indifferent to human behaviors. If you read the Abrahamic scriptures, you are not reading the life story of the simple first cause. It cannot play the role of God in Abrahamic practices. Spiritual naturalists seek to liberate the first cause from theism by embedding it in new systems of irreligious practices.

Dawkins suggests *naturalizing* the cosmological argument by identifying the first cause with some natural entity.[3] The *Natural Argument from Change* is a naturalistic version of the Thomistic Second Way. It runs like this: (1) Our universe is complex. (2) If any thing is complex, then the evolutionary principle (Sect. 1.1 in Chapter 5) says it has been generated by some evolutionary process that started out simple and climbed up through all lower levels of complexity. (3) Therefore, our universe has been generated by some cosmic evolutionary process. (4) Since this cosmic process starts out simple, it starts out with some nonempty set of simple things or thing. (5) Hence these simple *root things* exist. (6) Cosmic evolution resembles other evolutionary processes, such as the evolution of atomic, molecular, biological, or technical complexity. (7) When Dawkins discusses cosmic evolution, he thinks of it in biological terms.[4] Moreover, our universe is extremely complex. But the replication principle (Sect. 4.2 in Chapter 2) states that extreme complexity requires replicators. The biocosmic analogy implies that cosmic replicators make universes like spiders make webs or birds make nests. (8) So the process of cosmic evolution starts out with some set of simple self-reproducing cosmic replicators.

---

[2]CMI 77; GD 101–2, 184; BCD 420.
[3]GD 101–2.
[4]GD 98–9, 184–9.

1.2   A Root on Fire in the Earth

The naturalistic cosmological arguments justify the existence of some set of simple necessary things. They are the first cosmic replicators. Of course, this set might have only one member. And it is arguable that it does have only one member, so that there exists one single root thing. The *Argument for the Unique Ancestor* goes like this: (1) Suppose there are many *apparent root things*. (2) If there are many apparent roots, then they all share some common reproductive functionality. Each apparent root adds its own distinctive content to this common functionality. (3) But this distinctive content is some additional complexity. (4) Hence the common functionality is simpler than that of each apparent root. (5) Consequently, it is possible that there exists some *real root* which embodies only this simpler common functionality. Every apparent root is an offspring of this real root. (6) The Naturalized Cosmological Argument justifies this real root. It is the simplest of all possible things. And the cosmological arguments also imply that this simple object is also necessary. So there is a single simple necessary root thing. It is the ultimate ancestor of every complex thing. It is the first cause. It is the necessary ground of all contingent things, the ground in which they all are rooted.

Since Dawkins proposed that everything derives from a *single* first cause, spiritual naturalism agrees with the Argument for the Unique Ancestor.[5] It affirms that there is one root thing. This root is the initial cosmic replicator. It can be referred to as *Alpha*, the ultimate ancestor of all things. Thus Alpha is the ground of all concrete things. As the ground of all those things, Alpha explains them. And since Alpha is concrete, it explains them by causing them to exist. Alpha is the ultimate first cause; Alpha has *primacy*. Alpha is the *natural* first cause—it is the basis for the self-bootstrapping crane which brings all things into being. And Alpha is simple. Since it is simple, it contains no parts; it has no internal structure. Alpha is empty. But that might suggest, incorrectly, that Alpha is like a bag. Alpha is just a dot.

---

[5]GD 184.

Alpha is the seed of concreteness. It lies at the bottom of the ordinal spire, at the bottom of the *axis mundi*, buried in the earth. Struck by lightening from the Good, this seed bursts into flame. Alpha is a root on fire in the earth; it burns in the island of being-itself. Its light reflects the Good. As the simplest concrete thing, Alpha is the simplest universe. It is the origin of all other concrete things. Its simplicity does not prevent it from having the ability to self-replicate. Alpha has a nature. If Alpha lacks the functions required for self-replication, then there will not be any complex things at all. But since there are complex things, Alpha does have the functions needed for self-replication. Hence simplicity entails that the simplest universe is also the simplest animat. The simplest universe has no other cosmic content besides its animat.

The functions required for the self-replication of Alpha are simple. Since Alpha is simple, the function *make a self-copy* is also simple. This is the self-relation of Alpha; it is the reflexivity of Alpha; it is the root of all concrete reflexivity. Alpha causes other things to exist by replicating itself; its offspring then replicate themselves; they replicate again; and so it goes. So Alpha is the initial animat. As Alpha begets its successors, the world tree rises up out of Alpha. Thus Alpha is the ultimate source of the power which flows through the world tree. It is the ultimate source of the power which actualizes the possibilities in the world tree. The power which actualizes possibilities is axiotropy (Sect. 4.3 in Chapter 3), and Alpha is the hidden spring from which all axiotropy flows.

1.3   Buried in This Fertile Soil

The Arguments from Change reason backwards along causal chains to some simple first cause, which is the original concrete thing. However, since concreteness itself is derivative, those arguments were traditionally backed up by deeper arguments. So the Thomistic First and Second Ways were backed up by his Third Way, which uses the deeper categories of contingency and necessity. To say that a thing is *contingent* means that it depends on something else for its being. Its dependency entails that it *might not be*. Since its probability of existing is conditional, its self-probability is zero. To say that a thing is *necessary* means that it does not depend on anything else for its being. Its independence means it *must be*. It cannot be blocked from existing. Its self-probability is one. The Third Way motivates the Leibnizian cosmological argument.[6] For theists, these arguments ran to God. But we naturalize them.

The *Naturalized Argument from Contingency and Necessity* runs like this: (1) The Dawkinsian theory of complexity entails that every complex thing is a contingent thing. (2) Every contingent thing has an explanation. (3) The explanation for any contingent thing lies in some other thing. (4) Every class of contingent things is a contingent thing. (5) Let *nature* be the class of all contingent things. It is the collection of all the actual universes in the world tree along with all their parts. (6) Thus nature itself is a contingent thing. (7) Nature has an explanation. (8) The explanation for any class of things is not a member of the class. (9) So the explanation for nature is not any thing in nature. (10) If any thing is not in nature, then it is not contingent. (11) So the explanation for nature is not contingent. (12) If something is not contingent, then it is necessary. (13) Therefore, the explanation for nature lies in some necessary being which is not in nature. That is, the explanation for the class of contingent things lies in some necessary being which is not in that class. (14) If there are many necessary beings, then exactly one is ultimate. (15) Consequently, there exists one ultimate necessary being.

The ultimate necessary being is that which gives existence to all the beings among beings. The ultimate necessary being exists ontologically but not ontically. It is being-itself, which is the ground of being, the island floating in the abyss. Being-itself lies outside of nature in exactly the sense that it is not any thing among things. Does this mean that being-itself is supernatural? It does not. Dawkins uses the term *supernatural* to refer to complex agents which did not evolve. Being-itself is neither complex nor is it an agent. More generally, the supernatural goes back to the medieval great chain of being. If that chain were correct (it isn't), then being-itself would be *subnatural*. But our ontology does not rely on the medieval great chain of being. The location of being-itself in our system of images and concepts coincides with the location of the empty set, the initial book in the modal library, the first lifter, the first

---

[6]Leibniz (1697).

cause, the first animat, and the first universe. These are all ultimate by originality. While these original entities are not identical, they coincide in the structure of the Sanctuary.

2   Like Birds in This Tree

2.1   Cosmic Calendars

The power of being-itself erupts into concreteness at the initial universe. It is the root of the infinitely ramified world tree, which rises to the Good. As this tree branches into ever greater complexity, one of its paths runs from the initial universe to our universe. Our universe is surrounded by many other universes in the world tree. Although we can use our imaginations to travel to those other universes, our universe is our home. Since evolutionary processes design things, our universe is filled with artists making works of art. Every star is an atomic artist. Every planet is a molecular artist, and planets with life are biological artists. Our earth is filled with many technological artists: bees and birds and beavers all design artifacts—and there are human artists too. Dawkins often talks about scientifically-inspired art. Art often has an imitative function: art imitates nature. Imitative art participates in reflexivity. Scientifically-inspired art includes works in all media that illustrate the sublime structures of nature revealed by science.

Science reveals that our universe has a sublime time-scale.[7] To illustrate this scale, Carl Sagan developed his *cosmic calendar*. It maps the history of the universe onto the three-hundred and sixty-five days of the year. All human history occupies the last ten seconds of the last day. Dawkins makes a similar cosmic calendar using a line of books. Each page in each book records the history of a single year. To describe even the history of earthly life, the line of books would run from London to Scotland. As a work of art, this line of books resembles large-scale sculpture, like Robert Smithson's *Spiral Jetty*. It might even be laid out using

---

[7]Cosmic calendar (Sagan 1977: ch. 1). Dawkins line of books (UR 10–12). Open your arms wide (UR 12–13).

stone markers along some highway. Dawkins also uses a small work of dramatic or performance art to illustrate the great age of the universe. Open your arms wide. Your leftmost fingertip is the big bang, while your rightmost fingertip is the present. Humanity is the dust on your right fingertip.

Another way to commemorate cosmic evolution is the *cosmic walk*.[8] It dramatically celebrates the milestones in the evolution of complexity. The cosmic walk uses a large spiral laid out on the ground. It might be a logarithmic spiral, a form common in nature. The central point of the spiral refers to the big bang. Using some time scale, other points on the spiral are marked with the creatively significant alphas in our universe. These can include the first stars, the first life on earth, the first mammals, the first humans, and so on. The spiral continues beyond the present into the open future. Large unlit candles are placed at the marked points on the spiral. The cosmic walk requires three actors: a reader, a striker, and a walker. The walker starts at the center of the spiral. The reader begins the cosmic walk with a very short poem or declaration about the big bang. When the reader finishes, the striker hits a bell or gong, and the walker lights the candle at the center. The walker then proceeds along the spiral, stopping at each candle. At each candle, the reader narrates the text for its event. The gong is struck, the candle lit. The ritual continues until the walker passes out of the spiral into the open future. There are plenty of variations on this general script. The entire ceremony may be watched by an audience, who may also one by one walk the spiral after the candles are lit.

Many wheels roll along the cosmic spiral.[9] The largest of these are galactic: our solar system orbits the black hole at the center of the Milky Way. More locally, all the planets in our solar system orbit our sun. From our earthly perspective, the sun, the moon, and the visible planets trace out complex arcs across the sky. These movements serve as the basis for astrology. Dawkins replaces astrology with your set of *birthstars*. Your birthstars are the stars whose distance from earth is your age in light years. If you are forty years old, your birthstars are forty light years away.

---

[8] Taylor (2007: 249–252).
[9] Birthstars and existence spheres (UR 117).

Since it takes their light forty years to reach the earth, when you look at one of your birthstars, you are looking at an event that occurred when you were born. Conversely, the sphere forty light years in radius is your *existence sphere*. Here there are many opportunities for other works of art that integrate our lives into the cycles in the sky.

Your birthdays count your revolutions around the sun. But the sun is a star, and the stars fuse simpler elements into heavier elements. So it is entirely appropriate to correlate your birthdays with the processes of atomic evolution in the stars. These are your *elemental birthdays*.[10] Your *hydrogen birthday* is your first birthday, your *helium birthday* is your second birthday, and so on. If you live to be ninety-two, you will celebrate your *uranium birthday*. The elemental birthdays are a naturalistic way of correlating the microcosm of a human life with the macrocosm of the evolution of physical complexity. Your body is a crane which grows from the simplicity of your zygotic alpha to the complexity of your mature body. Elemental birthdays are one way of correlating your body-crane with all the cranes in the universe.

2.2   Beyond Religious Culture

Dawkins characterizes himself as a cultural Christian.[11] He enjoys religious (that is, Christian) works of art. These include religious works of music and painting, as well as literary works like the King James Bible. And they include works of performance art or drama like Christian rituals and ceremonies. He welcomes participation in religious weddings and funerals. He happily says religious grace at meals. He loves traditional religious Christmas celebrations. He wants Christian culture to continue.

For participating in these rituals, Dawkins was accused of hypocrisy.[12] He defended himself by pointing out that the rituals are meaningless

---

[10]Cleland-Host and Cleland-Host (2014).

[11]Cultural Christian (SITS 246; GD 383–7). Music and painting (GD 111; FH 117). King James Bible (GD 383–7; FH 111–14). Weddings and funerals (GD 387). Grace (BCD 29–31; FH 119–20). Christmas (SITS 246–51).

[12]Hypocrisy (BCD 30; FH 119–20). Incongruity (GD 111). Fictional (GD 51, 111; FH 117). Evil (GD chs. 7, 8, 9; etc.).

to him and that his participation is merely politeness. But he was also accused of incongruity: how can he enjoy these works of art if he regards their motivating doctrines as false? He correctly replies that he regards these religious works of art as fictional. Obviously, people can enjoy fictional works of art without believing that they are true. But Dawkins has also argued that religion is evil. Hence the charge of incongruity goes deeper: by perpetuating Christian culture, he is perpetuating evil. If he is right that Christianity is evil, then he ought to seek to replace its culture with some better culture.

For the sake of practical consistency, Dawkins ought to urge the development of alternatives to religious works of art.[13] He makes some efforts in this direction. Weddings, funerals, and naming ceremonies are all works of art. By now there are many alternatives to religious ceremonies. Dawkins himself planted a memorial tree for Douglas Adams. One time when he was asked to say grace before a meal, Dawkins was abruptly put on the spot to come up with a secular grace. He attempted to do this; however, he did not have enough time. Since Isaac Newton was born on 25 December, he proposes that we celebrate Newton Day as an alternative to Christmas. And of course Dawkins knows that Christmas itself comes from ancient pagan celebrations of the winter solstice, an event celebrated by many atheists.

Dawkins urges us to transform science into poetic or artistic ways of life. To live artistically on the earth is to participate in the artistry of nature. This natural artistry includes the cyclical music of nature. We participate biologically in the oscillatory music of the earth as it orbits the sun. The rhythms of that solar music are generated by the drum beats of the solstices and the equinoxes. Here again believing in Dawkins means building on Dawkins. It means replacing religious holidays, pageants, and festivals with works of performance art inspired by nature. Of course, to provide those dramatic works of art with content, it is necessary for us to use meaningful symbols. The ontology of spiritual naturalism (Chapter 6) provides many symbols for solar festivals. The sun itself symbolizes the Good. Spiritual naturalism provides symbolic

---

[13]Memorial tree (ADC 167). Secular grace (BCD 31). Newton Day (2007). Atheists winter solstice (Cimino and Smith 2014: ch. 4).

interpretations of the poetic elements of water, earth, air, fire, and light. We have made great use of trees and stars as natural symbols. The four compass directions, as well as the vertical directions on the *axis mundi*, are all available for poetic use. Natural shapes like circles and spirals can be used in dances, rituals, ceremonies, and even in sculpture and architecture. We believe in *rituals without dogma*: interpret the symbols as you see fit. As an alternative to religious culture, spiritual naturalists urge festivals on all the solar holidays.

2.3   Monuments for the Sun

Some spiritual naturalists may want to go beyond festivals. Sculpture and architecture are also artistic mediums. If cathedrals are religious architectural works, then naturalists can raise their own works of spiritual architecture. Many ancient cultures raised solar monuments, which marked the solstices and equinoxes. These monuments include Stonehenge and myriad other ancient stone works. They include the many solar calendars and sun daggers made by the Native Americans in the southwestern United States. They include many modern stone circles, like the Earth Clock in Burlington, Vermont. Some of these monuments are currently used for solar performance art.

But how to interpret these monuments? What do they mean? People have often raised monuments to give thanks. Many monuments have been raised to honor the soldiers who served or died in war. Monuments have been raised to honor military animals, such as horses and dogs who served or died in wars. At least one monument was raised to honor the laboratory mice who served in scientific experiments.[14] All these monuments were raised to commemorate, honor, and give thanks to lives that were offered in sacrifice for great values. They are monuments inspired by destructive conflict. Of course, evolution thrives on conflict.

At least one monument bears witness to the fact that evolutionary conflict can generate goods greater than its losses. The town of Enterprise, Alabama, raised a statue to give thanks to the boll weevil. The

---

[14]Sharp (2019: 119–120).

economy of Enterprise was dependent on cotton. One year, after boll weevils destroyed their cotton crop, the farmers of Enterprise used this destruction as an opportunity to switch to growing peanuts and other crops, and to diversify their economy. This diversification brought them considerable wealth. The destruction of their cotton drove them to greater creativity—it was an opportunity for successful self-surpassing. As an expression of gratitude, they raised a monument to the boll weevil in 1919. It's inscription reads: "In profound appreciation of the Boll Weevil and what it has done as the herald of prosperity this monument was erected by the citizens of Enterprise, Coffee County, Alabama."

Dawkins recognizes the value of gratitude (Sect. 1 in Chapter 9). However, he is not clear about how naturalists should express it. Can we give thanks to the sun? If people can raise monuments to give thanks to humans, dogs, horses, mice, and insects, then we can also raise monuments to give thanks to the sun. The sun is sacrificing itself, burning itself out, pouring its energies into the abyss. The sun is an artist. Its artistry includes its fusion of hydrogen into helium (a kind of atomic sculpture), but its artistry extends to all the creative processes driven by its thermodynamic gradients. The artistry of the sun includes the evolution of molecular and biological complexity on all its planets, including our earth. Evolution designs things. Every earthly species is an artistic product of solar creativity. Although the sun does not act purposively or teleologically, it does act algorithmically and teleonomically. Every convergent algorithm performs its operations for the sake of its finalities, for the sake of its ecstasies. The sun pumps entropy out of its solar system for the sake of every complex thing that emerges on its planets. The sun sacrifices itself for the sake of every evolved structure in its solar system. It sacrifices itself for every living thing on earth. Of course, it would be a serious error to worship or pray to the sun. It is nevertheless appropriate for us to give thanks to the sun. We can do this by raising solar monuments or having solar festivals.

2.4   Simulation of Parallel Universes

Our universe is surrounded by many others in the world tree. Further out, it is surrounded by dark universes. One task of art is to represent other possible universes (bright or dark). The variabilities of local structures in our universe enable us to represent distant possibilities. A string of DNA has four possible letters at each place. Other possible genomes represent other possible organisms. The cells in our retinas are like pixels that can vary in activation. Any retina defines an enormous class of possible images, which are images of possible visible objects. They represent those possible objects by structural correspondence—just like they represent the objects we see. We are like birds that sit in the world tree, looking out into other universes.

Dawkins says our brains run simulation software.[15] It takes our sensory inputs, and uses them to create a virtual reality universe. We live directly in this virtual reality (VR) universe. Your brain-built VR universe often corresponds truthfully to the universe inhabited by your body—that is, to your *local* universe. But that correspondence can shift to some other universe. It shifts when you experience an illusion. Your virtual reality then corresponds to *some non-local universe*. These non-local universes are other possible universes. We simulate other possible universes when we run "what if" scenarios, or while dreaming or fantasizing. Futurists typically depict other universes. Although it is not likely that our universe will evolve to an infinite omega point, the prophesies of Tipler and Kurzweil are surely *possible* (Sect. 5 in Chapter 2). The futuristic visions of transhumanists probably mostly portray other universes. And we simulate other universes using computer programs and video games. Dawkins' biomorphs program simulates some possible forms of life in some alien universe.

Fictional works of art take place in other universes. To say that painting is *fictional* does not mean that it has no truth. It means that it truthfully corresponds only to some scene in some *fictional universe*, which is some non-local universe. Fictional stories likewise represent

---

[15]Virtual reality universe (SG 56–60; 1988; ADC 11–13, 46; UR ch. 11). Other possible universes (SG 58–9; UR 312; 1993: 121–124). Dreaming or fantasizing (UR 282). Biomorphs (BW ch. 3; EE; BCD 363–94).

other possible universes. Sometimes they represent *alien universes*, which share little or no structure with our universe. For example, the *Lord of the Rings* universe is an alien universe. Other times they represent *parallel universes*, which can be overlaid on our universe. The adventures of Sherlock Holmes take place in a parallel universe. The Holmesian London can be overlaid on our London. So the Holmesian London is a *counterpart* of the London in our universe. The Harry Potter stories also take place in a parallel universe. The Potterian London can also be overlaid on our London. But the Potterian London corresponds less with our London, while the Holmesian London corresponds more. So the Potter universe is logically farther away from our universe, while the Holmesian universe is closer to us.

Dawkins says good science fiction should not portray its universe as a magical place.[16] He enjoys neither *The X-Files* nor Tolkien. He allows that science fiction may portray its universes as having different laws of nature. But he thinks that magic is lawless and disorderly, and good science fiction cannot portray its universes as lawless or disorderly. Dawkins is wrong to link magic with disorder—elsewhere he correctly recognizes that computer games can be lawfully magical. Magical universes are lawfully ordered (they have laws of magic). Consider massive multi-player online role-playing games (MMORGs). These are video games like *League of Legends* or *World of Warcraft*, which allow many players to socially participate in the same fictional universe. Even though they involve magic, these fictional universes still follow laws (otherwise, it would be pointless to play games in them). Their laws are embodied in computer code that generates the magical physics of the universe.

---

[16]Good science fiction (UR 28–9). Lawfully magical (1993: 108).

3   Shifting to Other Universes

3.1   From Fictional Universes to Religions

When you read some fictional story, or watch some fictional movie, your brain can simulate its universe. This simulation can become so intense that you get lost in the story. Getting lost in some fictional story often involves deep emotional involvement in its universe. Dawkins describes getting lost in fiction and being moved to tears.[17] When you get lost in some work of fiction, you lose your place in logical space. This means that you *shift* your position in logical space to the position of the fictional universe. Since books and plays and movies are technologies for shifting our simulations away from our universe to some non-local universe, they can be referred to as *modal vehicles*. New technologies, like video games and virtual reality, have produced new modal vehicles, which can shift our simulation software even more intensely.

Many people can read the same fictional books or watch the same fictional movies, and they can absorb these fictional works at similar times, so they collectively participate in the same fictional universe. When many people watched the *Lord of the Rings* movies around the same time, they collectively shifted to the LOTR universe. Likewise, when many people read the Harry Potter books around the same time, they collectively shifted to the Harry Potter universe. This collective shifting caused their mental simulations to become coordinated. This coordination produced *institutional facts* about those fictional universes. These institutional facts have objective truth, that is, truth that is independent of any single person. It is objectively true that Frodo carried the One Ring into Mordor, and that Harry Potter had a scar on his forehead. But institutional facts can support *institutional practices*. People can also collectively shift to those fictional universes by socially taking on the roles of the characters in those fictions. Through live-action role-playing (*larping*), you can assume the role of Harry Potter or Gandalf. To assume the role of some fictional character is to *channel* that character. You can

---

[17]FH 117.

channel some fictional character by playing that character in a stage-play or video game.

Widely shared works of science fiction and fantasy have inspired religions.[18] As illustrations of science-fiction religions, Dawkins mentions Scientology and Heaven's Gate.[19] But there are others. The *Starwars* movies inspired the Jediism religion. The movie *The Big Lebowski* inspired the Dudeism religion. The LOTR books inspired the Tolkien religions.[20] Some Tolkienists say he had visions of some mythic universe. One fictional religion, Pastafarianism, worships the Flying Spaghetti Monster.[21] R'amen. These fiction-inspired religions motivate a *modal theory of religion.*[22] The modal theory has three parts. First, religious texts portray fictional parallel universes (call them *religious*). Second, people use religious texts to collectively shift to these religious universes. Those texts generate institutional facts, which get reinforced by institutional practices.[23] Those facts and practices support moral norms. Third, people channel fictional characters in religious universes. Typically, they channel more positive versions of themselves. This channeling inspires positive emotions, like hope.

The modal theory of religion is *not* an anthropological theory. It is a philosophical theory, concerned with logical issues like meaning, reference, and truth. According to the philosopher David Lewis, to say there is *truth in fiction* means that there are some non-local universes at which the fiction can be told as fact.[24] Lewis believes there is truth in all religions. He believes in infinitely many deities.[25] Of course, all he means is that he thinks that there are other possible universes at which those religions are factual, and at which their gods exist. But Lewis is an atheist. The religious stories are not true at our universe, and their gods do not exist at our universe. The religious universes are not our universe—they

---

[18]Cusack (2010).

[19]UR 27–8.

[20]Davidsen (2014).

[21]Henderson (2006).

[22]Atran and Norenzayan (2004), Bloch (2008), and Bulbulia (2009).

[23]Wood and Shaver (2018).

[24]Lewis (1978).

[25]Lewis (1983: *xi*, note 4).

are far away in logical space. On these points, any atheist can agree with Lewis. So spiritual naturalists say that religions are fictions, and, like other fictions, there are universes in the modal library where they are facts.

Dawkins characterizes the Bible as fiction.[26] If all possible universes exist, then the Bible describes a class of universes. These Biblical universes, like all possible universes, are entirely natural. The Bible could be a script for a video game.[27] At Biblical universes, the simulation hypothesis is true (Sect. 2.2 in Chapter 4). The digital people trapped in that video game can tell the Bible stories as mostly fact. They use the term "God" to refer to the entirely natural designer-creator of their game. Although supernatural Biblical Gods do not exist anywhere, natural godlike game makers are possible, and so inhabit fictional universes in the modal library. However, since the Bible is so ethically defective, the Biblical universes are not actualized.[28] They are merely dark books in the metaphysical horror section of the modal library.

Our universe is not Biblical. On this point, both empirical science and rational moral philosophy entirely agree. Nevertheless, the Biblical universes can be overlaid on our universe: their Jerusalems are counterparts of our Jerusalem; their Pauls are counterparts of our Paul. People on our earth can use the Bible to collectively shift into those Biblical universes. Their collective shifting generates the facts of institutional Christianity, facts which get reinforced by Christian practices. Practices like communion enable people to vicariously participate in the Last Supper. These practices are live-action role-playing. Just as you can channel Gandalf by dressing up as a wizard and acting out some scene from LOTR, so you can channel a Biblical character by taking communion. You can pretend to participate in the Last Supper. To be sure, that Last Supper takes place, not in our universe, but in some other parallel universe.

[26]GD 51; FH 117.
[27]Moravec (1988).
[28]GD chs. 7, 8, and 9.

3.2 Religious Shifting to Fictive Universes

Dawkins often talks about the *memetic theory of religion*.[29] According to that theory, religions are complexes of *memes*. Memes are ideas that replicate across brains. Writers like Susan Blackmore and Daniel Dennett have done heroic work trying to use memes to understand religions. However, the scientists who study religions have not found memes useful: Atran sharply criticizes memetic theories of religion; Aunger says memetics has no substance; Tremlin dismisses memes in three pages. By the time of Barrett and Norenzayan, memes are ignored. Scientists don't take memes seriously. So if Dawkins has no other theories of religion, then his theory of religion is in serious trouble.

Fortunately, Dawkins also supports the modal theory of religion.[30] He links religion with fiction and simulation. He says that when children have imaginary friends, these are stable fictional objects. When a child entertains an imaginary friend, they are simulating a parallel universe. Dawkins suggests that gods resemble imaginary friends. He says religious experiences are glitches in our virtual reality software. Thus religions emerge from the simulation software of the brain. Religious glitches can afflict social groups. During the Our Lady of Fatima events, thousands of people apparently saw the sun behave anomalously. Dawkins says their visions were part of a shared hallucination; they were simulating similar fictional universes in unison.

The events linked with Our Lady of Fatima were said to be *miraculous*. On the literal interpretation, miracles take place in our local universe, and they violate its laws. But then they need some explanation. Literalists infer that, since some laws are violated, there must exist some agent sufficiently powerful to violate them. And if an agent can break the laws of nature, then it must be supernatural, in the sense of being superior to physical laws. So, granted the miracle, the literalist

---

[29]Religions involve memes (ADC ch. 3.2). Memes (SG ch. 11; EP 165–70; ADC chs. 3.1, 3.2; UR 302–9; GD 222–34; BCD 404–11; AT 324–6). Blackmore (1999: ch. 15), Dennett (2006), Atran (2002: ch. 9), Aunger (2006), Tremlin (2006: 151–154), Barrett (2007) and Norenzayan (2013).

[30]Imaginary friends and parallel universes (GD 389–94; see Mackendrick 2012). Glitches (UR 266–7, 282). Religions from simulation software (GD 112–7, 389–94). Our Lady of Fatima (UR 133–5; GD 117).

infers some supernatural agent. Dawkins correctly debunks this literalism.[31] Since this literalism is incorrect, either miracles are meaningless, or they require some figurative interpretation. Dawkins mocks figurative (symbolic) interpretations of miracles.[32] But mockery is not argument. Moreover, he has the resources to develop figurative interpretations. He says miracles are just glitches in our VR software. Moreover, since figurative interpretations are supported by the cognitive science of religion, he should have developed one. Spiritual naturalism does develop one—here I am building on Dawkins.

Miracles have a figurative interpretation. On the figurative interpretation, they are mental events that abruptly and involuntarily shift your location in logical space. A miracle throws you into some fictional universe. Of course, miracles happen only in your brain. They are glitches in your VR software. And, as should be obvious, they do not violate any natural laws—they are entirely natural shifts in your simulation software. But these glitch-miracles are mentally disruptive: they violently throw your mind into some other parallel universe. There is no mind-body dualism here—to say that your mind goes into some other universe just means that your brain simulates it. When you are immersed in fiction, you keep one foot in our universe. You retain your place in logical space. But glitch-miracles are disorienting, unsettling, and emotionally intense. They show that you can be *thrown out* of our universe. You land with both feet somewhere else. Both dreams and drug-induced hallucinations are glitchy. Consequently, the modal theory may explain why dreams and drugs are used in many religions.

Miracles (in the glitchy sense) teach an important lesson: you are only *contingently* located in our universe; you are not *essentially* here. Of course, you can learn this lesson if you study the metaphysics of possible universes. But when you study metaphysics, you only learn this lesson intellectually. Miracles *impose* this lesson on your body with the full force of experience—they are *revelatory*. They reveal the existence of other possible universes and the existence of your counterparts in them. They *transfigure* your body into the bodies of your counterparts.

---

[31] UR 134–5; GD 82–4.
[32] UR 183.

This transfiguration occurs only in the VR software of your brain. And, precisely because it is *software*, that software can become *identical* with the software running on your counterpart brains.

According to the modal theory, religions depict parallel universes in writing, painting, music, and so on. If that's all they did, they would just be art. But religions also use tools and techniques to forcefully shift people to religious universes. They reliably induce glitch-miracles in human brains. So the modal theory of religion says religions include technologies of modal transport. These are devices for mentally throwing you out of this universe and into some overlaid religious universe. These devices include the markers of sacred separateness. They include innumerable methods for inducing trances, dissociation, and other altered states of consciousness.

Religious technologies strive to generate shared glitch-miracles. Rituals work to shift many people at the same time into the same shared fictional universe.[33] To this end, religious rituals often involve shared behaviors. They involve many bodies that come together into the same place, and that move together in the same ways. Religious bodies collectively speak, sing, chant, sway, gesture, dance, and so on. Through shared behaviors, these bodies become entangled and entrained; they become coordinated, so that they all read in unison from the same fictional book in the modal library.

3.3   Religious Channeling

Dawkins says it's glorious to live in this world.[34] But most people live extremely difficult lives. For example, Dawkins says that science did not bring good news to Keats; on the contrary, it brought him news of his fatal tuberculosis.[35] Faced with this scientific bad news, Dawkins speculates that Keats sought solace in his fantasy universes. But science ultimately brings all of us bad news. This is the atheistic problem of evil (Sects. 5 in Chapter 1 and 3.1 in Chapter 6). Against the bad news of

---

[33]Seligman et al. (2008) and Willerslev et al. (2014).
[34]UR 1–6; FH 99.
[35]UR 27.

our universe, salvific religions claim they bring good news. They preach salvation from our universe.

A salvific religion declares that you can be redeemed from your identity with your flawed body and its flawed contexts. Christianity is salvific: it preaches salvation from the ills of our universe. Salvific religions have technologies that shift you to parallel universes where you are truly otherwise than negative. If you can be saved (and Christianity says you can), then you are truly otherwise than failure. Salvific religions affirm that you are truly otherwise than orphaned, abused, sick, poor, alone, ugly, despised, wretched, criminal, insane, worthless, futile, aged, mortal, and—and the mind reels. But none of this *essentially* belongs to you. Salvific religions distinguish between your apparent and real location in logical space. You *appear* to be located in our universe, you *appear* to be identical with your present body. But you are *really* located in some fictional universe in which you do not suffer any negativities.[36] You are *really* identical with some other better counterpart of your body.

Of course, this is escapism, or dissociation. Nevertheless, if parallel universes do exist, then it is true that you are only *contingently identical* with your present body; hence you are only contingently located in our universe. You don't need religion to believe that. This is partly a consequence of Turing universality of the human brain: since you are a universal computer, you can run a different program, you can be otherwise. Any organism that lacks this universality is bound much more tightly to its present self. This universality is closely linked with self-consciousness, which allows you to conceive of yourself, and therefore to conceive of alternative versions of yourself.

The modal theory of religion shows how religions generate hope. As long as some state of affairs is not logically impossible, you can hope that it will become actual. It may not be rational to believe in highly improbable outcomes. Yet hope begins where belief ends. You may believe that you will die of cancer; the probability may rise ever closer to certainty; still you may always hope that you will not. Since hope involves possibilities of things in this universe, it requires parallel universes. To say that your hope for some scenario is true means that there exists some parallel

---

[36]Zamulinski (2003).

universe in which that scenario occurs. Your thought "I hope I won't die of cancer" is true if and only if there exists some universe in which your counterpart does not die of cancer.

Religions provide technologies for enabling you to simulate some hoped-for counterpart, some better version of yourself. These hoped-for counterparts can be called *aspirational*. And the simulation of aspirational selves is *channeling*. By enabling people to channel aspirational selves, religions provide psychological benefits like the *illusion of control*.[37] The illusion of control can help people overcome depression, despair, and learned helplessness. Religions reinforce this channeling through ritual practices that cause socially organized groups to channel their aspirational selves together. So you can also channel a *socially aspirational* self. However, the aspirational counterparts of persons and societies might not exist in *morally better* universes—people often hope for *immoral* scenarios. Christianity shifts its groups to fictional universes where all their enemies are tortured forever in hell. Since hell is morally hideous, these Christian universes are dark universes.[38] Aspirations can be vicious.

Dawkins is constantly amazed that religions insist on falsehoods no matter how great the evidence against them.[39] But the modal theory of religion explains this. If you want to live in another universe, you have to deny this one. Religions train you to *refute* facts, and to affirm counter-factuals in their place.[40] Here is the young earth creationist Kurt Wise, channeling one of his counterparts in some fictional Biblical universe. He is lost in logical space. For him, *the earth* refers to some Biblical counterpart of our earth. Since he is not wrong about the age of *that earth*, no evidence about *our earth* will ever change his mind. Only by channeling some non-Biblical self can he be redeemed from his nescience. Liberation from religion does not come through better critical thinking skills. It comes from novel technologies of modal transport (Sects. 2–4

---

[37]Langer (1975).
[38]GD 360–2; SITS 250.
[39]GD 151–61, 232, 319–23; SITS 267, 309–13, 383; etc.
[40]ADC 138–41.

in Chapter 9), which generate new institutional facts. Those institutional facts can agree with science.

3.4   Artistic Channeling

Dawkins believes that science can inspire great art.[41] He says the artists who created great religious works of art could have created even greater secular works of art. He urges us to develop poetic science—science writers should use stimulating analogies, images, and metaphors. Poets might use science to create even greater poetry. He suggests that if Michelangelo had been commissioned by scientific institutions, he would have produced something at least as beautiful as the ceiling of the Sistine Chapel. He urges the composition of Goethean Hymns to Nature (Sect. 3.1 in Chapter 6).

Perhaps science is already inspiring great art.[42] Dawkins provides a list of scientists whose prose he thinks deserves the Nobel prize in literature. Consider the *Missa Charles Darwin*. Structured and sung like a medieval mass, its liturgy comes from Darwin's works. It is widely agreed to be a gorgeous composition (I've heard it, I agree). Thinking about the aesthetic potentials of science leads Dawkins to quote Sagan on the emergence of a superior future scientific religion. Thus Dawkins hopes scientifically-inspired art will motivate the emergence of his Einsteinian religion. This suggests that Dawkins himself might make some art—after all, he is very good with symbols. Surprisingly, when Dawkins deals with poetic science, he spends almost all of his discussion criticizing bad poetic science. He does offer a few positive uses of analogy in scientific discovery and exposition. But these are pedestrian rather than poetic. After lavishing such praise on poetry, it seems astonishing that Dawkins offers not one fruitful metaphor or symbol. Does he lack imagination? Here is one suggestion.

---

[41]Secular art (UR 24; GD 111; FH 108). Metaphors (UR 180, 233, 299). Greater poetry (UR 27). Sistine Chapel (GD 111; FH 108). Hymns to Nature (UR 24).

[42]Nobel prize (SITS 4). *Missa Charles Darwin* (Brown 2013). Sagan (SITS 80). Poetic science (UR ch. 8). Positive analogies (UR 186–7).

Dawkins writes about the deep beauty of crystals.[43] He uses them to illustrate the orderliness and rationality of the universe as described by science. So those inspired by his brand of poetic science might wear a crystal to symbolize the beauty, order, and rationality of the universe. It might be objected that this would expose those naturalists to misunderstanding: they might be mistaken for New Age fools. But why let the New Agers have all the symbols? Dawkins despises hijacking; it's time to take the symbols back. Can other superstitious symbols be reclaimed? Artists have created two science-inspired tarot card decks: the *Science Tarot* and the *Women of Science Tarot*.[44] Each deck beautifully interprets the cards in terms of scientists and scientific concepts. Like Brian Eno's *Oblique Strategies*, these decks can be used for creative problem-solving. You can also use them for describing hoped-for futures. On the one hand, Dawkins encourages scientifically-inspired symbolism. On the other hand, his constant criticisms suggest an overwhelming *fear* of symbols. If you're going to start to develop a system of poetic symbols, you might get lost in aspirational universes. But art takes risks.

Dawkins uses the *Dies Irae* to illustrate his hope for great scientifically-inspired music.[45] The *Dies Irae* is an old medieval chant describing the Christian last judgment—it's very name means *day of wrath*. Our earth did indeed suffer a day of wrath: the day the Chicxulub asteroid struck our earth and exterminated the dinosaurs and other species. Dawkins says that, if Verdi had known of this asteroid, and the mass extinction that followed, then he might have written a great scientifically-inspired *Dies Irae*. But why the *Dies Irae*? That song expresses the *hope for salvation* from the day of wrath. If Dawkins is right, then there was no hope for the dinosaurs. These reflections on art point back to the atheistic problem of evil (Sect. 5 in Chapter 1): if atheism does not provide salvation, then how can there be beautiful atheistic works of art?

When Dawkins suggests that scientific art might rival religious art, Hitchens disagrees.[46] Their disagreement suggests an argument: (1)

---

[43] ADC 43–6.
[44] Science Tarot (2019) and Women of Science Tarot (2019).
[45] UR 24.
[46] FH 108.

The greatness of art derives from its representation of the aspirations of our selves and societies. (2) Since religions describe aspirational universes, there can be great religious art. (3) However, since science can only describe our present universe, it cannot represent any aspirational universes. Hence there cannot be any great scientific art. Science cannot depict universes that satisfy our counterfactual emotional or moral demands—it is aspirationally blind; it has no *vision*; it does not *dream*. You cannot channel aspirational selves and societies through scientific illustrations of this universe. Of course, *science fiction* often portrays aspirational selves and universes. But as it portrays the flight from here to there, it turns into *transhumanism*. Transhumanism can look religious. However, rather than forcing us to bow down to any gods, it aims to raise us to godlikeness. It is an irreligious lifting. It is a naturalistic kind of *theurgy* (Sect. 3 in Chapter 9).

4   Atheistic Mysticism

4.1   Thrown into Ecstasy

Dawkins says many things arouse feelings of sacredness and holiness in him.[47] Standing among the giant redwoods in the Muir Woods, he feels the sacredness of those huge trees. He feels awed wonder when he looks up into the transcendental heights of the Milky Way, as well as down into the abysmal depths of the Grand Canyon. He says this awed wonder resembles religious experience, and makes him feel reverence. The fossil footprints of an ancient hominid in Tanzania are "another holy relic in my version of the sacred." The vaults in Kenya that store ancient hominid fossils have "a very strong and affecting sense of holiness." He says "you feel as though you are undergoing a religious experience when you are looking at the fossils in the Kenya National Museum." All these examples involve big things, big times, or big spaces.

---

[47] Sacred and holy (SS 135–7). Muir Woods (SS 136). Milky Way and Grand Canyon (2009b: 1:44–2:55; SS 137; SITS 1). Fossils (SS 136–7).

Profound experiences of the holy and the sacred are often associated with mysticism. At the opening of *The God Delusion*, Dawkins describes a mystical experience that one of his teachers had as a boy. He says that boy could have been him. When he was a boy, he was dazzled by the stars, "tearful with the unheard music of the Milky Way, heady with the night scents of frangipani and trumpet flowers in an African garden."[48] He says that nature often induces a mystical feeling in scientists, and that "I am only one of many who have experienced it."[49] Of course, I cannot possibly know whether Dawkins *really* had a mystical experience. Nevertheless, his words suggest he did.

And atheists *do* have mystical experiences, experiences which have nothing to do with God, and which often even more deeply confirm their atheism. Mystical experience was hijacked by being bound to God; but spiritual naturalism seeks to reclaim it for naturalism. Our Sanctuary includes naturalized mysticism. Six atheists who had mystical experiences include Nietzsche, John Dewey, John McTaggart, Bertrand Russell, Ursula Goodenough, and Alan Lightman.[50] But here I will very briefly mention three other cases of atheistic mysticism. The journalist Arthur Koestler described an atheistic mystical experience inspired by mathematics.[51] While reporting on the Spanish Civil War, he was accused of being a spy, and thrown into prison. While imprisoned, he wrote out the Euclidean proof that there are infinitely many primes. Contemplating this proof, he was overcome with ecstasy. His ego dissolved and he felt himself absorbed into ultimate reality. But Koestler rejected the identification of that ultimacy with God or any divine mind. He remained an atheist. The young and devoutly Catholic philosopher Pierre Hadot had atheistic mystical experiences.[52] His experiences were often aroused by looking into the starry sky or at mountains. They threw him into

---

[48]GD 31–2.

[49]TL 59–60.

[50]The mystical experience of Nietzsche is discussed in Gutmann (1954); that of John Dewey in Aisemberg (2008); that of John McTaggart in Dickinson (1931: 46, 92–98); that of Bertrand Russell in his (1967: 147); that of Goodenough in her (1998: 100–103); and that of Lightman in his (2018: 5–6).

[51]Koestler (1969: 428–430).

[52]Hadot (2011: 5–12, 75–78).

the immensity of nature. His ego dissolved and he experienced pure existence. These experiences were deeply disturbing to him, because they revealed that reality contains no God. They helped to drive him away from his youthful religion. Comte-Sponville had atheistic mystical experiences.[53] They were also aroused by looking into the starry sky. He experienced himself as being a part of a vast natural whole. He became aware of his participation in "the All." As his ego dissolved, he experienced "being itself".

4.2   Themes in Atheistic Mysticism

Five themes appear in the nine cases of atheistic mysticism mentioned above. The first theme is *structural insight*. Atheistic mystics gain some structural insight into existence. They become aware of their participation in some natural system of relations. They realize that there is a deep or hidden order in nature; or that the elements of nature are arranged in beautiful ways; or that they are nodes in some network of loving connections. So the second theme is *valuable connection*. Connection is love; but disconnection is suffering. Love binds all things together into the network. The third theme is *wholeness*. Atheistic mystics become aware of being contained by or being parts of some surrounding whole or encompassing totality. The fourth theme is *extremity*. Atheistic mysticism is aroused by sublime or vast aspects of nature. It reveals that the structure of reality is maximally inclusive. Nature is an unsurpassable whole. The wholeness of nature is open rather than closed; it is boundless. The fifth theme is *dissolution*. The egos and objects of the atheistic mystics often dissolve; during this dissolution, they become aware of being-itself or pure existence.

These themes suggest that some kind of *structuralism* is the best way to interpret atheistic mysticism. Here structuralism just means that objects are constituted by their positions in networks of relations.[54] Structuralism entails that nature is ultimately a connect-the-dots network. More formally, nature is a graph. Existing objects, including atheistic

---

[53]Comte-Sponville (2006: ch. 3).
[54]Shapiro (1997).

mystics, are subgraphs of this connect-the-dots network. To make this structuralism precise, I turn to set theory (Sect. 1.5 in Chapter 6). Set theory portrays nature as a connect-the-dots network in which the dots are sets and the connections are membership arrows. Sets are entirely constituted by their positions in the membership network. Set theory therefore depicts the world as a purely relational structure. The graph of the membership relation is absolutely infinite. It is that than which none more inclusive is logically possible. It includes every consistently definable structure. But this set-theory also includes the proper classes, which are unsurpassably great.

On the set-theoretic interpretation, atheistic mysticism begins with the awareness of being included in some bigger structure in some way. At first, this may be awareness of inclusion in ever-larger spatio-temporal wholes. But it rapidly turns into the awareness of participation in an ultimate relational network. If the set-theoretic interpretation of this network is right, the atheistic mystic becomes intuitively aware of being the origin of membership arrows that rise ever higher and expand ever further outwards in the network of sets. Atheistic mysticism turns into an awareness of infinitely long chains of inclusion relations. It finally turns into an awareness of inclusion in the absolutely infinite proper classes. Of course, this awareness is intuitive. It is the immediate consciousness of interconnectedness. It is *not* awareness of the axioms of set theory. You can be aware of spatio-temporal relations without knowing the formal structure of Minkowski space-time. You can be enrapt by the Golden Spiral without knowing its logarithmic equations. You can speak a language without explicitly knowing its grammar.

According to this ontology, the brain and body of the atheistic mystic are subnetworks of the world-network. They are subnetworks in the great connect-the-dots network of reality, the network of all the beings among beings. It is plausible that *the All* of Comte-Sponville refers to this network of beings. This network rises without bound to the highest heights of existence—it rises to the stars. When the brain of the atheistic mystic becomes intuitively aware of an all-inclusive wholeness, it becomes aware of its direct and unmediated participation in the network of beings. It becomes intuitively aware of its immediate participation in an ineffable and transcendental plenitude. It becomes intuitively aware of its direct participation in the All. But this awareness of the All leads to ego dissolution: it leads to the awareness of being-itself.

4.3   The Flight of the Solitary to the Solitary

Ordinary awareness is ontic: it is the awareness that some subject has of its objects. Both the subject (your ego) and its objects are beings among beings. It is an ontic fact that your ego exists, that its objects exist, and that your ego is aware of its objects. Following Quine (Sect. 1.4 in Chapter 6), this means that your ego and its objects are the values of bound variables. It is an ontic fact that (there exists some x)(x is your ego) and also that (there exists some y)(y is the object of your ego). Putting these together, it is an ontic fact that $(\exists x)(\exists y)$(your ego x is aware of its object y). During ontic awareness, your brain runs software that simulates those ontic facts and their logical elements. It simulates the variable x and its existential quantifier $\exists$. It simulates the y and its $\exists$. It simulates the awareness that the x has of the y. You live entirely in this virtual reality. And here your simulation is *veridical:* you are *truthfully* simulating those facts. Since the facts simulated by your brain are ontic, the powers of the x and y are balanced with the powers of their $\exists$s. Since the powers of those elements are balanced, your brain pumps equal amounts of neural energy into its simulations of each element.

Ontic awareness can become mystical, and our Sanctuary contains an atheistic analysis of mystical experience. This analysis is logical rather than psychological. As expected, since the Platonists were deeply interested in mysticism, this analysis is Platonic. Atheistic mysticism begins in the cave of physical awareness. It starts with the awareness of large networks of physical relations (like staring up into the starry sky). It is aroused by the vastness of the physical connect-the-dots network. This awareness shifts from awareness of objects to their structural interrelations. It shifts from the dots in the physical network to the arrows that connect them. The axiotropic fire runs through these arrows to make the blazing world tree. Hence mystical experience is often associated with fire. This fire appears to be *integrally omnipresent:* one flame is wholly present in every substructure of the physical connect-the-dots network.

Although you are surrounded by these blinding flames, you see through them.

As your mystical experience progresses, you become intuitively aware of the abstract set-theoretic network that lies behind the physical network. You see that things are bound into structures via the membership relation. But the membership network expands beyond our universe and beyond all possible universes. You still experience your own ego as a central point which radiates cognitive arrows towards its objects. As you become aware of ever larger structures, your awareness shifts from the sets to the membership arrows that bind them together. You become intuitively aware that *there exists a single relational power which binds all things together into the unified wholeness of existence.* All beings participate in membership in some way. Hence that relation is integrally omnipresent: one relation is wholly present in every subnetwork of the world-network.[55] It is wholly present in every part of the pleroma (Sect. 1.5 in Chapter 6).

As your mystical experience progresses, you experience your own participation in the membership relation. Your ego shrinks down to the minimality of the empty set. To say that your ego shrinks means that your brain puts less and less energy into its simulation of the x in the fact that $(\exists x)(x$ is your ego$)$. Its puts less energy into its construction of its sign for itself in its own virtual reality software. Your ego is starting to dissolve. At the same time, your object expands to the maximality of the pleroma. Your brain pumps more and more energy into the y in the fact that $(\exists y)(y$ is your object$)$. Mental energy is getting pumped out of your ego and into its object. It therefore rushes from your ego into its object. You experience this flux as both the expansion and dissolution of your ego into its object. Your awareness expands in every direction from your ego to its object. As it expands, it begins to surpass the ontic awareness of larger and larger structures, and you begin to see the stars. Your ontic awareness becomes intuitive participation in the transcendence of the stars. It expands towards the simulation of the All.[56]

---

[55]Plotinus (*Enneads*, 6.4–5).
[56]Comte-Sponville (2006: 165–168).

As your brain simulates the stellar ambiguities (Sect. 2.5 in Chapter 6), your awareness dissolves into the horizon between being-itself and non-being. You were experiencing the shrinking of your ego and the expansion of its object. But those ontic movements reveal an ontological shift. Your brain both minimizes the energy put into simulating your ego x and maximizes the energy put into simulating its object y. These changes affect its simulations of the $\exists$ to which each variable is bound. At both the extremes of minimality and maximality, the being of beings vanishes. As the ego x vanishes into minimality, only its $\exists$ remains. As the object y vanishes into maximality, only its $\exists$ remains. Your brain pumps all its mental energy into its simulation of the $\exists$s, and none into its simulations of the x and y. The powers of the x and y are no longer balanced with that of their $\exists$s. You experience the expansion of the $\exists$ as bliss and the contraction of x and y as terror. This is the ecstasy between non-being and being-itself.

As your mystical experience climaxes, your awareness shrinks to a point of pure wildness. You become agitated with glory and paralyzed with fear. Your $\exists$ vanishes into maximality while your x and y vanish into minimality. Your ego and its objects dissolve. The ontic fact that $(\exists x)(\exists y)$(the ego x is aware of the object y) becomes paradoxical. This is the flight of the solitary to the solitary.[57] With x and y gone, the solitude of $\exists$ relates through awareness to the solitude of $\exists$. Your brain simulates the desolation of being-itself, bereaved of its beings. This is *ontological awareness*. This awareness is *veridical*. By simulating the solitude of $\exists$, your brain *truthfully* signifies being-itself. Hadot writes that dissolution involves "the pure feeling of existing."[58] Comte-Sponville writes that in mystical ecstasy "What fulfills you then is not a particular state of being but being itself."[59] The content of this awareness is being-itself as the self-negation of non-being. And this is the ecstasy through which the abyss emptied itself of its negativity.

---

[57] Plotinus (*Enneads*, 6.9.11).
[58] Hadot (2011: 8).
[59] Comte-Sponville (2006: 165–168).

References

Aisemberg, G. (2008). Dewey's atheistic mysticism. *The Pluralist, 3*(3), 23–62.

Atran, S. (2002). *In Gods We Trust: The Evolutionary Landscape of Religion.* New York: Oxford University Press.

Atran, S., & Norenzayan, A. (2004). Religion's evolutionary landscape. *Behavioral and Brain Sciences, 27,* 713–770.

Aunger, R. (2006). What's the matter with memes? In A. Grafen & M. Ridley (Eds.), *Richard Dawkins: How a Scientist Changed the Way We Think* (pp. 176–188). New York: Oxford University Press.

Barrett, J. (2007). Cognitive science of religion: What is it and why is it? *Religion Compass, 1,* 1–19.

Blackmore, S. (1999). *The Meme Machine.* New York: Oxford University Press.

Bloch, M. (2008). Why religion is nothing special but is central. *Philosophical Transactions of the Royal Society B, 353,* 2055–2061.

Brown, G. W. (2013). *Missa Charles Darwin.* Performed by New York Polyphony. North Hampton, NH: Navona Records.

Bulbulia, J. (2009). Religiosity as mental time-travel: Cognitive adaptations for religious behavior. In J. Schloss & M. Murray (Eds.), *The Believing Primate* (pp. 44–75). New York: Oxford.

Cimino, R., & Smith, C. (2014). *Atheist Awakening: Secular Activism and Community in America.* New York: Oxford.

Cleland-Host, H., & Cleland-Host, J. (2014). *Elemental Birthdays: How to Bring Science into Every Party.* Solstice & Equinox Publishing.

Comte-Sponville, A. (2006). *The Little Book of Atheist Spirituality* (N. Huston, Trans.). New York: Viking.

Cusack, C. (2010). *Invented Religions: Imagination, Fiction and Faith.* Burlington, VT: Ashgate.

Davidsen, M. (2014). *The Spiritual Tolkein Milieu.* Dissertation, University of Leiden.

Dennett, D. (2006). *Breaking the Spell.* New York: Viking Penguin.

Dickinson, G. (1931). *J. McT. E. McTaggart.* Cambridge: Cambridge University Press.

Goodenough, U. (1998). *The Sacred Depths of Nature.* New York: Oxford University Press.

Gutmann, J. (1954). The 'Tremendous Moment' of Nietzsche's vision. *The Journal of Philosophy, 51*(25), 837–842.

Hadot, P. (2011). *The Present Alone Is Our Happiness: Conversations with Jeannie Carlier and Arnold I. Davidson* (M. Djaballah & M. Chase, Trans.). Stanford, CA: Stanford University Press.

Henderson, B. (2006). *The Gospel of the Flying Spaghetti Monster*. New York: Villard.

Koestler, A. (1969). *The Invisible Writing*. New York: The Macmillan Company.

Langer, E. (1975). The illusion of control. *Journal of Personality and Social Psychology, 32*(2), 311–328.

Leibniz, G. W. (1697). On the ultimate origination of the universe. In P. Schrecker & A. Schrecker (Eds.) (1988) *Leibniz: Monadology and Other Essays* (pp. 84–94). New York: Macmillan Publishing.

Lewis, D. (1978). Truth in fiction. *American Philosophical Quarterly, 15*(1), 37–46.

Lewis, D. (1983). *Philosophical Papers* (Vol. 1). New York: Oxford University Press.

Lightman, A. (2018). *Searching for Stars on an Island in Maine*. New York: Random House.

Mackendrick, K. (2012). We have an imaginary friend in Jesus: What can imaginary companions teach us about religion? *Implicit Religion, 15*(1), 61–79.

Moravec, H. (1988). *Mind Children: The Future of Robot and Human Intelligence*. Cambridge, MA: Harvard University Press.

Norenzayan, A. (2013). *Big Gods: How Religion Transformed Cooperation and Conflict*. Princeton, NJ: Princeton University Press.

Russell, B. (1967). *The Autobiography of Bertrand Russell* (Vol. 1). London: George Allen & Unwin.

Sagan, C. (1977). *The Dragons of Eden: Speculations on the Evolution of Human Intelligence*. New York: Ballantine Books.

Science Tarot. (2019). Online at sciencetarot.com. Accessed 3 May 2019.

Seligman, A., Weller, R., Puett, M., & Simon, B. (2008). *Ritual and Its Consequences: An Essay on the Limits of Sincerity*. New York: Oxford University Press.

Shapiro, S. (1997). *Philosophy of Mathematics: Structure and Ontology*. New York: Oxford University Press.

Sharp, L. (2019). *Animal Ethos: The Morality of Human-Animal Encounters in Experimental Lab Science*. Oakland: University of California Press.

Taylor, S. M. (2007). *Green Sisters: A Spiritual Ecology*. Cambridge, MA: Harvard University Press.

Tremlin, T. (2006). *Minds and Gods: The Cognitive Foundations of Religion*. New York: Oxford University Press.

Willerslev, R., Vitebsky, P., & Alekseyev, A. (2014). Sacrifice as the ideal hunt: A cosmological explanation for the origin of reindeer domestication. *Journal of the Royal Anthropological Institute, 21*, 1–23.

Women of Science Tarot. (2019). Online at shop.massivesci.com/products/ women-of-science-tarot-deck. Accessed 3 May 2019.

Wood, C., & Shaver, J. (2018). Religion, evolution, and the basis of institutions: The institutional cognition model of religion. *Evolutionary Studies in Imaginative Culture, 2*(2), 1–20.

Zamulinski, B. (2003). Religion and the pursuit of truth. *Religious Studies, 39*, 43–60.

8

Humanity

1   Human Animals

1.1   Genetically Programmed Survival Machines

Dawkins rejects mind-body dualism.[1] A human person is not a whole composed of a material body and an immaterial mind—we don't have any immaterial parts at all. Thus Dawkins is a *materialist* about human persons. Every human person is strictly (but only contingently) identical with his or her body: you *are* your body. Every human body begins with a fertilized egg, that is, with a zygote. The zygote contains a genetic program for the growth of the human body from its single-celled origin to its mature adult form. Our adult bodies are far more complex than our zygotes. Hence the genetic program that starts running in your zygote is an entelechy—it is a complexity-increasing program. Your growing body moils towards its adult form, which is its finality. Your growing body is a crane, which participates in all the upthrust of the universe.

---

[1]GD 209–10.

© The Author(s) 2020
E. Steinhart, *Believing in Dawkins*,
https://doi.org/10.1007/978-3-030-43052-8_8

As our bodies grow, our genes dictate their human architectures. Dawkins says our bodies are genetically programmed robots.[2] Human organisms, like all organisms, are survival machines built by genes for the sake of their self-replication. The organs of our bodies are tools used by our genes. So our bodies are technologies, they are complexes of purely material machines. Dawkins declares that biological organisms and artificial machines are functionally equivalent. Spiritual naturalism agrees with Dawkins on all these points: human persons are wholly material robots.

Robots are controlled by computers, and the computers that control human robots are their brains. Brains are networks of neurons, which transmit activity to each other across synaptic arrows. So the structure of the brain is defined by arrows of the form neuron → neuron. Since variable weights can be assigned to these arrows, the brain looks like one of the physical computers from Chapter 2.[3] Dawkins says the brain is a biological computer.[4] It is functionally equivalent to a finite Turing machine. This implies a computational theory of mentality: the mind is the computational activity of the brain. Minds are generated by brains as they run various neuro-genetic programs.

Genes establish the basic functional architectures of our brains and the mental algorithms running in our neural networks.[5] Those algorithms are extremely abstract and general. For instance, our genes ensure that our neural networks run pattern recognition algorithms and abstract learning algorithms. However, our genes do not precisely control our behaviors. Our genes only exert statistical influences on our behaviors. Consequently, while our brains are programmed by genes, they are not determined by them. Dawkins says we can over-ride our genetic imperatives.[6]

---

[2]Built by genes (SG 2; BW 3). Genetic robots (SG *xxi*, 19, 270–1; EP 15–17). Bodies are technologies (SG 24, 47). Machines are organisms (BW 1).

[3]The brain learns by varying its connection patterns and weights. This motivates some formal analogies between neurological and ecological computation (Pritchard and Dufton 2000; Watson et al. 2016). These analogies suggest that evolution learns.

[4]SG 48–62, ch. 4, 276–81; EP ch. 2; UR 286–7, ch 12.

[5]SG 58, 69, 197.

[6]SG *xiv*, 201, 331–2; ADC ch. 1.1.

1.2   Against Magical Mental Mystery

Dawkins is a *naturalist* about human persons. He demands naturalistic accounts of human intelligence, rationality, consciousness, agency, freedom, morality, and so on.[7] According to Dawkins, a naturalist affirms three axioms about humans: (1) human brains are computers; (2) human bodies are mechanical robots; (3) human bodies are programmed by their genes. Dawkins insists that the only alternatives to these three axioms about humans are supernatural. But supernaturalism is merely negative. It tells you that your brain isn't a computer—but it doesn't tell you anything useful about your brain. The naturalist can tell you that much mental illness comes from inflammation in the brain, which comes from auto-immune disorders in which immune cells attack neural tissue. While the naturalist can propose systematic treatments for mental illnesses, the supernaturalist offers you only further illness.

Dawkins provides naturalistic accounts of *mental representation, consciousness,* and *self-consciousness.*[8] Our brains run elaborate simulation software (Sect. 2.4 in Chapter 7). As this VR software becomes more complex, *consciousness* emerges from the activity of the brain. Dawkins suggests that consciousness emerges when the brain runs simulations that start to include itself. As your brain represents its environment, it represents itself as doing the representing. Hence your awareness becomes conscious awareness. Since conscious awareness emerges from self-simulation, it includes self-consciousness. Of course, this is reflexivity. Consciousness and self-consciousness emerge as the brain maximizes reflexivity. Dawkins says we all live in virtual realities. If you are aware of some object, your ego-simulation is aware of some object-simulation.

Dawkins provides a naturalistic account of the human *self* (or ego) as a unified agent.[9] As your brain simulates itself in its environment, it appears to itself in itself. The unified self is just the *sign of the brain* in

---

[7] EP 25–6; GD 34–5.

[8] Simulation software (SG 57–60). Virtual reality (UR ch. 11). Consciousness (SG 50; ROE 156–8). Include itself (SG 59, 278).

[9] Sign of the body (UR 283–4; SITS 2–3). Selfish cooperators (UR 308–9). War of all against all (UR 309; SITS 2–3). Buddhism self is illusory (Rahula 1974: ch. 6; Dennett 1993; Blackmore 2011).

its self-simulation. For Dawkins, the *unified self* is an illusion, which can be explained through cooperation. Dawkins says that unified bodies emerge as warring genes congregate into groups of selfish cooperators. Our brains support many interacting patterns of neural activity. They cooperate with each other in order to better compete. Here Dawkins echoes the Platonic idea of the self as a city. But a kind of Hobbesian sovereign emerges from this neural war of all against all. Hence the self emerges from the cellular *agon* in the brain—it is an emergent product of neurological *arete*. Of course, your illusory self is a real pattern in your brain. To say that your self is an illusion means just that your brain is naturally deceived about itself. Your brain thinks it is a substance which is *immaterial*, which *endures eternally*, and which has a *freedom* that transcends physics. In reality, your brain (and body) have none of those features. They are illusory. These points about the self follow Hume and Nietzsche. They are developed by Daniel Dennett. They agree with Buddhist ideas about selfhood. The psychologist Susan Blackmore shows how meditative techniques from Zen Buddhism can cut through these illusions of selfhood.

Dawkins provides a naturalistic account of *free will*.[10] Although Dawkins does not discuss rival philosophical theories of free will, it is easy to see that he is a *compatibilist*. Compatibilists say that freedom means the absence of coercion. You behave freely as long as your behavior emerges from your own bodily programming. If your behavior is being directed only by your genes and your brain (that is, by the logic encoded in your own body), then you are behaving freely. Freedom is not opposed to determinism; it is opposed to coercion. Both robots and humans act with equal freedom. The chess-playing program Deep Thought plays chess *freely* because its moves come from its own nature (from its programming) and not from something else. What else would freedom be? Those who oppose compatibilism are known as *libertarians*. But they have never provided any account of libertarian free will. It is an unsolvable mystery.

---

[10]SG 270–7.

Dawkins provides a naturalistic account of *goals* and *purposes*. Purposes emerge as organisms develop feedback loops.[11] The theory of feedback loops is a part of the more general theory of *dynamical systems*. A dynamical system has many possible *states*, which make up its *state space*. And it has some *operator* which transforms states into states. Each transformation is an *arrow* from this state to that state. These arrows can be used to define a *landscape of attractiveness*. If the arrows in some region of the state space all point towards some state, then that state is an *attractor*. If they all point away from some state, it is a *repeller*. Gravity provides a helpful analogy: valleys are attractors while mountain peaks are repellers. Gravitational arrows point down from the tops of mountains to the bottoms of valleys. Or consider a computer program which takes some input number and repeatedly divides it in half. Since the value of the repeatedly divided number converges to zero, the attractor for this program is zero. The attractors of any dynamical system are its finalities. These finalities emerge from purely mechanical operations. So the goals of organisms are their attractors.

1.3   A String of Counterparts

We can agree that Richard Dawkins in 1947 *is the same person as* Richard Dawkins in 2017. And we can agree that Dawkins has changed over those seventy years. Young Dawkins in 1947 differs from Old Dawkins in 2017. So what does it mean to be the same person? There are two main theories about personal persistence.

The first main theory is known as *endurantism*. Endurantists believe that human persons are enduring things. An enduring thing remains the same thing despite its changes. So endurantists believe in *identity through time*: Young Dawkins is identical with Old Dawkins. They believe that there exists exactly one thing, namely, Richard Dawkins, which remains invariant through all its changes. They do not believe that Young Dawkins is one thing while Old Dawkins is another thing. So

---

[11]SG 50–1.

what is this single thing which remains invariant through seventy years of change?

Many endurantists say that, when it comes to human persons, the one thing that remains invariant through change is an immaterial mind. However, Dawkins denies that there are any immaterial minds. So it cannot be his immaterial mind that keeps its identity during the course of his life. For Dawkins, human persons are bodies. So what about bodies? Our bodies constantly change their parts. New atoms enter as old atoms leave. But different parts make different wholes. As old atoms leave and new atoms enter, your old body changes into a new body.[12] Since those two bodies are composed of distinct atomic parts, they are two distinct bodies. Your old body is not identical with your new body. So, for the case of Dawkins too, it cannot be his material body that keeps its identity during the course of his life. Of course, what holds for the case of Dawkins holds for all human persons, and, indeed, for all earthly organisms.

From these considerations, it follows that endurantism is false. Complex material things that change their parts over time cannot remain identical through time. Organisms do not remain identical through time. And there are good reasons to think that even simple things like quarks and electrons cannot remain identical through time. For human persons, identity through time is an illusion, much like the illusion of the single unified self. It is an illusion of self-consciousness.[13] Personal *persistence* is not personal *identity*. To say that Young Dawkins is the same person as Old Dawkins does not mean that Young Dawkins is identical to Old Dawkins. On the contrary, they are two distinct persons. So what does it mean to say that they are the same?

The main alternative to endurantism is *perdurantism*. Perdurantists say that persisting things are both spatially and temporally extended. Persisting things are *four-dimensional space-time wholes*. Just as a loaf of bread is a series of distinct slices, so a persisting four-dimensional thing is a series of distinct three-dimensional slices. Just as a movie is a series of instantaneous photos, so a persisting thing is a series of instantaneous

---

[12]GD 415–6.
[13]SITS 2.

stages. The stages are bound together into a four-dimensional whole by various kinds of continuities. It is plausible to say that if some set of stages composes a persisting thing, then there must be some continuous flow of information across all those stages. Later stages must carry information about earlier stages.

For the perdurantist, to say that some earlier person is the same person as some later person just means that the earlier and later persons are both 3D parts of the same 4D whole. So to say that Young Dawkins is the same person as Old Dawkins just means that there exists some *four-dimensional life* which contains Young Dawkins as one of its earlier stages and Old Dawkins as one of its later stages. The sameness comes from the fact that both 3D stages are parts of the same 4D life. But (unless you are Dawkins) there isn't any 4D life that contains stages from your life and stages from the life of Dawkins. So you aren't the same person as Dawkins. You are two distinct persons.

A three dimensional person is a stage of a four dimensional life. The distinct persons in the same life are all *counterparts* of each other. So your life is a series of persons who are all counterparts. You are your own present counterpart. But you have past counterparts in earlier times. To say that you *were* asleep last night means that you have a *past* counterpart who *is* asleep. And you have future counterparts in later times. To say that you *will be* older means that you have a *future* counterpart who *is* older.

1.4   Blessed by This Present Light

Dawkins describes his subjective experience of time using the classic metaphor of the moving spotlight.[14] The spotlight of presence moves along some timeline. All the moments behind it are in the darkness of the past, while all the moments ahead of it are in the darkness of the future. Only the present moment shines with brightness, only the present sits in the spotlight. Of course, the trouble with this metaphor is that the spotlight itself has to be moving *in time*. The metaphor refutes itself.

---

[14]UR 3, 312; GD 404.

Dawkins recognizes that his picture of time may be just another subjective illusion.[15] It is indeed illusory: time is not a spotlight moving in some bigger time. Nevertheless, it is necessary to explain this illusion. Why think there is some spotlight of presence? Start with the four-dimensional space-time of our universe. The fourth dimension is just the gigantic line of time. This line is *eternal* in the sense that it is *timeless*. Eternity is not an endless time, eternity is mathematical timelessness. The only alternative to the spotlight is that the entire timeline is illuminated in a timeless eternity. Since the entire timeline is eternally illuminated, every moment of this timeline is eternally illuminated.

Since the entire timeline is eternally illuminated, every 3D body in your 4D life is eternally illuminated. Nevertheless, every 3D body has *three different relations* involving time. Every 3D body in your life *sees* only its present self and its environment; it *remembers* its past bodies but it does not see them; it *imagines* its future bodies but it does not see them. It is blind to its past and future bodies and hence to their environments. To your present body, the present *appears* bright while the past and future *appear* dark. Your 3D bodies use temporal operators to express their situations. They use tensed language. Each of your 3D bodies truly says "I *am* bright *now* but my *earlier* bodies *are* dark *now*" and "I *am* bright *now* but my *later* bodies *are* dark *now*." These tensed statements and the facts they represent are *time-indexed*. They express relations between your temporal counterparts. Thus "My earlier bodies are dark now" means your past selves appear dark to your present self; it does not mean your past selves are dark. Time-indexed facts about light and darkness are consistent with the eternal illumination of all moments. Eternal facts are not time-indexed. The moving spotlight is an illusion like the bentness of a stick in water. There is no contradiction between "the stick looks bent to me" and "the stick is straight." Likewise there is no contradiction between "the spotlight seems to move to me" and "the spotlight does not move."

Although this explanation is physically adequate, it is not rationally adequate. It does not explain the origin of the eternal light. The metaphor of the luminous present originates in Platonic models of time

---

[15]UR 3; SITS 329.

and eternity. Platonists said the Good shines like the sun. For spiritual naturalists, the Good shines with holy light. This light illuminates all and only the cosmic forms in the treasury. The cosmic forms in the treasury lie eternally in the light of actuality while the cosmic forms not in the treasury lie eternally in the darkness of unactualized possibility. The spotlight of presence emerges when the sets in these cosmic forms are tied together with lengths of physical string. Each length of physical string is an instance of the *is-present-to* relation, so that actualized sets become present to themselves and to each other (Sects. 2.2 in Chapter 5 and 5.3 in Chapter 6).

Dawkins says we are lucky to be actual, lucky to awaken in the light of day.[16] Of course, luck plays an important role here. However, if our actuality were *entirely* due to luck, then the probabilities on causal arrows in our universe would be random. But they are non-randomly skewed towards complexity. Hence our actuality is not entirely lucky. Moreover, our universe itself did not appear by chance (Chapters 4 and 5). Our universe is actual because its form lies in the best of all possible worlds. Its form *deserves* to be actual. But the Good entails that if any cosmic form deserves to be actual, then it is actual. The Good itself is the source of the light shining on our universe. Our universe is a blazing shrine; it reflects the light of the Good back to itself.

Many bright books (cosmic forms) contain human bodies while many dark books also contain human bodies. Dawkins discusses the difference between bodies that lie eternally in the night of unactualized possibility and bodies that lie eternally in the day of actuality.[17] Bodies that lie in darkness remain asleep while bodies in the day of actuality have the capacity to be awake. When they are awake, when their eyes are open, they reflect the universe back to itself. Speaking of our capacity to reflect the universe in vision, Dawkins says that we are "hugely blessed."[18] At every moment of time, if you are awake, if you are conscious, your present body receives the light of the Good and reflects it back in vision. To receive this light is to be awestruck by the wonder that anything exists

---

[16]UR 3–5, 312–13; GD 404.
[17]UR 3–5.
[18]UR 5.

at all; it is to be *blessed* by the Good; it is to be *graced* by the Good. The grace of the Good burns in your body. It is a light you can shine into the world.

2   Evolutionary Ethics

2.1   Genetic Egoism

The replicators have an ideal finality: make self-copies without end. They aim at the ecstasy of infinite self-replication. The philosopher Max Black argues that ideals generate obligations: an agent *ought* to do something if and only if it is the most effective way for the agent to realize its ideal finality; so your genes ought to do whatever maximizes their self-replication.[19] This leads to a *reproductive imperative* at the level of the gene: always follow whatever strategy maximizes self-replication. Selfish genes obey an ethics of *genetic egoism*: use all other things as means to the end of your reproductive success. If organisms directly inherit their imperatives from their genes, then organisms will also obey reproductive imperatives. As a genetic egoist, you ought to strive for genetic success. Your self-interest is the self-interest of your genes.

But genetic self-interest is blind. Self-replication has no foresight, so it cannot learn from its mistakes. Hence self-replicators (that is, genes) prefer strategies which offer short-term success over strategies which offer greater long-term success.[20] As an illustration, consider aggression. Evolution seems to favor aggression.[21] It even appears to *morally justify* aggression: nature is a war of all against all; since we are natural, we ought to act naturally; therefore, we are justified in seeking our own gain by all means possible. It seems to justify the equation of ethical value with competitive success: the winners are morally superior while the losers are morally inferior. It seems to imply that tyrants and robber barons shine most brightly with *arete*.

---

[19]Black (1964).
[20]SG 200.
[21]SG ch. 5.

Genetic egoism appears to entail social Darwinism (Sect. 4.2 in Chapter 3).[22] And social Darwinism seems to amount to the ethical doctrine that "might makes right"; or, as Thucydides put it "the strong do what they will and the weak suffer what they must." It says we ought to chose social policies which reward competitive success. Social Darwinism appears to justify unrestrained economic and unrestrained nationalistic aggression. It appears likewise to justify racist evils, such as ethnic cleansing and vicious eugenics. Dawkins correctly says these implications of Darwinism are morally revolting. He opposes social Darwinism. He says that, to derive our moral values, we should invert the apparent values of Darwinism. Having evolved intelligence, and therefore ethical self-awareness, we can oppose the selfish imperatives of our genes.

2.2   From Selfish Genes to Altruistic Organisms

However, it is not clear why Dawkins opposes Darwinism in morality and politics. After all, he often argues that selfish *genes* do not entail selfish *organisms*.[23] He often argues that *selfish* genes lead instead to *altruistic* organisms. Genetic egoism drives us to behave according to the Golden Rule. Much of *The Selfish Gene* argues that genetic egoism drove the evolution of cooperation. His chapter on The Roots of Morality in *The God Delusion* is devoted to the Darwinian roots of altruism and morally positive cooperative behavior.[24] He says in that chapter that there are four Darwinian roots of altruistic behavior: (1) Organisms act altruistically to benefit their genetic kin. By benefitting their kin, they further propagate their own genes. (2) They act altruistically for the sake of reciprocity. They do good now to obtain reciprocal benefits later. (3) They act altruistically in order to obtain a good reputation. (4) They act altruistically to produce hard-to-fake signals of fitness and power.

---

[22]Against social Darwinism (ADC 10–2; SITS 34, 275). Invert social Darwinism (SG *xiv*, 2–3). Oppose our genes (SG 200–1).

[23]GD 246–7.

[24]GD ch. 6.

All human morality has emerged from evolution by natural selection.[25] Where else would it have come from? It did not come from God. Evolution gave us four Darwinian reasons to be altruistic. But it also gave us *three moral gifts:* "the gift of understanding the ruthlessly cruel process that gave us all existence; the gift of revulsion against its implications; the gift of foresight." Thanks to evolution, we have a *universal moral algorithm*. This is our *moral instinct*. It encodes an *evolutionarily stable strategy* for human animals. The local environmental features of specific human groups lead to local specializations of the universal moral algorithm. It becomes specialized into local conventions or customs. Given some local constraints, the universal moral algorithm generates conventional behavioral dispositions. Accordingly, if you are in this situation, then it is customary for you to perform this action.

Customary actions are often regarded as *morally correct* actions: if it is *customary* for you to perform some action in some situation, then you *ought* to perform that action in that situation. This definition of moral correctness emerges from the genetic imperative: genes ought to do whatever they have to do in order to maximally self-replicate; once genes have become bound up in human bodies, the universal moral algorithm encodes the genetic strategies for maximal self-replication; hence, as vehicles carrying human genes, we ought to do what the moral algorithm tells us to do.

Nevertheless, Dawkins correctly denies that customary actions are always morally right actions. He argues that the changing moral climate (the changing moral "Zeitgeist") shows that custom has been wrong.[26] The moral instinct in the past has lead societies to embrace slavery, sexism, racism, nationalism, fascism, religious fanaticism, and a vast host of self-destructive evils. But the moral instinct has a second-order drive towards increasing goodness. The moral algorithm contains a second-order algorithm for re-writing its own first-order rules. We have made moral *progress* away from those ancient evils towards greater modern

---

[25]Not from God (OTTR 397–8; GD ch. 7). Gifts (ADC 12). Algorithm (GD 254–8).
[26]GD ch. 7.

goods.[27] Of course, this also involves progress away from religiously orga-nized societies and towards secular societies. It is *morally better* for a society to base its decisions on science instead of superstition. To make these claims about moral progress, it seems that Dawkins must assume some ideal morality toward which customary morality progresses. There must be some ideal moral standard, some ideal source of moral value, or *ideal moral algorithm*.[28]

2.3 Objectively Existing Values

According to Dawkins, we *can* rebel against genetic egoism.[29] And since customary morality emerges from genetic egoism, we *can* surpass customary morality. But he also says that we *ought* to rebel against genetic egoism. And so we *ought* to surpass customary morality. More-over, one of the main points of *The God Delusion* is that we also *ought* to surpass the moralities of ancient religions. Of course, if we ought to surpass those three inferior moralities, then there must be some ideal morality which we ought to work towards. Dawkins presupposes that this ideal moral code exists. He also presupposes that it is morally supe-rior to the codes derived from genetic egoism, the codes derived from ancient religions, and the morality of custom.

While Dawkins often suggests that there does exist an ideal moral code, he also says it is not his job to find it.[30] Still, even if he has not explicitly developed an ethical theory, he has written quite a bit about ethics. And when his ethical remarks are systematized and completed, a defensible ethical theory does emerge. Clearly, given his rejection of the supernatural, and his commitment to the justification of beliefs through evidence, Dawkins must be some kind of *ethical naturalist*. He clearly believes that *truth is an objective value*. It is objectively morally wrong to lie, and objectively morally right to tell the truth. He is a realist

---

[27]GD 304.

[28]OTTR 398; GD 289, 298–308.

[29]Can rebel (SG 200–1). Should rebel (SG *xiv*). Surpass customs (GD ch. 7; SITS 250).

[30]Not his job (GD 306). Objective truth (ROE 31–2; UR 21, ch. 6; ADC chs. 1.2, 1.7; GD 18–9, 319–20; GSE ch. 1; SITS 7, 22–3). Objective beauty (UR 118).

about cognitive values (such as the truth-values of theories). He rejects all forms of cognitive relativism. Some theories really are truer than others. Dawkins also clearly believes that *beauty is an objective value*. The objective value of beauty is a persistent theme in *Unweaving the Rainbow*. He also believes that religious and superstitious ways of life are ugly. He is a realist about aesthetic values—some ways of life and belief systems really are more beautiful than others.

The fact that Dawkins is a realist about cognitive and aesthetic values supports the thesis that he is also a realist about moral values. The fact that he rejects cognitive relativism suggests that he also rejects moral relativism. The fact that he criticizes genetic egoism, morality based on religion, and the morality of custom, also suggests that he believes in objectively superior moral ideals. And the fact that he affirms that some things are sacred or holy also suggests he believes in objective values generally.[31] On the basis of these four facts, the Dawkinsian texts support moral realism. They entail that there are objectively real cognitive, aesthetic, and moral standards or value-scales. Of course, as usual, I do *not* claim that Dawkins himself is a realist about these values. I claim *only* that his writings support that realism. Dawkins makes much less sense if those values are merely subjective. Believing in Dawkins means trying to make the most sense out of his writings. So believing in Dawkins means building on Dawkins: spiritual naturalism includes realism about cognitive, aesthetic, and moral values.

3   On Natural Duties

3.1   Emergent Norms

According to our ontology (Sect. 1 in Chapter 6), non-being negates itself while being-itself affirms itself. Hence being-itself adopts two policies towards itself. It expresses them through the purely logical activity of its dyadic self-relation. The first policy is that being-itself minimizes self-inconsistency. The second policy is that it maximizes self-consistency.

---

[31] SS.

But the policies that being-itself adopts for itself serve as norms for all the beings that emerge from it. If those policies are not norms, then those beings do not emerge; but there are beings among beings. Being-itself *forbids* every being from adopting any strategy which introduces self-inconsistency. Being-itself *obligates* every being to adopt only strategies which maximize self-consistency.

According to our ontology, truth is better than falsehood (Sect. 4.2 in Chapter 6). Since truth is better than falsehood, self-consistent strategies are better than self-contradictory strategies. Therefore, acting according to a strategy that contradicts itself is wrong; but acting according to a strategy that is maximally self-consistent is right. Every being *ought* to follow self-consistent strategies and every being *ought not* to follow self-contradictory strategies. It is *obligatory* to follow self-consistent strategies and it is *forbidden* to follow self-contradictory strategies. Since being-itself generates all the beings, its policies hold universally. So if any being adopts a policy which remains self-consistent when universalized, it is obeying being-itself. But if it adopts a policy which introduces self-inconsistency, then it is disobeying being-itself. By disobeying being-itself, it refutes its own being, and therefore destroys itself.

As a purely logical principle, the *categorical imperative* sorts policies into the obligatory and the forbidden. It says beings should obey being-itself and they should not disobey it. Hence policies which are self-consistent when universalized are obligatory, while those which are self-inconsistent when universalized are forbidden. Consequently, the categorical imperative emerges from the reflexivity of being-itself. It regulates the evolution of complexity. Start with mathematics. Mathematical existence is pure self-consistency. Physical existence supervenes directly on the mathematics. Hence the laws of physics in every possible universe are consistent. As things evolve into complexity, they can come into conflict. Conflicts are not inconsistencies; however, they are images of inconsistency. As things evolve into complexity, some of them become agents. Any computer that runs an entelechy is an agent; hence every crane is an agent. Many simpler universes contain stellar and planetary cranes. Our universe contains at least one crane, namely, our earthly crane, which generated replicators. When they emerged, reflexivity became explicit. It began to concentrate itself into life.

The normative power of being-itself acts in every agent. It expresses itself in the axioms of our deontic logic. The first axiom of our deontic logic states that if some agent can maximize value, then it ought to maximize it. Since this axiom requires only agency, it applies to all cranes in all universes. It applies to all celestial computers in our universe or other universes. It applies to all organisms. Since any organism can maximize value, it ought to maximize it. Organisms have the duty to maximize it. As organisms diversify, their duties can come into cooperation and into conflict. The second axiom of our deontic logic states that any rational agent strives to do its duty. All organisms participate at least in the structural rationality of existence. So they strive to do their duties. They strive to maximize value in their own ways. These strivings come into cooperation and conflict. They drive Darwinian evolution.

Many qualities evolve slowly from primitive simplicity to glorious complexity: agency, duty, intelligence, consciousness, reasoning, purposiveness, morality. They grow by slow accumulation from bacteria to humans, then into possible transhumans and godlike superhumans. On any evolutionary account, these must all emerge in primitive ways in early life. Game-theory generates axiological demands which emerge very early in evolutionary history. As these qualities grow in complexity, more complex forms of self-relation emerge. Hence the categorical imperative becomes more explicit.

3.2   Applying the Categorical Imperative

As any celestial computer does its duty, universal Darwinism entails that replicators evolve on the surfaces of its planets. These evolve into cells. Any cell has some tiny degree of agency—it is an agent whose maxim or policy is encoded in its genes. Every possible mutation of its genome is an alternative policy. Thus cells adopt genetic policies towards themselves. As a purely logical principle, the categorical imperative applies to cells. A cell should not adopt a genetic policy which kills it. It should adopt a genetic policy which enables it to live and reproduce. And cells strive to do their duties: they develop software for correcting mutations to their genetic codes. Any free-living cell has the duty to reproduce to infinity. However, as cells congregate into social groups called bodies, their reproductive duties become constrained. Cells still adopt genetic policies towards themselves. But now they also adopt those policies towards other cells, as well as towards the society of cells in which they live. The categorical imperative applies to cells in bodies. A cell should not adopt a genetic policy which kills its body. It should adopt a genetic policy which enables its body to flourish.

The application of the categorical imperative to cells in bodies can be illustrated using cancer. Among the alternative genetic policies of any cell in some body are some that lead to unconstrained reproduction—they are cancerous. Should a cell adopt a cancerous policy or not? The categorical imperative states that an agent ought to adopt only those maxims that can be universalized without self-negation, that is, without self-destruction or self-refutation. There are two ways that a cell in a body can universalize some genetic policy. The first way universalizes it by extending it to all the offspring of the cell. The second way extends it to every cell in the body. On either way, if any cancerous policy were universalized, the body would die; but then its cells would die; therefore, when cancerous policies are universalized, they lead to biological self-negation. Any cell which adopts a cancerous policy eventually kills itself. Because they are suicidal, cancerous policies violate the categorical imperative. Every cell *should not* adopt a cancerous policy. Cancerous policies are forbidden, while policies of controlled reproduction are obligatory. Natural selection reinforces the categorical imperative: it finely tunes cells to strive to perform their obligatory policies. Cells evolve circuits to prevent cancerous growth. Tragically, cells often fail to do their duties.

The logical version of the categorical imperative applies to any agent which depends for its survival on the flourishing of its social network. It therefore applies to many non-human animals. Of course, Kant would have disagreed: he denied that non-human animals have duties. On this point, Kant is just plain wrong. More recent philosophers have outlined natural theories of value which allow non-human animals to be agents with duties.[32] Social insects are agents with duties.[33] If some bee finds

---

[32]Foot (2001: ch. 2), Bekoff and Pierce (2009), and Rowlands (2012).
[33]SG ch. 10.

nectar and returns to the hive, then it *ought* to do an accurate waggle dance to show its hivemates the way. Social mammals are agents with duties. Every wolf who shares in the bounty of the common feast has an *obligation* to hunt with the pack. Free-riders fail to do their duties. But we are neither bees nor wolves. If ethics emerges from the teleonomic flourishing of species, then humans will have distinctively human ethics. Perhaps there are some asocial and irrational animals for whom *arete* is blind brutality—if so, they are not humans. Human *arete* includes the virtues of rationality and sociality.

Dawkins says that thinking about right and wrong is the business of moral philosophers.[34] So which moral philosophies are best supported by his theories of value? He briefly considers Kantian foundations for morality. Two considerations show that the Dawkinsian theory of value supports Kantian ethics. The first is that Dawkins values truth very highly; but valuing truth leads to Kantian ethics. The second is that Dawkins use game-theory to present the evolution of cooperation. Kantian ethics rules out uncooperative strategies in games. The idea behind Kantian ethics is the categorical imperative: always act so that the maxim of your action can be willed as universal law. The maxim of your action is some principle you follow when you act.

For example, if you refrain from stealing something, you are acting according to the maxim that stealing is wrong. Your maxim is part of a larger moral strategy, which involves the concepts of property and owner-ship. But suppose you decide to steal. You are now following a maxim which says that stealing is permitted. If this maxim were universalized, so that everybody could steal whenever they wanted to, there would be no property. But then it would be impossible to steal. Stealing depends on the concept of property. Since stealing refutes its own assumptions, it is practically self-contradictory. Likewise lying presupposes truth-telling and thereby contradicts itself. And sexual infidelity presupposes fidelity, and thereby contradicts itself. The categorical imperative implies that you ought to keep your promises and obey your contracts. Dawkins correctly understands that it rules out parasitic strategies in games: if everybody in

---

[34]Moral philosophers (OTTR 397; GD 265; SITS 271). Kantian foundations (GD 264–5). Strategies in games (SG ch. 12).

a game acts as a selfish parasite, there won't be hosts left to parasitize.[35] So adopting parasitism as a moral maxim is self-refuting. It is anti-reflexive.

According to the categorical imperative, to say that a maxim is wrong means that it is anti-reflexive. It participates in the self-negativity of the abyss. Since maxims permitting murdering, stealing, sexual infidelity, breaking promises, breaking contracts, and refusing to cooperate are all self-negating, they are all wrong. But maxims which are self-affirming are right. For example, the *golden rule* says do unto others as you would have them do unto you. The golden rule can be universalized. When universalized, it is self-affirming. Other self-affirming maxims say to keep your promises; honor your contracts; refrain from murdering, stealing, and cheating; and strive to cooperate.

3.3   From Rationality to Natural Goods

The Dawkinsian theory of value also supports the *natural goods* approach to ethics (an approach which is consistent with Kantian ethics). The Dawkinsian texts support biological approaches to ethics; but the natural goods approach is biological. According to this approach, organisms naturally rank their own states from worst to best. This is an objective ranking. It follows from your nature as a living thing that health is objectively better than sickness. This ranking is typically based on the natural *capacities* or *capabilities* of the organism. Organisms naturally seek those states in which they fully manifest their capabilities, or in which they perform their functions well. Likewise they naturally strive to avoid those states in which they cannot manifest their capabilities, or in which they perform their functions poorly. Most birds have the natural ability to fly. Health is good because it enables a bird to fly. Injury or being caged is bad because it prevents a bird from flying. The natures of organisms objectively define the conditions of their ideal existence, the conditions in which they flourish.

We flourish when we maximally exercise all our capabilities; we naturally strive to flourish; flourishing is an ideal goal. The philosopher Max

---

[35]GD 265.

Black moves from facts to obligations like this: to say that you ought to *be* something means that something is an ideal end for you. Thus you ought to *be flourishing*. To say that you ought to *do* some act means that act is an optimal means to an ideal end. Here an optimal means is one that is most effective or which maximizes the probability that you will end up at your ideal goal. If eating a balanced diet is part of an optimal means to the ideal goal of flourishing, then you ought to eat a balanced diet. It is a matter of logic that if you ought to do something, then you ought to do whatever that requires. You ought to be healthy; being healthy requires eating nutritious food and exercising; so you ought to eat nutritious food and exercise. You ought to flourish; but humans are social animals; so your flourishing requires the flourishing of humanity; therefore, humanity ought to flourish. And you ought to do whatever is necessary to help humanity flourish. Likewise, since humans are animals embedded in an ecosystem, the flourishing of humanity requires the flourishing of the earthly ecosystem. So you ought to help the ecosystem flourish.

The fact that we are rational animals links the Kantian approach to ethics with the natural goods approach to ethics. Since we are rational animals, we flourish when we are being rational. Being rational is naturally good for humans. On the one hand, strategies which are self-consistent are rational; therefore, acting according to them is a kind of health which enables you to realize one of your natural goods. For a naturally rational animal, such as a human animal, it is a natural duty to act rationally. You are obligated by your own nature to act rationally; but this means you are obligated by your own nature to adopt only self-consistent strategies. On the other hand, strategies which are self-contradictory are irrational; therefore, acting according to those strategies is a kind of illness which prevents us from realizing one of our natural goods. For a naturally rational kind of animal, such as humans, acting irrationally is a natural evil. Thus, as a naturally rational animal, it is your natural duty to not act irrationally. You are forbidden by your own nature to adopt self-contradictory strategies. If you act irrationally, you act in contradiction to your own nature. Since you are a rational animal, you must follow the categorical imperative in order to flourish. It follows from the facts about your own nature that you ought to follow the categorical imperative.

As a scientist, Dawkins is interested in measurement. One advantage of the natural goods approach is that human flourishing can be measured. The natural goods approach to ethics leads to the *capabilities approach* to human welfare.[36] Since we have the basic biological capability to eat, we ought to have food. Hunger and famine are obviously bad for humans. So we can measure the extent to which people do have food. We can also measure other medical indicators of welfare such as longevity. But people need other things, like security from violence, to flourish. So we can measure the rates of crime in a society. We can measure other dimensions of social health. These ways of measuring personal and social health have been compiled into *human development indexes*.[37] These indexes aim to quantify human flourishing.

3.4   Maximize Beauty!

Dawkins believes that the universe is orderly, rational, and beautiful. By truthfully describing it, we participate in its order, rationality, and beauty. To do science is to truthfully describe the universe. By mentally simulating a beautiful rational thing (the universe), we become *rationally beautiful*. This is a deeper beauty than merely sensory beauty (Sect. 2.3 in Chapter 4). The rational beauty of the universe is a cosmic value. Consequently, by doing science, we participate in this cosmic value. Participating in this cosmic value makes our lives valuable—it makes life worth living.

Thus Dawkins writes that, by doing science, we can arouse in ourselves a feeling he calls "awed wonder."[38] He says it ranks among the most valuable of all possible human experiences. Its intensity can equal the most powerful emotions aroused by art. Most importantly, he says that feeling this awed wonder "*is truly one of the things that makes life worth living.*" So this awed wonder provides an aesthetic justification for life. Dawkins says our big brains enable us to "construct a working model of the universe

---

[36] Sen (1993) and Robeyns (2005).

[37] Burd-Sharps et al. (2008) and Deneulin and Shahani (2009).

[38] Awed wonder (UR x). Life worth living (UR x, my italics). Before we die (UR 312). In the first place (UR 313, my italics). Inspiration (GD 394–420).

and run it in our heads before we die." As we face our deaths, we can say that building these models explains *"why it was worth coming to life in the first place."* This implies that what makes life worth living is running a scientific model of the universe in your brain. Doing science may give you consolation or inspiration. But doing science has deeper meaning: it is the purpose of life. This is the old Platonic imperative (Sect. 6 in Chapter 1): *Maximize vision!* To run a model of the universe inside of your head is to participate in the self-representation or reflexivity of the universe (Sect. 2.5 in Chapter 3). Thus value emerges from reflexivity. To maximize reflexivity is to maximize value.

Dawkins says that cognitive participation in the rational beauty of the universe makes life worth living. If this is interpreted narrowly, then only scientists have lives worth living; however, people who are not scientists do have lives worth living; hence the narrow interpretation is wrong. It is therefore necessary to interpret this broadly, as saying that *being rationally beautiful makes life worth living*. This echoes Nietzsche's claim that "only as an aesthetic phenomenon are existence and the world justified."[39] It suggests that *maximizing participation in rational beauty* is an ideal goal for humans. Our highest capability is our ability to participate in rational beauty. Rational beauty includes but surpasses the shallow beauty revealed by the senses.[40] It includes the deep beauty revealed by mathematics. But this leads to an aesthetic imperative: *we ought to maximize human participation in rational beauty*. Being beautiful is more valuable than pleasure or happiness. The utilitarians said that we ought to adopt strategies which maximize happiness for the greatest number of people for the longest time. But a Dawkinsian may say that we ought to adopt strategies which aim to maximize beauty for the greatest number of organisms for the longest time. This motivates replacing the utilitarian greatest happiness principle with the *greatest beauty principle*: an action is *right* insofar as it facilitates the universal participation in rational beauty; it is *wrong* insofar as it hinders the universal participation in rational beauty.

---

[39] Nietzsche (*The Birth of Tragedy*, sec. 5).
[40] UR 63–4.

Maximizing rational beauty clearly requires maximizing other goods. Being beautiful requires being healthy; but being healthy requires access to nutritious food and good medical care; access to good food and good medicine require social and economic justice. Being beautiful requires participation in economic beauty. Participation in economic beauty requires having meaningful work, in decent labor conditions, without exploitation or economic injustice. Working in a factory is ugly; working in unjust conditions is ugly; working as an exploited laborer is ugly. Being beautiful requires participation in social beauty. Participation in social beauty requires security, fairness, justice, and social harmony. Social ugliness includes crime, violence, war, corruption, racism, sexism, and so on. Being beautiful requires participation in intellectual beauty. But this participation requires education. And it requires all the intellectual virtues of truth-telling. Living beautifully generally requires participation in virtue. Vice is ugly. Living beautifully requires self-enhancement. Greater beauty is greater complexity. Participation in beauty requires participation in challenges involving competitive excellence.

In his early writings, Dawkins said science cannot discover ultimate moral premises; he implies there cannot be any science of moral obligation.[41] However, in his review of *The Moral Landscape* by Sam Harris, Dawkins wrote that Harris made him change his mind. Harris made him realize that science has a lot to say about the foundations of morality. On the natural goods approach, science can help us understand our natural goods; it can help us understand what it means to flourish. It can likewise help us to understand what we need to do to flourish, and therefore what we ought to do. If our universe does have ideal possible histories, then we ought to make progress towards them. Perhaps these ideal histories move to omega points (Sect. 5 in Chapter 2). This leads to a *greatest progress principle*: an act is right insofar as it moves our history towards the ideal history; it is wrong insofar as it moves our history away from the ideal.

To figure out the policies that we ought to adopt, Dawkins sketches a method. Since his sketch is incomplete, spiritual naturalism fills it out. Filling out this method is part of building the Sanctuary for Spiritual

---

[41] OTTR 397; ADC 34.

Naturalists. When it is filled out, the method has four steps. (1) The first step involves simulations. Dawkins often appeals to computer simulations in his discussions of evolution. So we can use computers to create detailed simulations of multiple possible futures. (2) The second step involves ranking alternative futures using the concept of natural goodness (such as the capabilities approach, and naturalistic concepts of human flourishing). (3) The third step involves using computers to find the policies that maximize the probability of moving our present towards the best possible futures. We can digitally simulate the outcomes of alternative policies.[42] We adopt those that move us most effectively towards the best possible futures. For example, if maximizing rational beauty is right, then we adopt those policies which maximize rational beauty. If maximizing the idealness of our history is right, then we adopt those policies which maximize ideality. (4) The fourth step involves evaluating the effectiveness of the policies. Are they working? If not, we need to go back to the earlier steps and start over. We test and iterate. This is the experimental method applied to human self-evolution. This test-and-iterate method is sometimes called *hacking*. Thus spiritual naturalists use *hacker methodology*.

4   The Illusion of Identity

4.1   The Immortal Gene

When it comes to persistence, Dawkins contrasts genes with diamonds.[43] He notes that a diamond can persist for millions of years. The diamond persists as the same arrangement of atoms. Following the language used previously (Sect. 1.3 in this chapter), the diamond persists by *enduring*. A material thing endures as long as it consists of the same atoms arranged in the same way. Thus diamonds generally endure for very long times. If you think of a gene in concrete terms, then you think of it as a particular material thing. Concretely speaking, a gene is just a specific

---

[42]UR 312; SITS 71, 191.
[43]SG 34–5.

constellation of atoms. But genes do not have the stabilities of diamonds. Genetic constellations endure only for a short time. When its atomic constellation falls apart, the gene dies—it ceases to exist.

However, since genes are replicators, they can make copies of themselves. Of course, a copy is never identical with its original. Suppose some original gene X fissions into two copies Y and Z. Symmetry entails that if there is any identity between the original and its copies, then X is identical to Y and X is identical to Z. Substitution of identical terms tells us that Y is identical to Z. But that is false. When a gene from some parent organism produces a copy of itself in some offspring organism, that offspring copy is *not* identical with its parental original. For genes, as for everything else, *there is no identity through replication*. Hence there is no genetic identity across generations of offspring. So this distinction between the gene and its copies is not trivial. A gene does not literally *survive* in its copies. To survive is to continue to endure as the same stuff. However, when a gene makes a copy of itself, the information or pattern encoded in the gene gets transmitted to its copy. The pattern in the copied gene is identical with the pattern in the original gene. But the pattern is not a material thing—it is a form.

Since genes are replicators, they do not persist like diamonds. If diamonds persist by enduring, then genes persist by *perduring*. A thing perdures through its *counterparts*. Each copy of a gene is a counterpart of its original. The parental original is a *past counterpart* of each offspring copy and each offspring copy is a *future counterpart* of its parental original. But these counterparts are not identical. So while a gene does not survive in its copies, it is *recreated* in its copies. A *genetic chain* is a series of genes in which each next gene is a copy of its previous gene. The chain is a temporally extended *process* in which the specific genes are instantaneous *stages*. If we think of time as the fourth dimension, then the chain of genetic copies is a 4D process of 3D stages.

A genetic chain can persist for hundreds of millions of years. It can potentially extend into the future forever. However, while the chain can persist for a very long time, its constituent genes do not persist for very long. If the chain is immortal, it does not follow that its genes are immortal. When we say that the gene is immortal, we are speaking figuratively. We are figuratively attributing a property of the chain of genes to its parts (so the figure is part-whole metonymy). It does not harm to talk like this, so long as we remember that we are speaking figuratively rather than literally. Along any chain, the abstract pattern of the original gene moves from copy to copy. A chain of genetic copies is a stream of information—it carries an abstract genetic form down through the ages. And, since counterparts are not identical, the original gene X can be a counterpart of its offspring Y and of its offspring Z. So genetic reproduction can create a potentially endlessly branching tree of counterpart offspring.

4.2   The Soul Is the Form of the Body

The Dawkinsian conception of genetic immortality can be extended to anything that can be replicated. Dawkins believes that *memes* are mental replicators living in brains.[44] Memes can also have replicator immortality: they can persist across potentially endless sequences of counterparts. All the genetic information in your body can be copied into a clone. But all the memetic information in your brain can also be copied into a new brain, or into a computer. Indeed, all the information in your body can be copied into a new body, or into some functionally equivalent robot. Just as material genes are mortal, so material bodies are mortal. But just as patterns of genetic information can be immortal, so patterns of bodily information can be immortal.

Although our bodies store self-descriptions in their genes, Dawkins correctly writes that genes do not completely define persons.[45] You also store self-data in your brain. You would need to have a perfect memory to store an exact biography. Most of us augment our naturally faulty memories with written or photographic records. Much of the information about our lives is stored in external memories. Consider the total system of information that defines you. This total system is your *life-script*. It is a script because your life is a four-dimensional process. Your life-script completely and perfectly describes your entire personal history.

---

[44]SG ch. 11; etc.
[45]UR 1.

Anything which duplicates your life-script is an exact replica of your entire temporally extended life from birth to death.

Your life-script has two parts. Its first part is your nature or essence. Your essence encodes the invariant aspects of your life. Your essence is analogous to a computer program or algorithm. Most of your genetic information is in your essence. The second part of your life-script is your destiny. It encodes all the variable aspects of your life. It is like a string of inputs to your life-script. Your life-script generates your life by running your essence on your destiny, by running your life-program on your string of life-inputs. But the term life-script is awkward. Is there a better term? There is. This concept of the life-script is very close to the old Aristotelian concept of the form of the body.[46] But Aristotle referred to the form of the body as the *soul*. So your life-script is your soul. This is an entirely physical conception of the soul.[47] There is no reason to allow religions to hijack the soul—spiritual naturalists reclaim the soul from religions.

It might appear that a Dawkinsian cannot accept any concept of the soul.[48] But that is false. Dawkins distinguishes between two senses of the soul, which he refers to as *soul-1* and *soul-2*. Soul-1 is the immaterial mind; it is a supernatural substance that can exist without the body. We agree with Dawkins that these souls do not exist. But soul-2 is just the mental functionality of the human body; it is the mind that emerges from the body. Dawkins embraces the concept of soul-2. To these we can add *soul-3*. Soul-3 is just the modernized Aristotelian form of the body. Soul-3 is the life-script. The concept of the soul as a life-script is just the concept of the gene extended to the whole organism. It is a biological conception of the soul. A Dawkinsian can embrace that.

4.3   How Recreation Beats Survival

The illusion of the unitary self and the illusion of personal identity work together to produce the illusion of mind-body dualism. They produce

---

[46] Aristotle (*De Anima*, 412a5–412b21).

[47] Steinhart (2014: ch. 3).

[48] Against soul-1 (GD 35; SITS 5, 212–3). For soul-2 (SITS 5, 212).

the illusion that the mind is somehow distinct from its body. The body ages, while the mind remains self-identical. The mind itself remains free from the ravages of time. If you get Alzheimer's, your brain changes; nevertheless, according to dualism, your mind remains unchanged. The illusions of the unitary self and personal identity thus produce the *illusion of survival*. Your body dies. But since your mind remains self-identical, despite the changes of your body, your mind survives bodily death. This really is an illusion. Bodies do not survive death; but minds are activities of bodies; hence minds do not survive death.[49]

Although minds and bodies do not survive death, life after death requires neither bodily nor mental survival. You can deny survival while affirming life after death. If there is no survival, there can still be life after death. To say that *you will live again after you die* means exactly that you have a future counterpart who does live after you die. And there are many naturalistic ways you can have a future counterpart who lives after you die. Genetic technology is making progress. Dawkins argues that by 2057, you will be able to clone your body.[50] He then raises this question: will your clone "be a resurrection of your conscious being, a reincarnation of your subjectivity?" Your clone will just be an artificial monozygotic (identical) twin. Since twins do not share the same mind, your clone will not be a revival of your mentality.

Of course, your future clone won't have your memories—it won't be psychologically continuous with you. But phrases like "a resurrection of your conscious being" and "a reincarnation of your subjectivity" are far from clear. Reincarnations traditionally have their memories wiped clean. And the Christian doctrine of the resurrection *of the body* never said much about the continuity of the mind. Cloning from a genomic book would probably satisfy St. Paul's description of the resurrection (in 1 Corinthians 15: 35–57). Your future clone will not be the same person as you in the old psychological sense of personal identity. But if the perdurantists are right, then identity is not what matters. Biological objects persist through replication. There is a clear biological sense in which you will live again by means of your clone. Your clone will be

---

49 ADC 13; GD 220–1, 398–404; SITS 3, 268.
50 ADC chs. 2.5, 3.4. Reincarnation of your subjectivity (SITS 215).

your future counterpart. And, through it, you will live again after you die. Cloning might not be personally satisfying to people brought up in the culture of mind-body dualism. But it does illustrate the distinction between survival and life after death. And it is strange for Dawkins to worry about the survival of consciousness and subjectivity. After all, Dawkins argued that the conscious self is an illusion (Sect. 1.2). He would be more consistent with his own theory of human animality if he said that any new body generated by the same genes enables the same person to manifest itself again in a new way.

5   Life After Death

5.1   Promotion into Your Next Life

Dawkins frequently affirms that there is no life after death.[51] His argument appears to go like this: (1) All the old religious theories of life after death are false. (2) If all those old theories are false, then all theories of life after death are false. (3) Therefore, there is no life after death. The argument fails at the second step. It's like saying that if all the old religious theories about the origin of life are false, then all theories of the origin of life are false. Obviously, they are not. Dawkins replaces religious creationism with scientific evolution. He should have argued that false religious theories of life after death can be replaced with more plausible scientific theories of life after death. The example of cloning suggests that life after death does not require religion. Spiritual naturalists reclaim life after death. We liberate it from religious hijacking by making naturalistic theories of life after death. At least two plausible naturalistic theories of life after death are supported by things Dawkins says about persons and universes.[52]

The first Dawkinsian type of life after death comes from the simulation hypothesis. Dawkins admits that the simulation hypothesis might be true (Sect. 2.2 in Chapter 4). But simulation permits human persons

---

[51]ADC 13; GD 21, 221, 397–401; FH 24; SITS 3, 268; etc.
[52]Steinhart (2014).

to be recreated after death. We might be *promoted* from the simulation into the surrounding reality of the simulators.[53] Suppose some cosmic engineers design and create a simulated universe. It runs on a sidereal computer in their laboratory. They watch as life evolves inside of their simulated universe. Whenever any intelligent organism appears, they start to record the details of its life. They record the structure of its body and its biography. They record its soul, which is just a gigantic data file in their sidereal computer. After the organism dies, they promote it into their own surrounding universe. They use their robotics skills, and their records of the old organism, to make a new body for it. The old organism that previously lived inside the computer now lives outside of it as a new organism. They have recreated the old organism (in fact, this has actually been done). But the new organism is not identical with the old organism. The new organism and the old organism are counterparts. This idea of promotion resembles the ancient idea of the *resurrection of the body*. Promotion is an example of the distinction between survival and life after death. Promotion is consistent with a naturalistic conception of human persons. It is likewise consistent with everything Dawkins says about persons and universes.

5.2   Revision into Your Next Life

The second Dawkinsian type of life after death is *revision*.[54] Revision follows from the Dawkinsian theory of complexity. The *Argument for Revision* goes like this: (1) Our universe is extremely complex. (2) The evolutionary principle (Sect. 1.1 in Chapter 5) states that every complex thing comes from some simpler version of itself. (3) So our universe comes from some simpler version of itself—it is the offspring of some less complex parent. (4) But the evolution of complexity follows Dennett's principle of the accumulation of design, which states that design is mostly copied from simpler things to more complex things. (5) Hence the design of each offspring universe is mostly copied from the design of its parent. (6) But the designs of wholes depend on the designs of

---

[53]Moravec (1988: 152–153) and Steinhart (2014: ch. 5).
[54]Steinhart (2014: ch. 7).

their parts. If the design of the next whole is mostly copied from the last whole, then the designs of the next parts are mostly copied from the last parts. Parts correspond to parts. (7) So it's likely that your present life is an improved version of some past life in the parent of our universe. You have a past counterpart in the parent of our universe. Likewise it's likely that every offspring of our universe will contain an improved version of your life. You will have better future counterparts in the next universes. Of course, an improved version of your life is a minimally more intrinsically valuable version of your life. Improvement implies increases in complexity, reflexivity, beauty, and so on. Improvement increases functional excellence. It increases the *arete* that emerges in the strife-torn *agon*. It need *not* increase happiness—we are *not* utilitarians.

The Dawkinsian theory of complexity supports a very strong case for revision. (8) Since the complexity of any whole depends more intensely on the complexities of its more complex parts, the probability of that some past part is copied into some future universe increases with the complexity of that part. Humans are extremely complex things. Hence it is extremely likely that we will be copied into the cosmic offspring of our universe. (9) Since more complex things are more fragile, the probability that some past part is very accurately copied into some future universe increases with the complexity of that part. If the structure of a rock is only approximately copied across universes, then the result is likely to be another rock, or at least something with similar complexity. However, if the structure of an organism (like a human) is not extremely closely copied from one universe to the next, then the result is likely to be a very dead and very simple thing. More precisely, if the structure of your genes and your brain are not extremely closely copied, your complexity will be *entirely* lost. (10) Therefore, as the complexities of things increase, it becomes more likely that their structures will be accurately copied from each universe to its offspring. This copying implies only that each future counterpart includes all the complexity of its past counterpart. It permits complexity to be increased from past to future counterparts. Intrinsic value can (and must) increase. Obviously, the premises of this long argument are speculative. But I have spent the entirety of this book arguing for them. And they all follow directly from things that Dawkins himself says. The Sanctuary includes revision.

Revision follows from the thesis that animats improve their universes. Every animat improves its universe in every possible way. Universes have part-whole structures. A very simple universe just contains some simple things. A more complex universe has simple things which serve as parts of more complex wholes: the simple parts fuse into some wholes. An even more complex universe has some simple things which fuse into first-level wholes; those first-level wholes fuse into second-level wholes. As universes gain complexity, they gain more levels of part-whole structure. It is a law of logic that better wholes are composed of better parts (arranged in better ways). As parts get improved, their wholes also get improved. Thus improvement bubbles up from the simplest parts to the most complex wholes. If any part is improved, then its better version must be compatible with some better version of its whole. For if the new version of the part were not compatible with any better version of its whole, then that part would have been made worse rather than better. Hence every improvement of every part of some whole is a part of some improvement of that whole. It follows that, if some whole is improved in every possible way, then every part of that whole is improved in every possible way. An example may help to clarify this reasoning.

Suppose a universe contains exactly one simple organism. This organism has a head H and a tail T. Since this organism is the original, its head can be called H_0 while its tail is T_0. Now suppose that there are two ways to improve the head. These are H_1 and H_2. And the two ways to improve the tail are T_1 and T_2. Improved organs come together to make improved organisms. Each better version of an organ can enter into at least one better version of its organism. For suppose some new version of the tail could only combine with some degraded version of the head to make a degraded version of the organism. If that were the case, the new tail would be worse tail rather than a better tail. Hence any better tail can enter into at least one combination that makes a better organism. The logic generalizes: any better part can enter into at least one combination that makes a better whole. So the improvements of the organism are composed of improvements of its organs. There are eight ways to improve the organism: (H_0, T_1), (H_0, T_2), (H_1, T_0), (H_1, T_1), (H_1, T_2), (H_2, T_0), (H_2, T_1), and (H_2, T_2). The only combination that is not an improvement is just the original combination (H_0, T_0).

Since these combinatorial ideas are purely logical, they can be universally generalized. They apply to any wholes composed of any parts. They scale up from the organs of organisms to entire universes.

You are obviously *not identical* with any of your future counterparts. But Dawkins denies the importance of identity. Likewise, there is *no survival* from the old universe to the new universe—there is only recreation. Each life in the old universe has a counterpart in the new universe. Each later counterpart is a revision of some previous life. Revision across universes illustrates the distinction between survival and life after death. It is consistent with a naturalistic conception of human persons. If the argument for revision is sound, then you have future counterparts in future universes in the world tree. Since these universes are very similar to ours, they can be overlaid on our universe. Since they are better versions of our bright universe, they are bright universes. Since they realize the logic of hope, they are aspirational. Through transformational practices (Sect. 4 in Chapter 9), you can mentally identify yourself with some better future counterpart. Through rituals of hope, you can mentally shift into some aspirational universe.

5.3 Animatic Design Constraints

Since every animat improves itself in every possible way, it improves its universe in every possible way. The maximization of value entails that all these ways are *Pareto optimal*. As animats produce new universes from old universes, they produce all the Pareto optimizations of their old universes. Four rules define these Pareto optimizations. These rules ensure that the value of the old universe (the old whole) is optimally increased in the new universe (the new whole). Here are the rules: (1) The first rule is that every part in the old whole must have at least one new version of itself in the new whole. Hence no value is lost by absence. The new version of the old part is a *counterpart* of the old part. (2) The second rule says that distinct parts in the old whole must have distinct counterparts in the new whole. Hence no value is lost by erasure of uniqueness. (3) The third rule says that no part in the old whole can have a less valuable counterpart in the new whole. The values of the parts are never decreased. (4) The fourth rule says that at least one part in the old whole must have a more valuable counterpart in the new whole. The value of at least one part must be increased. Of course, as animats generate improved wholes from improved parts, they must also ensure that these new parts fit together coherently. The parts must be mutually compossible. They must fit together either as cooperators, competitors, or neutral parties.

These rules ensure that old lives are revised into better lives in new universes. They entail that lives are immortal in the same sense that genes are immortal. To see how these rules lead to genetic immortality for lives, consider some old universe which will be surpassed in every possible way. This is the original universe running on some original animat. This old universe contains the three lives A, B, and C. The animat which ran those lives figures out that each can be improved in three ways. These are the *seeds* of each life. So the seeds of A are A_1, A_2, and A_3; the seeds of B are B_1, B_2, and B_3; and those of C are C_1, C_2, and C_3. Suppose that seeds with the same numbers are compossible, while those with distinct numbers are incompossible. Hence there are three sets of compossible seeds. These sets are cosmic *clutches*. The first clutch is $\{A_1, B_1, C_1\}$; the second clutch is $\{A_2, B_2, C_2\}$; and the third clutch is $\{A_3, B_3, C_3\}$. Each of these clutches satisfies all the rules for Pareto optimality: each old thing has at least one counterpart; distinct old things have distinct new counterparts; no things are made worse; at least one thing is made better. Since these clutches satisfy the Pareto constraints, they are *viable*. Each viable clutch can serve as part of the cosmic script for an improved version of the original universe. So the animat which ran the original universe produces three offspring which run three improved cosmic scripts based on these clutches of lives.

Each life in the parent universe is therefore surpassed by a future counterpart in the next offspring universe. For example, the life A is surpassed by A_1, by A_2, and by A_3. So the original life A is followed by the better lives A_1, A_2, A_3. These in turn will be followed by their own improved lives in later universes. Animatic creativity generates an infinite progression of these successor lives. The rules for infinite animatic creativity imply that every infinite progression of successor lives is surpassed by an infinite limit life. So immortality proceeds into the infinite. Of course, these lives are not identical with each other—each next life is a future counterpart of its previous life. This is exactly like the genetic immortality described by Dawkins. Lives, like genes, are immortal through their counterparts. But this logic transfers directly to souls. Lives are generated by running souls. As old lives get revised into better lives, so old souls get revised into better souls. Souls have a kind of genetic immortality.

This genetic theory of life after death has some resemblances to ancient theories of *reincarnation*. Souls get copied across universes. However, since there are more differences than similarities, it isn't appropriate to refer to this theory as reincarnation. This genetic theory of life after death has more in common with the theory of *rebirth* in Theravadic Buddhism.[55] Theravadic Buddhists don't believe in an immaterial soul. They reject mind-body dualism. The theory of genetic life after death also rejects mind-body dualism. It is a kind of *naturalized rebirth*. But the danger remains: using ancient terms links the new theory with too many old and unintended connotations. It is better to use a new term. It is better to describe this new theory as *revision*.

References

Bekoff, M., & Pierce, J. (2009). *Wild Justice: The Moral Lives of Animals*. Chicago: University of Chicago Press.

Black, M. (1964). The gap between "is" and "should". *Philosophical Review, 73*, 165–181.

Blackmore, S. (2011). *Zen and the Art of Consciousness*. Oxford: Oneworld.

Burd-Sharps, S., Lewis, K., & Borges Martins, E. (2008). *The Measure of America: American Human Development Report 2008–2009*. New York: Columbia University Press.

Deneulin, S., & Shahani, L. (2009). *An Introduction to the Human Development and Capability Approach*. Sterling, VA: Earthscan.

Dennett, D. (1993). *Consciousness Explained*. New York: Penguin.

Foot, P. (2001). *Natural Goodness*. New York: Oxford University Press.

---

[55]Rahula (1974: 34).

Moravec, H. (1988). *Mind Children: The Future of Robot and Human Intelligence*. Cambridge, MA: Harvard University Press.

Pritchard, L., & Dufton, M. (2000). Do proteins learn to evolve? The Hopfield network as a basis for the understanding of protein evolution. *Journal of Theoretical Biology, 202*, 77–86.

Rahula, W. (1974). *What the Buddha Taught*. New York: Grove/Atlantic.

Robeyns, I. (2005). The capability approach: A theoretical survey. *Journal of Human Development, 6*(1), 93–114.

Rowlands, M. (2012). *Can Animals Be Moral?* New York: Oxford University Press.

Sen, A. (1993). Capability and well-being. In M. Nussbaum & A. Sen (Eds.), *The Quality of Life* (pp. 30–53). New York: Oxford University Press.

Steinhart, E. (2014). *Your Digital Afterlives: Computational Theories of Life After Death*. New York: Palgrave Macmillan.

Watson, R., et al. (2016). Evolutionary connectionism. *Evolutionary Biology, 43*, 553–581.

9

Spirituality

1   On Gratitude

1.1   From the Sublime to Gratitude

Dawkins says that we are not merely *lucky* to be alive, but that we are *blessed* and *privileged* to be alive[1] When he observes the sublime aspects of nature (its depths of space, time, and detail), he says he experiences powerful positive emotions. These emotions drive him to something very close to *worship*. Of course, he does not mean worship of any personal deity—he only means he experiences a kind of gratitude: he is grateful to exist in this glorious universe. Elsewhere, Dawkins says the natural processes which brought us into being are "wondrous, amazing, and they are cause for us to give thanks." This gratitude, unfortunately, motivates an argument against atheism.

The *No Thanksgiving Argument* against atheism goes like this[2]: (1) Atheists justifiably feel urges to give thanks for their lives and good

---

[1]Blessed and privileged (UR 1–5; ADC 12). Worship (2009b: 1:44–2:55; SITS 269). Glorious universe (FH 99). Cause to give thanks (2010: 27:01–17; 2009: 1:44–2:55).

[2]Bishop (2010) and Colledge (2013).

© The Author(s) 2020
E. Steinhart, *Believing in Dawkins*,
https://doi.org/10.1007/978-3-030-43052-8_9

fortunes. (2) However, since atheists do not believe in any divine persons, they have nobody to give thanks to.[3] (2) So their urges to give thanks are frustrated. (3) But the satisfaction of these urges is required for an emotionally or even ethically full life. (4) Therefore, atheists cannot live ethically or emotionally full lives. This is the atheistic problem of evil (Sect. 3.1 in Chapter 6).

There are at least two responses to this argument. The first response tries to argue for *intransitive gratitude*: it is possible to be *grateful for* something without being *grateful to* somebody.[4] Dawkins endorses intransitive gratitude. But giving thanks is distinct from mere gratefulness. To give thanks requires something *to which* thanks is given. So this response seems inadequate—it offers at most an incomplete satisfaction. The second response suggests that we can give thanks to impersonal things. Perhaps we can thank providence or fate. Dawkins says evolution *blesses* us with *gifts*. So perhaps we can thank impersonal evolution for its gifts. Hence the second response looks more carefully at the range of objects to which thanks can properly be given.

1.2   Computational Benefactors

Humans routinely give thanks to soldiers or others who have sacrificed their lives. Being deceased, they neither know they were thanked nor can they reply. Hence giving thanks to something does not require that it can reply with "You're welcome." Humans can and do exchange thanks with domestic animals.[5] Cats bring us gifts; dogs behave in ways that seem grateful. Although exchanging thanks does require high intelligence, philosophers generally do not regard domestic animals as persons. Hence giving thanks to something does not require that it is a person of any kind.

It is arguable that intelligent machines can give gifts to humans. A human lying in an intensive care ward in a hospital may be kept alive by

---

[3]Nobody to give thanks to (2010: 27:01–17; SITS 243–5).
[4]Intransitive gratitude (Lacewing 2016; Dawkins, SITS 245). Thank fate (SITS 243). Blesses us with gifts (ADC 12–13).
[5]O'Hagen (2009).

vigilant computers, which intelligently control much medical machinery. Upon recovery, she might justifiably give thanks to their programmers. But the computers might have gained their skills through learning. As computers grow more intelligent, they gain autonomy.[6] It seems likely that they will soon become entirely responsible for their own actions. At that point, we cannot praise or blame any human programmers (a fact which already causes legal problems with autonomous vehicles, such as self-driving cars). As artificial intelligence makes progress, computers themselves will give us gifts, and it will be increasingly appropriate for us to thank them. Moreover, as they gain intelligence, they will thank us in return. These intelligent machines may resemble animal minds. However, they will not be animal minds. They will have only mindlikeness. If this reasoning is correct, then thanks-giving requires only some degree of mindlikeness.

Dawkins says that biological evolution is a computer.[7] It runs a Darwinian algorithm. It has memory and perhaps it even learns. It moves with non-random directionality up Mount Improbable. Dawkins portrays evolution as a mindlike designer (Sect. 3 in Chapter 3). He says it makes a specific kind of progress: lineages of organisms accumulate adaptations that make them better at their ways of living. They accumulate *beneficial* adaptations. Evolution sharpens the teeth of the predator; it toughens the hide of the prey. It makes both the cheetah and the gazelle faster and more agile. It increases the functionalities of hands and eyes and brains. It makes progress in a value-laden sense. It maximizes the competitive excellences of organisms. A Stoic would say that it maximizes *arete*. It should thus come as no surprise that Dawkins says evolution has given us gifts. He even says these gifts are *blessings*. Evolution has blessed us with big brains. Consequently, we should be able to give thanks to evolution.

---

[6]SG 51–2; EP 24–7.
[7]Evolution computes (CMI 72, 326; UR 238–45; ADC 12). Memory (UR 257; GSE 405–8). Evolutionary progress (ADC 208, ch. 5.4). Value-laden progress (BW 305; CMI 85–100; CMI 132–6; AT 681–9). Blesses us with gifts (ADC 12).

On the basis of what Dawkins says about evolution and gift-giving, it is plausible to say that we can and should give thanks to the biological crane for all our adaptations.[8] Since evolution has blessed us with eyes, we can thank it for our eyes. Since it has graced us with powerful brains, we can thank it for our brains. We can thank it for our abilities to reason and to act morally. But the biological crane depends on other cranes. So if we can give thanks to it, then we can also give thanks to those other cranes. Since the physical crane has blessed us with our beautiful habitable earth, we can give thanks to it for our home. We can thank our sun for pumping entropy out of our solar system (Sect. 2.3 in Chapter 7). Our universe has many virtues which facilitate the evolution of complexity. According to the cosmological reasoning in earlier chapters, cosmological evolution gave our universe those virtues. We should give thanks to cosmological evolution. We can thank all the cranes. If our theory of agency is correct (Sect. 2.4 in Chapter 3), then we can compose Goethian Hymns to Nature. We can agree with the Stoics that the sun does its duty, and we can say "Thank you, sun."[9] Moreover, our theories of moral obligation entail that we *ought* to give thanks—we *should* thank all the cranes.

Spiritual naturalists affirm that we can and should give thanks to cranes. Since these cranes emerge from the laws of nature, we should thank them too. Like the Stoics, we should thank the impersonal rational order of nature. We should thank the Logos. Since all our benefits ultimately derive from the skewing of probabilities towards the Good, we should thank the ultimate skewer. Like the Platonists, we should thank the Good. People already say "Thank goodness!" Unfortunately, theists will try to hijack this thanks-giving by turning it into worship. Dawkins urges us to resist theistic hijackers. We need to liberate thanks-giving from its bondage to theism. We must invent poetic rituals of gratitude. And, in exactly the *philosophical* sense of the term *paganism*, the sense that comes from the best in the Stoic and Platonic traditions, these rituals will be pagan. Of course, if we can and should give thanks to cranes, then

---

[8] Blessed with eyes (CMI ch. 5). Blessed with brains (ADC 12). Blessed with reason and morality (AT 681–2). Blessed with Earth (UR 4–6).
[9] Epictetus (*Discourses*, 3.22.5–6).

it seems like we can and should also give criticisms. If we can praise them, then we can blame them. We can curse them for all the poorly designed features of our bodies.

1.3 Evolutionary Providence

Perhaps it will still be objected that evolution involves far too much wickedness for it to be a legitimate object of thanks-giving.[10] On this objection, it is not appropriate to give thanks to any power which produces so much misery and destruction. The objection says that evolution is not *providential*—but only a providential power deserves thanks. Here it is interesting to return to the theistic problem of evil: despite the fact that God has total control over the earth, and that the earth contains enormous misery and destruction, theists maintain that God is good. The strategies used to defend the goodness of God can be applied to defend the providence of evolution. Parallel to the old theodicies, Dawkins says there will be evolutionary theodicies—call them *evodicies*.[11] They will defend evolution.

Dawkins writes that there is "a kind of grandeur in nature's serene indifference to the suffering" in evolution.[12] This is a Stoic point. Nature, like the Stoic Zeus, exhibits equanimity (*apatheia*). Evodicies will argue, along with the Stoics, that evolution is providential. And, from the perspective of Stoic value-theory, suffering does not refute providence. Stoic value-theory entails that *utilitarianism is false*. Here is an unsound argument: (1) If evolution is providential, then it maximizes happiness. (2) It does not maximize happiness. (3) Therefore, it is not providential. The argument is unsound because its first premise is false. Providence does not entail maximizing happiness. Nor does it imply that this is the best of all possible universes. Universes (and their parts) are beings; but every being is surpassed by better beings. Providence merely means that evolution does the best it can do given its constraints.

---

[10]SG 284; ROE ch. 4; ADC ch. 1.1; GSE 390–402.

[11]GSE 390–5, 401.

[12]GSE 401; see ROE 96, 133.

Dawkins gives a *Comparative Argument* for the goodness of our planet[13]: (1) Our planet has both positive and negative features. Its positive features include its fertility and its beauty. They include all the intrinsically valuable entities generated by evolution, such as rational moral human animals. Its negative features include its many disvalues: deserts, wastelands, slums, crime, poverty, tyranny, and on and on and on. (2) To assess the value of our planet, it is necessary to compare it with others. Dawkins says that if we survey the other known planets, our earth turns out to be excellent. (3) Thus our earth, even with all its disvalues, is intrinsically superior to other planets. Life is valuable; intelligent life even more valuable; rational moral intelligent life is extremely valuable. Yet many planets fail to possess those values. (4) They fail to have the values of earth because evolution does not run on them. (5) Therefore, planets on which evolution runs are far more valuable than those on which it does not run.

The Comparative Argument motivates an argument that the suffering produced by evolution is necessary for the greater goods of complex life. This is the *First Greater Goods Argument*. It goes like this: (1) If evolution on earth doesn't work the way it does (with its horrific toll of suffering), then complex life will not exist on earth. (2) Universal Darwinism shows that this result generalizes from earth to all planets. Evolution is *necessary* for the production of complex life. (3) So, if evolution does not work the way it does, there will not be any sentient morally responsible organisms at all. (4) At most, the universe would only contain bacteria, or just rocks. (5) A universe without sentient life is far worse than one with sentient life. Indeed, sentient life is the greatest good a universe can contain. (6) Therefore, the suffering needed to produce sentient life is justified. It is the price paid for the greater good of sentient life. Moreover, this price is worth paying: sentient life is *extremely* more valuable than the alternatives.

Here is the *Second Greater Goods Argument* for defending evolution against the charge of evil: (1) All the waste and suffering associated with evolution is *necessary* to raise up any complex ecosystem. (2) The values manifest by any complex ecosystem are greater than those destroyed in

[13]UR 4.

raising it. Thus *Unweaving the Rainbow* stresses over and over again that just to be alive is blessing, despite the fact that life involves suffering. (3) But suffering which is necessary for the achievement of a greater good is not evil. (4) Therefore, evolution is not evil; on the contrary, evolution is good. It might be objected that the toll of suffering in both Greater Goods Arguments is not necessary: it could be lower. The reply is that evolution in fact moils to minimize suffering: less fit organisms suffer more; but since evolution optimizes fitness, it also minimizes suffering. It is not possible to produce complex sentient life with less suffering.

Plotinus once gave an *Argument from Impermanence to Goodness.*[14] It goes like this: (1) Since animals are not eternal abstract objects, it is necessary for them to disintegrate and die. (2) Disintegration and death are painful. (3) So it is necessary for animals to suffer; it is not possible to eliminate the suffering of disintegration and death. (4) However, either that suffering can be wasted or it can be used to create good. (5) A universe in which it is used to create good is better than one in which it is wasted. (6) Our earth shows that it is used to create greater good: since life eats life, the suffering of disintegration and death is used for the evolution of greater life. (7) Thus our earth shows how suffering can be converted into goodness. (8) More generally, out of the necessity of painful death, evolution brings the goodness of more life.

The Argument from Impermanence to Goodness leads to a more general argument that evolution moils to convert suffering into goodness. Here is the *Benevolent Emergence Argument* that evolution is good: (1) Evolution moils to produce organisms which cooperate. Out of selfishness, it brings altruism. Evolution moils to produce organisms which are rational and moral. It even moils to produce godlike organisms. More generally, evolution moils to bring goodness out of evil. (2) But anything which moils to bring goodness out of evil is good. (3) Therefore, evolution is good. These arguments entail that evolution is providential. They entail that the biological, physical, and cosmic cranes are providential. Once more, this does *not* entail that our universe is the best of all possible universes. Our universe can be improved; every universe can be improved; cosmic evolution entails that they will be improved.

---

[14]Plotinus (*Enneads*, 3.2.15).

2   Stoic Spirituality

2.1   From the New Atheism to Stoicism

There are at least four ways in which Dawkinsian ideas resemble Stoic ideas. The first way concerns the rationality of nature. Following Heraclitus, the Stoics affirmed the rational order of nature, which they referred to as the Logos. The Logos is a dynamic rationality, which actively works to structure the universe. Dawkins affirms this rationality.[15] For the Stoics, the Logos is not something we make up, it is not a human projection into nature. It has objective, mind-independent reality. Dawkins argues against the cultural relativism that reduces science to local mythology.[16] But those arguments presuppose the objectivity of the Logos. So spiritual naturalism can include a modernized Logos. The Logos is an impersonal and unconscious rationality, like the rationality of a system of mathematical axioms or an algorithm.

The second way that Dawkins resembles the Stoics concerns the spiritual status of the Logos.[17] The Stoic Logos doctrine shaped the religious rationalism of Spinoza, which shaped the religious naturalism of Einstein. Dawkins admires Einsteinian religion, which is our spiritual naturalism. It regards the rational order of nature as the proper object of positive spiritual emotions (awe, reverence, wonder, admiration, adoration, praise). Dawkins says that he is religious in the Einsteinian sense. He says "Science can be spiritual, even religious in a non-supernatural sense of the word." Like the Stoics, Dawkins regards the rational order of nature (the Logos) as the proper object of positive spiritual emotions. He has a naturalistic conception of sacredness and holiness. When he endorses Einsteinian religion, Dawkins resembles the Stoics. Of course, the Dawkinsian theory of the Logos does not entirely agree with the Stoic theory. The Stoics sometimes referred to the Logos as God. But Dawkins

---

[15]UR *xi*, 151; TL 73–4; SSSF.
[16]ROE 31–2; ADC chs. 1.2, 1.7; SITS 23–4.
[17]Einsteinian religion (TL 58–64; GD 33–40). Spiritual in a non-supernatural sense (ADC 27; see SITS 269). Science as a religion (ROE 33). Sacred and holy (SS). Logos is not God (GD 33–41). Misleading (ADC ch. 3.3; TL 59).

denies that the rational order of nature is God. God is the object of Abrahamic religious practices. It would not be appropriate to worship the Logos. And Dawkins worries that thinking of science as a religion can be misleading. Still, the Logos is spiritually significant.

The third way Dawkins resembles the Stoics concerns fate.[18] The Stoics were determinists. They thought that the cosmos unfolds according to the inexorable lawful power of the Logos. We must accept our destinies. Since we must accept our destinies, we ought to learn to love them. This affirmative attitude is *amor fati*, love of fate. Dawkins often affirms a similar attitude. After discussing the ways that an evolutionary view of life may seem pessimistic, he says "There is deep refreshment to be had from standing up full-face into the keen wind of understanding." He praises "tough-mindedness in the debunking of cosmic sentimentality." After discussing the theory that the universe will end in emptiness, he writes: "the eternal quietus of an infinitely flat nothingness has a grandeur that is, to say the least, worth facing off with courage." Dawkins values painful truths above comforting illusions. Such an evaluation entails facing reality with Stoic courage. Dawkins praised John Diamond for facing death with Stoic bravery. We should not complain about having to die. On the contrary, since we are among the lucky few who have been born, we should look at death as a *privilege*. These thoughts indicate an attitude similar to *amor fati*. However, Dawkins never really develops the theme of *amor fati*. He does not give us anything like Russell's "A Free Man's Worship," or a Stoic theory of virtue, or a Dionysian-Nietzschean theory of value. So spiritual naturalism added these theories of virtue and value.

The fourth way Dawkins resembles the Stoics concerns cosmology. The Stoics argued for the biocosmic analogy. They argued that universes resemble self-reproducing organisms. Dawkinsian principles also support the biocosmic analogy (Sects. 1–3 in Chapter 5). On the one hand, that analogy can give you some hope: much as genes pass from organism to organism, so our lives pass from universe to universe (Sect. 5 in

---

[18]Wind of understanding (ADC 13). Cosmic sentimentality (UR *ix*). Facing off with courage (AK 188). No comforting illusions (ADC 13; UR *ix*; GD 20, 394). John Diamond (ADC ch. 4.4). Death is a privilege (UR 1; GD 404–5).

Chapter 8). On the other hand, that analogy entails the utter destruction of every valuable thing. Both the Stoics and modern physics argue that our universe will end in desolation. Although we may build glorious transhuman civilizations, the second law of thermodynamics will wipe them out. All complexity and value will be destroyed. Grief increases along with entropy. Our universe is burning up. Dawkins recognizes this total destruction, and he also recognizes that it drives people to despair.[19] Entropy is grief.

Your *locality* is the universe which contains your body. It is your location in logical space. The Stoics developed techniques for adapting to the negativities of locality. But Stoicism is not religious. Like Buddhism or Confucianism, it is a philosophical way of life.[20] Since Stoicism does not require worshipping any God, many atheists have turned to Stoicism in order to cope with adversity. There are many Stoic self-help books.[21] Dawkins recognizes that atheists need coping techniques.[22] Since Dawkinsian ideas often resemble Stoic ideas, it is appropriate to add Stoic technologies of consolation to the Sanctuary for Spiritual Naturalists. By adding these techniques, I am building on Dawkins. But these Stoic techniques go deeper than mere comfort. They were intended as *technologies of salvation*. Does this sound too Christian? The Christians got many of their ideas from the Stoics. Salvation can be naturalized.

2.2   Stoic Technologies of Salvation

According to the Stoics, we suffer from many forms of immaturity. We are born in physiological immaturity—we are defenseless against bodily harms. We are born in cultural immaturity—we cannot speak to summon aid from others. And we are born in cognitive immaturity—we do not know what hurts us and what helps us. As we grow up, we become more and more able to defend ourselves in many ways. But the Stoics

---

[19]SG *xiii*; UR *ix*.
[20]GD 59; Hadot (1995).
[21]Irvine (2009), Robertson (2015), Pigliucci (2017), and so on.
[22]GD 21, 394–8.

recognized that life always entails facing adversity. We are fragile animals. We face illness, injury, personal failure, loss of possessions, loss of social status, loss of family members, and loss of our own lives. We become victims of disease, crime, institutional injustice, war, natural disasters, old age, and ultimately death.

One form of immaturity, namely, *spiritual immaturity*, stands out for the suffering it brings. If we are spiritually immature, then adversities lead to negative emotions. The negative emotions include anger, rage, craving, lust, greed, ambition, obsession, addiction, jealousy, envy, disappointment, bitterness, vengefulness, pride, vanity, slavishness, hatred, anxiety, fear, superstition, depression, despair, sloth. These lead to vicious behaviors and violent actions which are harmful or destructive to self and others. The negative emotions involve the loss of self-control. Your will becomes captured by something outside of your control. We become slaves to alien powers. These negative emotions also express themselves in evil social and political institutions.

Fortunately, many types of immaturity have natural remedies. We naturally grow to physiological maturity. We naturally acquire cultural maturity (we naturally learn the language and customs of our culture). We are naturally inclined to cognitive maturity (we are inclined by natural curiosity to learn skills and gain knowledge). This natural inclination needs to be artificially assisted through educational institutions. We are naturally obligated to gain spiritual maturity. This natural obligation needs to be artificially cultivated through spiritual education. We have the natural capacity to respond well to adversity. But this capacity needs to be cultivated by spiritual training. We need to learn good habits for responding to adversity.

The ancient Stoics proposed a variety of training exercises. These are *spiritual practices* or *technologies of the self*.[23] These are technologies of self-relation, techniques of reflexivity. They involve learning to control your emotional responses to adversities. They must be practiced, so that the proper responses become habitual. One technique is *Daily Meditation*. Each morning, you remind yourself that you will face adversities and you rehearse the Stoic response. Every evening, you record in your

---

[23]Foucault (1988).

journal how well or poorly you handled the troubles of the day. Another technique is the *Premeditation of Adversity*, in which you remind yourself that you will lose all that you have. You tell yourself that your coffee cup will break because it is fragile; that your family members will die because they are mortal; and that you will become ill and die because you too are mortal. A third technique is *Voluntary Adversity*, in which you deliberately cause yourself hardship. This might include fasting, or a week spent backpacking and sleeping on the hard ground. A fourth technique is the *Contemplation of Exemplars*. According to this technique, you reflect on how people in the past have responded well to adversity. For the ancient Stoics, the death of Socrates was a primary example of courage in the face of an unjust death. But the Stoics also used as models Roman Senators who defied the Emperor for the sake of justice, despite exile and death.

The goal of Stoic spiritual training is to become virtuous. The goal is to transform your dispositions, so that, even in the face of excruciating hardship, you can maintain your rational self-control and emotional equilibrium. The goal is to become an ideal person, a Stoic *sage*. The sage is serene in the face of adversity, undisturbed even in the face of death. The Stoic sage has the positive emotions (*eupatheia*). These include peace of mind, fearlessness, invulnerability, freedom, contentment, joy, serenity, cheerfulness, clarity of will, steadfastness, resoluteness, honesty, magnanimity, and so on. These positive emotions go hand in hand with the virtues. For the Stoics, these included the four cardinal virtues of wisdom, courage, justice, and temperance. There is a very real sense in which the Stoic sage has been *saved* or *redeemed* from bondage to locality. But this is an entirely naturalistic concept of salvation.

The goal of Stoic training is a kind of detachment from locality, an indifference to your participation in our universe. Stoicism produces freedom from locality. You will encounter negativities. But when you suffer from negativities, your emotional responses are not necessary; on the contrary, they are contingent on your decision. You are free to respond in many emotional ways. You are not mentally bound to our universe; on the contrary, you can always shift to other possible universes. Stoic techniques dissolve your bondage to locality. Stoicism shows you that you are not essentially identical with your local self. You are not essentially located in this possible universe. Stoicism shows you that you are only contingently identical with your local self. It teaches you that you can at least emotionally shift to an alternative universe. If our universe is surpassed by better future universes, then you are surpassed by more virtuous future counterparts. You can emotionally shift into those better universes by mentally simulating (by channeling) your more virtuous future counterparts. Stoicism teaches you how to shift into fictional future universes which differ only emotionally from our local universe. However, it does not shift you into alien universes with other physical laws.

2.3   Mindfulness Meditation

Mindfulness meditation is a spiritual practice which aims at a kind of mental well-being. It comes from Buddhism. Dawkins thinks of Buddhism less as a religion and more as a philosophical way of life. And Buddhism can be secularized.[24] So mindfulness meditation can be part of an irreligious spirituality. And while mindfulness meditation did not originate with Stoicism, it is similar in many ways to Stoic techniques of mental self-discipline. So I will include mindfulness with Stoic spirituality. Dawkins agrees that meditation may be a beneficial kind of mental discipline.[25] However, although he tried it, it didn't produce any benefits for him. But here we are building on Dawkins. Among the other Four Horsemen of the New Atheism, Sam Harris advocates meditation. He argues that it can be part of an irreligious spirituality. Harris focuses on meditation in his book *Waking Up: A Guide to Spirituality Without Religion.*[26]

Harris focuses on Buddhist breath-meditation.[27] The basic idea is easy to understand. You sit in a comfortable position. You close your eyes and start breathing. As you breathe, you concentrate on your breath. You allow thoughts and feelings and sensations to emerge, pass, and fade away. If you find yourself getting distracted, you gently return your focus

---

[24]Batchelor (1997).
[25]Dawkins (2004c).
[26]Harris (2014). Cited here as WU.
[27]WU 34–40; ch. 4.

to your breathing. This can be very hard to do at first. But over time, with practice, you can get better. By focusing on your breath, your mind begins to free itself from its constant inner turmoil. It becomes clear. And the experience of having a unified permanent self starts to disappear—the ego fades away.

Dawkins says that the unified permanent self is an illusion. Meditation helps you break free from that illusion by extinguishing your ego.[28] It destroys the illusion of the unified permanent self and awakens you from the dream in which you experienced yourself as an ego. You no longer identify yourself with your thoughts, emotions, or sensations. You are no longer in bondage to them. Through meditation it is possible to become pure selfless consciousness. You float free from your local self. You stop channeling any local self burdened with its local negativities.

According to Buddhism, meditation can help you face adversity, and help relieve suffering. Suffering arises from the bondage of consciousness to its ego.[29] The ego is bound up with a universe in which it has many troubles. By meditating we destroy the illusion of the ego; by destroying that illusion, we stop suffering. Meditation (like Stoic practices) reduces pain, anxiety, rumination, and depression. Since meditation detaches you from your local self, you gain a profound freedom. Harris says "If you are injured and in pain, the path to mental peace can be traversed in a single step: simply accept the pain as it arises, while doing whatever you need to do to help your body heal."

Both meditation and Stoic training aim at detachment. Your ego binds you to your local body, in local relations with local things. So your ego binds you to those things. But as meditation dissolves your ego, it also dissolves your bondage to your local body and its local relations to local things. You cease to be attached to those things. You no longer experience your self as being constituted by its relations with local things. You no longer cling to them nor do you crave them. You can be in love without identifying yourself with the love or clinging to your beloved. You cease to be emotionally enslaved by local things. You no longer fear losing

---

[28]Extinguish the ego (WU 31, 37, 82). No longer identify (WU 45, 101, 102, 140). Pure selfless consciousness (WU 37, 102, 123, 125). More familiar with it (WU 199).

[29]Stop suffering (WU 45, 48, 123, 171). Reduces pain (WU 35, 39, 121–2). Profound freedom (WU 49). Help your body heal (WU 149).

them. Emotional bondage expresses itself as anxiety, fear, jealousy, anger, grief, and depression. As you become detached, those negative emotions become weaker and do not last as long.[30] Through mindfulness, you learn to respond better to adversity. Negative stimuli hurt you less.

Meditation relieves you of the negativities that come from your locality.[31] It relieves you of your bondage to the negativities of your local body and the negativities of its local relations to local things. Meditation also produces positive effects. Through meditation, you can ultimately arrive at a kind of well-being that cannot be disturbed by any negative stimuli. You can achieve Stoic equanimity, and a smooth flow of life. Meditation can improve cognitive functions like learning and memory. It can help with emotional self-regulation. Moreover, it has health benefits. It reduces stress, improves immune function, and helps your cardio-vascular system. It has shown promise in treating eating disorders and addictions. And, as Harris often points out, you do not have to take these claims on faith. They have been verified by scientific studies. The goal of meditation is to enlightenment—a kind of pure freedom from all attachment. The enlightened meditator resembles the Stoic sage. They have been *saved* or *redeemed* from bondage to any location in logical space. Here again this salvation is naturalistic.

Despite its claims about producing selflessness, meditation might be criticized as extremely self-centered.[32] You are, after all, just focusing on a process going on inside your own body. But Harris reminds us that Buddhism values ethical living with others. He says that if we awaken from the illusion of the ego, then we will become more ethical. Meditation can make you more accurately aware of the emotions of others. It can increase empathy and compassion. It arouses and sustains pro-social dispositions. It motivates you to reduce the suffering of others. It makes you want to help others flourish. Meditation can be done socially: many people can collectively dissolve into selflessness and worldlessness. They

---

[30]Weaker emotions (WU 98–100). Adversity (WU 149). Noxious stimuli (WU 121).

[31]Imperturbable well-being (WU 39, 44, 48, 124). Equanimity (WU 38). Learning and memory (WU 35). Self-regulation (WU 35, 47). Health benefits (WU 122).

[32]Ethical living with others (WU 30). Pro-social dispositions (WU 122). Help others flourish (WU 206).

can collectively detach from their local selves in their local universe. They can float free together.

By weakening the illusion of the ego, mindfulness helps you become detached from your local self. Your brain stops the simulation of its ego. It therefore stops defining your location in logical space. You cease to be attached to your local body in our universe. Nevertheless, meditation does not kill you: you still exist. And meditation does not destroy your self: you are not unconscious. What, then, does it do? It alters the balance of powers in the ontic fact that grounds your awareness (Sect. 4.3 in Chapter 7). This ontic fact is $(\exists x) (\exists y)$ (your ego x is aware of its object y). Mindfulness minimizes the powers of the beings x and y. It shrinks your ego to its simplicity and its objects to their simplicities. You participate more directly in the self-consistency of the One, which you experience as blissful peace. And the One, which is being-itself, has the freedom to develop in all possible ways. Consequently, meditation resembles mystical experience: both meditation and mysticism minimize the powers of beings. Nevertheless, they are not the same. During mystical experience, the power of the existential quantifier $\exists$ becomes maximized, in terrifying ways. But meditation aims to produce the bliss of minimizing the x and y without the terror of maximizing the $\exists$. It does not always succeed. Any alteration of the balance of powers in the ontic fact of awareness involves risks. Meditation can induce full-blown mystical terror—even psychosis.[33]

3   Platonic Spirituality

3.1   Striving for Godlikeness

Plato says you should strive to become as godlike as possible.[34] You ought to climb up out of the cave towards the sun, to climb up the divided line towards the Good. By the time of Plotinus, the godlike becomes the unsurpassable. So, if you strive for godlike excellence, then

---

[33]Dyga and Stupak (2015) and Baer et al. (2019).
[34]Plato (*Theaetetus*, 176a5–b2).

you strive for unsurpassable personal excellence. However, this excellence is not a degree of excellence; on the contrary, it is an unsurpassable series of surpassable degrees of personal excellence. Such a series is an *ideal person*, so you should strive to be an ideal person. To strive to be an ideal person is to strive to pass through every surpassable degree of personal excellence. Therefore, if you should strive to become as godlike as possible, then you should always strive to surpass yourself. You should always work to transform your current *arete* into greater *arete*. Of course, this Platonic duty generalizes: every human animal ought to become as godlike as possible. Likewise every human community— every family, city, and nation—should strive to become a community of godlike animals relating to each other in godlike ways. The entire human species ought to become as godlike as possible. Every family, city, nation, and even every earthly species, should strive to become their ideals. All these ideals are transcendental objects. They are stars so high they are heightless.

The Platonic duty entails that, for every excellent quality, you should strive to have more of it. But what are these qualities? To be human is to be a rational social animal. To become as godlike as possible implies that you should aim at *godlike animality, godlike sociality*, and *godlike rationality*. Godlike animality includes all possible degrees of excellence of all possible animal functionalities. Focus on godlike health: to strive for godlike health is to always strive to be healthier. This striving minimizes sickness, injury, and bodily dysfunctionality of any kind. It maximizes youthfulness, longevity, and bodily functionality of every kind. Godlike sociality includes godlike justice and godlike virtue. Godlike rationality includes godlike intelligence, along with godlike degrees of all the cognitive virtues. It includes the godlike emotionality outlined by the Stoics. It maximizes intensity while minimizing passion.

To become godlike is the Platonic goal. But what is the Platonic method? The Platonic method is the *hê telestikê technê*, the craft of self-surpassing.[35] To present this method, Plotinus uses two artistic analogies. Plotinus first uses the analogy of *self-sculpting*: you stand to yourself as

---

[35]Craft of self-surpassing (Johnston 2008; Dillon 2007, 2016). Plotinus on self-sculpting (*Enneads*, 1.6.9). On self-tuning (*Enneads*, 1.4.16, 2.3.13, 4.4.8, 4.7.8D).

a sculptor to their statue; your task is to perfect your statue. Plotinus second uses the analogy of *self-tuning*: you stand to yourself as a musician to their lyre; your task is to finely-tune your lyre to play beautiful music. The Plotinian craft of self-tuning is iterative: (1) You play your lyre. (2) You listen for the ways that the performance of your lyre can be improved. (3) You tune your lyre to improve its performance. (4) You repeat back at step one. By repeatedly tuning your lyre (or sculpting your statue), you surpass your current degree of excellence towards some superior degree. Following the Platonic imperative to become as godlike as possible, you should pursue this iterative method as far as you can.

After Plotinus, later Platonists use the term *theurgy* to refer to the craft of self-surpassing.[36] Theurgy aims to make you godlike. It is the craft of self-divinization. Of course, ancient theurgy was mostly magical nonsense. But theurgy contains the seeds of science. The theurgists aimed to reveal and optimize the *numbers of the body*. They aimed to understand the body as a system of mathematical patterns, as a system of codes and programs. As theurgy becomes naturalized, it puts ever more science and technology into the Plotinian iterative method. It uses modern science and technology to carry out the Plotinian self-sculpting and self-tuning. As theurgy is naturalized, it turns into *techno-theurgy*. The Plotinian iterative method turns into the experimental method. When we practice techno-theurgy, we use the experimental method to make our bodies and our communities as godlike as possible. We use science and technology to aim at godlike animality, godlike sociality, and godlike rationality.[37] This theurgical concept of self-experimentation was more fully developed by Nietzsche. After him, it enters transhumanism. The experimental method becomes the *hacker methodology*. Thus self-experimentation becomes self-hacking. Techno-theurgy is self-hacking which aims to make humanity as godlike as possible. Ultimately, it aims at the Good. Since it aims at the Good, techno-theurgy always uses science and technology in ethical ways.

---

[36]Theurgy (Shaw 2014). Numbers of the body (Steinhart 2019). Nietzschean self-experimentation (Bamford 2016).
[37]Harris (2007).

The Dawkinsian conception of *good spirituality* fits well into the Platonic framework of techno-theurgy. To get to the Dawkinsian conception of good spirituality, we need to pass quickly through his conception of *bad spirituality*. Dawkins usually criticizes popular spirituality for its abuse of science.[38] He says it is often fraudulent, even criminal. When Dawkins condemns popular spirituality, he often refers to medical quackery: *bad spirituality* goes with *bad medicine*. Bad medicine fails experimental tests (or it fails to subject itself to testing). If bad spirituality goes with bad medical practice, then *good spirituality* goes with *good medicine*. Good medicine is the system of practices which has passed rigorous scientific testing. It uses the experimental method to try to find solutions to medical problems (to cure illness, to maintain health, and so on). Good medicine uses the experimental method proposed by Plotinus for tuning your lyre or sculpting your statue. Just as good medicine uses the experimental method, so good spirituality also uses the experimental method. Hence the Dawkinsian conception of good spirituality follows the Plotinian conception. Both agree that good spiritual practices help us increase our degrees of excellence. Both agree that good spiritual practices are those that survive the ordeal of experimental testing.

3.2   Striving for Godlike Animality

Techno-theurgy aims to raise us all to godlike animality. To strive for godlike animality is to always strive to have greater physiological excellence. Just as there are many levels in the Platonic divided line, so there are many levels of this *medical techno-theurgy*. Platonic spirituality includes a first level of medical techno-theurgy. The first level strives for normal human physiological excellence. Here the norm is defined in purely statistical terms (tailored for sex, age, etc.). The first level of medical techno-theurgy is often called *therapy*. Therapy aims to use science and technology to raise all humans to normal degrees of physiological excellence. It aims to cure disease, to repair injury, and to overcome disability. Dawkins approves of therapy.

---

[38]Criticizes popular spirituality (ADC chs. 1.2, 1.6; SITS 22–4). Abuses science (UR 37, 115, 146, 188–9). Criminal (1995c). Testing (ADC 180, ch. 1.6).

Platonic spirituality includes a second level of medical techno-theurgy. This second level strives to raise all humans to elite human physiological excellence. This eliteness is the best any natural human body can perform on some task. The second level of medical techno-theurgy is sometimes called *optimization*. Optimization aims to help you run as fast as the fastest human runner; to see as well as the best human eyes can see; to climb mountains as well as the best human mountaineer; to play any game or sport as well as the best human player; to master any skill as well as the best human performer; to live as long as the oldest human; to be as healthy as the healthiest human.

Optimization is already done without ethical controversy. To see this, consider aging. If human history defines human normality, then it is normal for aging humans to get cataracts, lose their teeth, lose their fertilities, get cancer, heart disease, Alzheimer's, and so on. But medicine strives to prevent or reverse all those negativities. Thus medicine is already deeply committed to the process of human optimization. Likewise it is normal for humans to be vulnerable to infections. Our normal and natural immunological powers leave us vulnerable to smallpox, syphilis, malaria, and so on. But vaccination optimizes human immunological function. Dawkins seems to approve of optimization. Building on Dawkins, we include optimization in the Sanctuary.

Platonic spirituality includes a third level of medical techno-theurgy. This third level strives for the highest humanly possible degrees of physiological excellence. This third level is sometimes called *enhancement*. Enhancement aims to raise all humans to the highest possible degrees of human physiological excellence. It aims to enable our bodies to perform every possible human function as well as the best possible humans. Enhancement surpasses optimization because the best local humans are almost certainly not the best possible humans. Enhancement produces ideal human bodies. Of course, the paths to ideality form a richly ramified tree. There are as many ways to be ideal as there are stars in the sky. Enhancement almost certainly requires genetic modification of humans into transhumans. A transhuman body can perform every human function at least as well as every elite human body. And it can perform at least one human function better than every elite human body. Are transhuman bodies possible? Dawkins thinks superhuman aliens actually exist.[39] Hence the biological library contains species far more excellent than us. It is therefore plausible that many paths of genetic alterations run from humans to many kinds of physiologically superior transhumans.

There are four ways we could genetically change into transhumans.[40] The first way goes through continued evolution by natural selection. Dawkins is skeptical of this way. The second way goes through future eugenics. Eugenics involves using coercive political power to selectively breed humans for enhanced traits. Selective breeding of humans is possible. However, Dawkins opposes eugenics. Spiritual naturalists follow Dawkins on this point: we permit only ethical ways to enhance humans; since eugenics is coercive, it is not ethical. The third way is through future voluntary selective breeding. This is happening already with preimplantation genetic selection of embryos during in vitro fertilization (IVF). It happens when sperm donors or egg donors are selected on the basis of intelligence tests or other achievements. Dawkins offers an argument by analogy: it is not wrong to train children for valuable traits; but breeding for a trait is morally analogous to training for that trait; therefore, it is not wrong to breed children for valuable traits. However, Dawkins says there may be crucial moral disanalogies between breeding and training. Dawkins offers another argument: it is not wrong to reject IVF embryos with negative traits; if it is not wrong to reject embryos with negative traits, then it is not wrong to select embryos with positive traits; therefore, it is not wrong to select embryos with positive traits. But Dawkins does not clearly commit to positive selection. Here spiritual naturalism goes beyond Dawkins: there are ethically permissible and even ethically obligatory ways to select embryos for positive traits.

The fourth way to enhancement runs through genetic engineering.[41] Dawkins seems to approve of this way. He suggests it would not be unethical to produce a human-chimp hybrid using genetic engineering.

---

[39]ROE 151–61; IA 96–7; GD 98–9.

[40]Skeptical of future natural selection (Dawkins 1993b). Selectively breeding humans (GSE 37–9; SITS 36–8). Against eugenics (SITS 37). Breeding is analogous to training (Dawkins 2006b: 300; SITS 38). Selection for positive traits (SITS 36–8).

[41]Genetic engineering (BW 73–4; 1998b; ADC 28–9; GSE 304). Human-chimp hybrid (ADC ch. 1.3). Precautionary principle (1998b; see 2006b).

Of course, that would be enhancing chimps, and the enhancement would involve amplifying features crucial to personhood. If that is not unethical, then it would not be unethical to enhance humans by amplifying features crucial to personhood. But when it comes to genetic engineering, Dawkins appeals to the precautionary principle. Here we agree with Dawkins: when it comes to genetic enhancement, we need to proceed with great care. Platonic spirituality welcomes all and only *ethical* ways of enhancing human genetics. Dawkins often urges us to rebel against genetic short-sightedness. Using our foresight for the ethical enhancement of human genes is probably the best way to rebel against the tyranny of our genes.

Platonic spirituality includes a fourth level of medical techno-theurgy, which strives to raise humanity above the maximum possible human physiological excellence. This fourth level can be called *divinization*. Divinization takes us from the transhuman to the superhuman. It strives for Olympian degrees of physiological excellence. The Olympian deities did not age, did not die, did not suffer from any illnesses or incurable injuries. Divinization is ethically controversial. However, an argument can be made that it is just as ethical as any other medical practice. The argument runs like this: (1) If medicine is ultimately successful, then you won't die of any injuries or illnesses, nor will you die of aging. And if it is ultimately successful, then you will not suffer from any negativities of aging. (2) So, if medicine is ultimately successful, you will have the physiological excellence, youth, and health, of an Olympian deity. Thus medicine ultimately aims at divinization. (3) But all the practices of medicine are ethical practices. And practices that aim at higher degrees of excellence are even more ethical. (4) Therefore, divinization is an ethical practice. Just as there were many Olympian deities, so there are many ways to become divine. Divinity is multiple. Perhaps divinization takes us beyond carbon-based life into silicon-based life. It transforms us into cyborgs or super-organic robots. Dawkins often discusses the possibility that humans will be surpassed by silicon super-intelligence.[42] Techno-theurgy now joins up with the futurism of Kurzweil. We ought to surpass

---

[42]BW 157–8; ROE 160–1.

ourselves until the whole universe wakes up. Thus Platonic spirituality contains an infinite series of techno-theurgical levels.

3.3   Striving for Godlike Sociality

Techno-theurgy aims to raise us all to godlike sociality.[43] To strive for godlike sociality is to always strive to have greater moral and political excellence. Just as there are many levels of medical techno-theurgy, so there are many parallel levels of *moral techno-theurgy*. According to Platonism, existence has the duty to maximize reflexivity. Since you exist, you also have this duty. You have the duty to maximize moral reflexivity. Moral reflexivity is the core virtue of *rational self-control*. Rational self-control is also known as executive function. Spiritual Platonism includes *moral therapy*. Moral therapy uses science and technology to raise humanity to the level of normal moral functionality. It aims to equip us with at least normal degrees of virtue.

Spiritual Platonism also includes *moral optimization*. Moral optimization aims to make us all as virtuous as the most virtuous current humans. But what are the virtues? The core virtue of rational self-control manifests itself in many specialized virtues. The Stoics listed four special virtues: wisdom, temperance, justice, and courage. These can be filled out by turning to Buddhist theories of virtue. The transhumanist James Hughes lists the ten main Buddhist virtues as (1) generosity; (2) proper conduct; (3) renunciation; (4) wisdom and insight; (5) energy, diligence, vigor, and effort; (6) patience, tolerance, forbearance, acceptance, and endurance; (7) truthfulness and honesty; (8) determination and resolution; (9) loving-kindness; and (10) equanimity and serenity.[44] Platonic spirituality seeks scientifically precise definitions of these virtues.

Moral therapy and moral optimization include the self-application of *spiritual technologies*. By applying these technologies to our own bodies for the sake of the Good, we help to maximize moral reflexivity. All spiritual technologies involve tools and techniques for increasing rational self-control and other virtues. Spiritual technologies include

---

[43]Steinhart (2019).
[44]Hughes (2013).

Buddhist and Stoic self-help techniques. Newer spiritual technologies include psycho-active molecules, electrical brain-stimulation, magnetic brain-stimulation, and so on. But all currently available spiritual technologies are slow, ineffective, unreliable, or unsafe. Spiritual research programs aim to develop better spiritual technologies. These spiritual research programs includes *virtue engineering* and *moral enhancement*.[45] As we apply these technologies more widely and intensely to ourselves, we gain greater self-control and greater virtue. We become morally and politically godlike.

Spiritual Platonism includes *moral enhancement*. It aims to raise us all to the highest humanly possible level of virtue. It aims to transform us into *morally ideal humans*. Here too ideality is multiple: there are many ways to be morally ideal. The paths to moral ideality diverge into an ever-expanding tree of possible future morally ideal transhumans. The ancient Greeks and Romans referred to morally ideal humans as *sages*. Sages were extensively discussed by both the Stoics and the Platonists. Sages have the highest possible degrees of human self-control. They do not suffer from the moral vices. They do not suffer from emotional distress. Since their minds are less emotionally disturbed, they are more sensitive to reasons—they are more rational. Since they are more rational, they are more sensitive to their moral responsibilities and ethical duties. They are more accurately aware of what they ought to do. Likewise they are more capable of doing what they ought to do. Of course, sages are not unemotional—the Stoics emphasized the value of the positive emotions. The positive emotions help motivate sages to do their duties. Hence the sages are more motivated to do what they ought to do. Since they are more aware of their duties, more capable of doing them, and more motivated to do them, they are more likely to do their duties.

Moral enhancement also includes the self-application of spiritual technologies. Since self-control has a well-defined genetic basis, moral enhancement probably requires genetic engineering. Dawkins suggests that humanity would be greatly improved if we had one hundred

---

[45] Froding (2013).

clones of Carl Sagan.[46] Beyond moral enhancement, spiritual Platonists posit an endless series of degrees of further moral excellence. Moral
techno-theurgy aims to raise us through these degrees and into greater
godlikeness. Futurists like Kurzweil argue for an endlessly rising series of
degrees of bodily power (including the powers of robotic bodies). As we
gain those greater degrees of bodily power, we will encounter ever greater
moral problems, and to solve them we will need ever greater degrees of
moral self-regulation. We will need ever greater degrees of virtue.

3.4   Striving for Godlike Rationality

Techno-theurgy strives for godlike rationality. To strive for godlike rationality is to always strive to have greater cognitive excellence. Just as
there are many levels of medical and moral techno-theurgy, so there
are many parallel levels of *cognitive techno-theurgy*. Spiritual Platonism
includes *cognitive therapy*, which strives to raise humanity to the level of
normal rationality. Dawkins frequently argues that human irrationality
causes enormous harms. He devotes much of his writing to revealing and
opposing these harms. Both *Unweaving the Rainbow* and *The God Delusion* are book-length diatribes against irrationality. Spiritual naturalism
includes cognitive therapy.

Cognitive therapy begins with helping children to participate in the
magisterium of reason (Sect. 2 in Chapter 1).[47] Dawkins wrote *The
Magic of Reality* to help children to participate in that rational magisterium. More generally, he argues that science education should help
children to cultivate reverence for their own rationality and for the rationality of nature. It should help them to cultivate cognitive virtues and
spiritual values. Children cultivate these virtues and values by *doing*
science, *doing* mathematics, and *doing* moral reasoning. Of course, children sometimes grow up to pursue science as a vocation. Since science

---

[46]SITS 80.

[47]Cultivate cognitive virtues and spiritual values by doing science (UR chs. 1 & 2; SITS 266–
73), by doing mathematics (BW 74), by rational moral philosophy (SITS 271). Spiritual way
of life (SITS 21–41; FH 21–5). Irrational dispositions (SITS 267). Science is spiritual (ADC
27; SITS 5) as is writing about it (SITS 5).

aims to reveal the sacred truth of nature, it is a spiritual way of life. And science demands ethical self-transformation. You need to learn the scientific method. But that method involves cognitive self-discipline. You need to root out your own irrational dispositions. You need honesty. Analogously, Dawkins argues that doing good science writing is also spiritual practice. Good science writing can help many people see the beauty, order, and rationality of nature. It can help free them from their superstitious delusions. It therefore helps lift people to higher levels of rationality.

The practice of science can be more deeply understood in terms of the Platonic value of reflexivity. Some things in nature reflect nature. They are parts of nature which represent or mirror the whole of nature. These mirrors include our theories of nature. As science makes progress, these mirrors grow more accurate and more comprehensive. They rise through ever higher degrees of reflective excellence. They reflect the *light of nature* back into nature. But that light is the holy light of the Good. By reflecting that holy light, these scientific mirrors shine with holy light. Better scientific theories shine more brightly. Thus more scientific theories of nature are holier. They are holier than religious theories. For example, Darwin's *Origin of Species* is holier than Genesis. It follows that meditation on the meaning of science is a spiritual practice. Dawkins illustrates this in his meditation on the last paragraph of Darwin's *Origin.*[48]

Spiritual Platonism also includes *cognitive optimization*, which aims to make us all as rational as the most rational actual humans. Cognitive optimization (like moral optimization) requires the self-application of many spiritual technologies. Spiritual Platonism includes *cognitive enhancement*, which aims to raise us all to the highest humanly possible level of rationality. It aims to transform us into *cognitively ideal humans*. Here too ideality is multiple: there are many ways to be cognitively ideal. The old concept of the sage includes this cognitive ideality: sages are enlightened; they are as rational as humanly possible. Like medical and moral enhancement, cognitive enhancement almost certainly requires genetic modification of our bodies. It takes us beyond optimal human

---

[48]GSE ch. 13.

rationality into transhuman rationality. Beyond cognitive enhancement, spiritual Platonists posit an endless series of degrees of further cognitive excellence. Cognitive techno-theurgy aims to raise us through these degrees. Futurists (like Tipler and Kurzweil) have defined omega points with infinite degrees of cognitive excellence. Beyond those infinities, there are greater infinities.

4   Meturgy

4.1   Meturgical Shifting

Believing in Dawkins means building on Dawkins. It means extending his theories. I have argued (in Sect. 3 in Chapter 7) that his best theory of religion is the modal theory, which views religions as technologies for helping people to channel aspirational counterparts. By channeling those selves, they mentally and socially shift to aspirational universes. Spiritual naturalists are interested in *irreligious* ways for people to shift to aspirational universes. We are interested in technologies for irreligiously shifting socially organized groups of people into shared aspirational universes—we seek technologies that can provide the benefits of religion without the negative side-effects.

For new technologies, it will be useful to coin a new term. I propose the term *meturgy*. It comes from the Greek *met-urgy*, meaning change-working. It is similar in form to *theurgy*, the practice of making humans more godlike.[49] Meturgy is the practice of designing, creating, and using technologies to irreligiously shift groups into shared aspirational universes. These are the later universes in the world tree. Meturgy helps humans to participate in the flow of axiotropy through the branches of this tree. You can be carried away. Meturgy is irreligious in at least the sense that it does not involve faith, dogma, or authoritarianism. It rejects spiritual servility in favor of spiritual sovereignty. You do not bow down to any lord or master. Meturgy rejects worship and prayer. Meturgy does not try to shift all people to the same aspirational universe. If dogma

---

[49]Shaw (2014).

works to fix your modal location to a single universe, then meturgical practices are rituals without dogma. Meturgy affirms many ways to be a better person in a better universe. You can channel any better self you want. There are many stars in the sky. Three recent cultural movements stand out for their meturgical potentials.

The first recent meturgical movement includes *fire circles*. Fire circles involve drumming, dancing, and chanting around a central fire.[50] People dance all night long, and rest during the day. A typical fire circle might go on for three days. Dancing and drumming are traditional ways to arouse the natural energy of your body. The Stoic or Platonic interpretations are straightforward. Since your body-energy ultimately comes from the One, rituals involving dancing and drumming are ways to arouse the power of the One in your body. The central fire symbolizes the holy fire that animates the world tree. The cyclical patterns of rhythmic drumming and dancing represent the cycles of nature, including the cycle that transforms every universe into its successors. You can affirm your participation in the turning wheels of nature by reproducing them with your own movements. By dancing to the drums, you turn the wheel. Fire circles are sacred rituals intended to produce self-transformation. As you dance, you symbolically turn yourself into your aspirational counterparts. By channeling the transformative fire-energy into your own body, you transcend yourself. Fire circles often use alchemical symbolism: as the alchemists sought to transform lead into gold, so you seek through ritual to transform yourself into some godlike self.

4.2   Ecstatic Dance

A *rave* is a dance festival in which electronic music plays a central role. Raves also typically involve computer-generated imagery. The music and imagery at raves are designed to facilitate altered states of consciousness. As they dance, ravers enter ecstatic *hyper-arousal trances*. During their trances, ravers often experience a profound energy flowing through their bodies; their ego-boundaries dissolve; they see that all things are

---

[50]Winslade (2009).

connected and unified; they feel that this same energy flows through all things.[51] These trances resemble mystical experiences of being-itself. But aren't raves just drug-fueled dance parties? Raves can be less hedonistic and more spiritual.

Many writers have stressed the spiritual aspects of raves.[52] The anthropologist Robin Sylvan interviewed hundreds of ravers, and they were eager to present raving as a new spiritual movement. Despite much coaching from Sylvan, they almost always refused to refer their ecstatic experiences to God. Raving has more in common with older pagan rituals. Many aspects of raving can be interpreted using Platonism. The Platonists argued that things are unified through their participation in the One. Hence the powerful feelings of cosmic unity induced by raving come from the One. Since the One is immanent in all things, ravers are ultimately energized by its power. When ravers experience the flow of universal energy, they are of course just experiencing the extreme activation of their nervous systems. But this extreme activation symbolizes the flow of axiotropy through the world tree. It symbolizes the flow of fire-energy.

Ravers often report *pronoia*, the feeling that reality is out to help you.[53] This pronoia comes from the providential beneficence of the Good. Axiotropy flows from the One to the Good. Its flow carries you towards the Good. It transforms you from your actual self into your better future counterparts. Thus Platonism makes sense of the rave experiences of unity, universal energy, and pronoia. Rave dancing produces therapeutic benefits. The ravers interviewed by Sylvan said that raving helped them overcome anxiety, depression, and destructive behaviors. Raving made them more compassionate. It gave them hope, confidence, and courage. Arousing the universal energy oriented them towards positive social values, expressed in the rave ethic of PLUR (Peace Love Unity Respect). Thus raving can shift us towards godlike sociality.

---

[51] Sylvan (2005: ch. 3).

[52] Takahashi and Olaveson (2003), Gauthier (2004), St John (2004), Sylvan (2005), and Redfield (2017).

[53] Pronoia (St John 2004: 3). Therapeutic benefits (Hutson 2000).

Raves can be well-structured rituals.[54] Many raves have included ritual elements, such as altars, opening and closing ceremonies, ritual purification of the musical equipment and dance space, and so on. Many ravers told Sylvan that they channeled divine energies and deities when they dance. The psychologist Audrey Redfield also interviewed many ravers. One raver reported to Redfield that, when she danced, "I would feel like I would turn into a certain deity, … like some kind of ancient goddess." These accounts suggest that raving is a modernized version of the old Platonic practice known as *theurgy*. Theurgic rituals aimed to enable humans to channel divine currents of energy through their bodies and thus to become avatars of deities. Of course, while you cannot literally channel any deity, you can shift into an aspirational universe in which your counterpart is godlike or goddesslike. Through ecstatic dancing, you detach your identity from your local body, and this detachment helps you overcome the negativities of locality. Raves are transformational rituals. But they are rituals without dogma. Every raver is free to channel his or her own future divine selves. Raves do not try to shift all their participants to the same location in logical space. On this point, they differ from theistic religions. They are meturgical rather than religious. Sadly, classical rave culture is long gone. After reaching its peak in the 1990s and 2000s, it has gone into decline. For spiritual naturalists, classical rave culture is an existence proof—it proves that meturgies are possible. We seek new types of spiritual raves.

4.3   Burning the Man

The third recent meturgical movement includes *transformational festivals*, which have appeared in hundreds since the turn of the millennium. They express the meturgical shifting of the self and our earth. They include visionary art. They aim to transport their participants into aspirational universes. Of course, they suffer from many problems. Since they involve costly commitments of time and money, they are mainly attended only

---

[54]Well-structured rituals (Redfield and Thouin-Savard 2017). Ritual elements (St John 2004; Sylvan 2005). Channeling (Sylvan 2005: 84–95). Ancient goddess (Redfield 2017: 71). Theurgy (Shaw 2014). Decline (Anderson 2009).

by those who are economically privileged. And they can veer away from rationality into New Age nonsense. But spiritual naturalists seek ways of refining transformational festivals into spiritual technologies of modal transport.

The oldest and most well-known transformational festival is *Burning Man*.[55] Burning Man takes place in the Black Rock Desert in Nevada. The Black Rock Desert is a flat and desolate place—its entropy is maximum. It sees extreme temperatures, strong winds, and fierce dust storms. There are few places on earth where the stars shine as brightly in the sky. Burning Man involves about 70,000 participants. Arriving in the desert at the end of August, they build a temporary town, called Black Rock City. Pumping entropy out of the desert, they build artistic glories of complexity. Burning Man has well-defined spatio-temporal boundaries. To cross them is to shift from the *default world* into a sacred space-time. This crossing detaches you from your default self. You can socially channel alternative selves. You are free to wander through logical space. After a week in the desert wilderness, the burners clean up. Their goal in cleaning up is to *leave no trace*. The City vanishes into maximum entropy. Of course, like any human institution, Burning Man suffers from all the usual afflictions. It is open to corruption by money and fame. It is almost certainly unsustainable. For spiritual naturalists, Burning Man is an existence proof: meturgical paths to social spirituality are possible. Burning Man has replicated itself around the earth—there are hundreds of *regional burns*.

Burning Man celebrates artistic creativity. During the year, burners make art which they bring to the playa. Burning Man clearly express the Dawkinsian joy of awakening to see this world. For Dawkins, beauty is a central spiritual concept. Black Rock City is an oasis of art and beauty in a vast hostile landscape. It resembles Sagan's Pale Blue Dot sitting in the midst of an endless inhospitable space. The beauty gathered in the desert is precious, fragile, and rare; it is like life itself, and human life especially—it is *sacred* beauty. Burning Man illustrates the Dawkinsian metaphor of islands of viability in great seas of sterility. The pilgrimage

---

[55]Doherty (2004), Gilmore (2010), and Harvey (2017).

to the Black Rock resembles the evolutionary production of rare oases of sacred value in a vast desert of valuelessness.

Burning Man is a temporary utopia. It aims to illustrate an ideal way of forming a community. It is governed by ten general ethical principles. Burning Man is founded on a *gift economy*. The Man has often been lit by a fire which is kindled from the sun. This solar power obviously drives the evolution of life on earth. And the sun *freely gives* this power away. It generates beauty by squandering itself. The fire kindled from the sun symbolizes the eruption of energy in the big bang. It symbolizes the Stoic *pyr technikon*, the rational fire-energy which animates all things. And, if Platonism is correct, then the sun refers to the Good, and the fire kindled from it refers to the holiness which brings all concrete things into being. Dawkins says we ought to be grateful. Burning Man is a festival of gratitude, in which making art is giving thanks. To create art, especially ephemeral art, is to engage in useless squandering. It *sacrifices* time, energy, money, and other resources. This sacrifice is a sacred act of thanks-giving.

The Man stands in the center of Black Rock City. The Man (officially genderless) is a large wooden structure which outlines an indefinite human. On the penultimate night of the festival, the Man burns. His arms raised, he is lit on fire. Burning the Man is accompanied by shouting, dancing, and other acts of celebration. There are many ways to interpret the Man and his conflagration. For the Stoics, the entire universe replicates itself in an infinite cosmic cycle. Each cycle begins and ends with the dissolution of all things into the holy fire. Out of this fire, each new universe, like a Phoenix, is reborn. On this Stoic interpretation, burning the Man ritually enacts the cosmic cycle. The burning Man is Zeus in his absolute desolation. It is Zeus bereft of all the beloved complexity of his universe, serenely grieving for his lost beings, passing undisturbed through his own death and into the next cosmic cycle. And if the animatic replicators exist, the Man is every animat, too. Assuming that you, too, will be replicated in some future universe, the Man is you. When he burns, he dies; but he will be born again next year. When the Man's arms are raised, they are raised in victory. He will be victorious over death; he will rise in the next annual cycle. The Man will literally be resurrected. More intimately, the Man is your old self. Watching the

Man burn, you watch your old self symbolically consumed by flames. From the ashes of your old self, your new self will rise again. You have the freedom to reshape your life.

On one interpretation, the construction of Black Rock City symbolizes the evolutionary process in which the holy fire concentrates itself into sacred beauty. But Dawkins frequently reminds us that all value is produced through conflict. And if the construction of the Man symbolizes the evolutionary process, then the Man symbolizes all those who have passed victoriously through natural selection. If nature is red in tooth and claw, then the Man is stained with blood. The Man has accumulated all the negativities of actuality. He has accumulated all the negativities of life from the first replicator all the way up to humanity. These negativities are recorded in *the Temple*. The Temple is an elaborate wooden structure, whose form is taken from sacred architecture world-wide.[56] The Temple serves a special ceremonial purpose: burners decorate it with inscriptions, texts, photos, or other mementos. These express grief, loss, or struggle with adversity. They express the negativities of actuality. They express all that is lost in the strife-torn *agon*. On the last night of the festival, the Temple burns.

References

Anderson, T. (2009). Understanding the alteration and decline of a music scene: Observations from rave culture. *Sociological Forum, 24*(2), 307–336.

Baer, R., Crane, C., Miller, E., & Kuyken, W. (2019). Doing no harm in mindfulness-based programs. *Clinical Psychology Review, 71*, 101–114.

Bamford, R. (2016). Nietzsche on experience, naturalism, and experimentalism. *Journal of Nietzsche Studies, 47*(1), 9–29.

Batchelor, S. (1997). *Buddhism Without Beliefs: A Contemporary Guide to Awakening*. New York: Penguin.

Bishop, J. (2010). Secular spirituality and the logic of giving thanks. *Sophia, 49*, 523–534.

---

[56]Pike (2005).

Colledge, R. (2013). Secular spirituality and the hermeneutics of ontological gratitude. *Sophia, 52,* 27–43.

Dillon, J. (2007). Iamblichus' defense of theurgy: Some reflections. *The International Journal of the Platonic Tradition, 1,* 30–41.

Dillon, J. (2016). The divinizing of matter: Some reflections on Iamblichus' theurgic approach to matter. In J. Halfwasse, T. Dangel, & C. O'Brien (Eds.), *Soul and Matter in Neoplatonism* (pp. 177–188). Heidelberg: University of Heidelberg Press.

Doherty, B. (2004). *This Is Burning Man.* New York: Little Brown.

Dyga, K., & Stupak, R. (2015). Meditation and psychosis: Trigger or cure? *Archives of Psychiatry and Psychotherapy, 3,* 48–58.

Foucault, M. (1988). Technologies of the self. In L. Martin, H. Gutman, & P. Hutton (Eds.), *Technologies of the Self: A Seminar with Michel Foucault* (pp. 16–49). Amherst: University of Massachusetts Press.

Froding, B. (2013). *Virtue Ethics and Human Enhancement.* New York: Springer.

Gauthier, F. (2004). Rave and religion? A contemporary youth phenomenon as seen through the lens of religious studies. *Studies in Religion, 33*(3–4), 397–413.

Gilmore, L. (2010). *Theatre in a Crowded Fire: Ritual and Spirituality at Burning Man.* Berkeley: University of California Press.

Hadot, P. (1995). *Philosophy as a Way of Life: Spiritual Exercises from Socrates to Foucault* (M. Chase, Trans. & A. Davidson, Ed.). Malden, MA: Wiley-Blackwell.

Harris, J. (2007). *Enhancing Evolution: The Ethical Case for Making Better People.* Princeton, NJ: Princeton University Press.

Harris, S. (2014). *Waking Up.* New York: Simon & Schuster.

Harvey, S. (2017). *Playa Fire: Spirit and Soul at Burning Man.* San Francisco: HarperElixir.

Hughes, J. (2013). Using neurotechnologies to develop virtues: A Buddhist approach to cognitive enhancement. *Accountability in Research, 20,* 27–41.

Hutson, S. (2000). The rave: Spiritual healing in modern western subculture. *Anthropological Quarterly, 73*(1), 35–49.

Irvine, W. (2009). *A Guide to the Good Life: The Ancient Art of Stoic Joy.* New York: Oxford University Press.

Johnston, S. (2008). Animating statues: A case study in ritual. *Arethusa, 41*(3), 445–477.

Lacewing, M. (2016). Can non-theists appropriately feel existential gratitude? *Religious Studies, 52,* 145–165.

O'Hagen, E. (2009). Animals, agency, and obligation in Kantian ethics. *Social Theory and Practice, 35*(4), 531–554.

Pigliucci, M. (2017). *How to Be a Stoic: Using Ancient Philosophy to Live a Modern Life*. New York: Basic Books.

Pike, S. (2005). No novenas for the dead: Ritual action and communal memory at the temple of tears. In L. Gilmore & M. Van Proyen (Eds.), *AfterBurn: Reflections on Burning Man* (pp. 195–214). Albuquerque: University of New Mexico Press.

Redfield, A. (2017). An analysis of the experiences and integration of transpersonal phenomena induced by electronic dance music events. *International Journal of Transpersonal Studies, 36*(1), 67–80.

Redfield, A., & Thouin-Savard, M. (2017). Electronic dance music events as modern-day ritual. *International Journal of Transpersonal Studies, 36*(1), 52–66.

Robertson, D. (2015). *Stoicism and the Art of Happiness*. New York: McGraw-Hill.

Shaw, G. (2014). *Theurgy and the Soul: The Neoplatonism of Iamblichus* (2nd ed.). Kettering, OH: Angelico Press.

St John, G. (Ed.). (2004). *Rave Culture and Religion*. New York: Routledge.

Steinhart, E. (2019). Spiritual naturalism. In R. Nicholls & H. Salazar (Eds.), *The Philosophy of Spirituality* (pp. 312–338). New York: Brill.

Sylvan, R. (2005). *Trance Formation*. New York: Routledge.

Takahashi, M., & Olaveson, T. (2003). Music, dance, and raving bodies: Raving as spirituality in the Canadian rave scene. *Journal of Ritual Studies, 17*(2), 72–96.

Winslade, J. (2009). Alchemical rhythms: Fire circle culture and the pagan festival. In M. Pizza & J. Lewis (Eds.), *Handbook of Contemporary Paganism* (pp. 241–282). Boston: Brill.

Additional Resources

Additional resources which help to further explain the concepts and arguments in this book can be found at my website:

www.ericsteinhart.com

These resources include both texts and videos.

Glossary

The Glossary contains words used here in specialized ways, that is, as terms of art. It does not include technical terms in philosophy, mathematics, or science. Terms in bold appear in the Glossary. Section numbers refer to the place where the term is first introduced or first defined in detail.

abyss
: The abyss (Sect. 1.3 in Chapter 6) symbolizes **nothing**, non-being, or negativity. The abyss is the **Zero**. The abyss is pure self-inconsistency, which by negating itself forbids itself. As self-inconsistency, it is negative self-relation, which is **anti-reflexivity**. By negating itself, the abyss generates existence or **being-itself**. The abyss is symbolized by **water**. Black holes in our universe are physical symbols of the abyss.

accumulation of design
: The principle of the accumulation of design (Sect. 4.2 in Chapter 2) comes from Daniel

Dennett, and states that **complex** things copy most of their structure from simpler antecedents. **Cranes** accumulate design.

actuality   Actuality (Sect. 1.4 in Chapter 8) is a property of possible structures. A possible structure is actual when it gains the **reflexivity** of presence. This emerges as the **spotlight** of temporal presence, which appears in self-conscious awareness. The totality of actual structures is the **day**, which is the **world tree**. The **animats** do the work of bringing possible universes into actuality by knitting their parts together with relations of presence. But the ultimate explanation for actuality is the **Good**, which **skews** the probabilities of arrows so that animats actualize the **books** in the **treasury**.

actual world   The actual world (Sect. 5 in Chapter 6) is the class of all actual **universes**, which is identical with the **treasury**. It is the **day**, which is illuminated by the **light** of the **Good**. It has the form of an infinitely ramified tree, which is the **world tree**. The root of the world tree is the cosmic **alpha**, the simple original universe justified by the **cosmological argument**. Every universe in the world tree is surpassed by better universes.

agent   An agent (Sect. 2.4 in Chapter 3) is any self-powered thing that uses a **maxim** to change some value. Maxims are policies; that is, they are **algorithms** or programs. Thus agents are **computers**. The simplest and most primitive agents in our universe are stars. Stars run maxims encoded in their nuclear reaction probabilities. Organisms are agents running maxims encoded in their genes.

Agency evolves into rational moral agency. Persons (human or otherwise) are extremely complex and derivative agents.

air

Air (Sect. 2.4 in Chapter 6) symbolizes the system of abstract objects. These are purely mathematical objects like sets or numbers. They are stratified into the **sky**. Air is one of the five elemental powers, along with **water**, **earth**, **light**, and **fire**. These elements can be used symbolically in **meturgical** rituals or practices.

algorithm

An algorithm (Sect. 2.3 in Chapter 2) is a program for a **computer**. Algorithms are assignments of probabilities to **arrows**. A divergent algorithm wanders while a convergent algorithm runs **teleonomically** to some **finality**.

alpha

An alpha is some original thing. All the **libraries** contain alphas, which are the original **books** in those libraries. Since simple things are **actual** before complex things, the alphas are the first things to be actualized. The big bang is the alpha of our universe. The first atoms are the atomic alphas. The first universe in the world tree is the alpha universe. The empty set (the first set) is the alpha of sets.

animat

An animat (Sect. 2.1 in Chapter 5) is a cosmic replicator. Animats produce offspring which climb higher on the cosmic **Mount Improbable**. They are the replicators which drive cosmic evolution. Animats can be metaphorically thought of as cosmic organisms or cosmic computers. Every actual universe is actualized by its animat.

anti-reflexivity

Anti-reflexivity (Sects. 5.2 in Chapter 3, 1.5 in Chapter 6, 3.2 in Chapter 8) is negative self-relation. It is the privation of self-consistency, that is, the absence of **reflexivity**. It is self-negation, self-inconsistency, self-contradiction, self-conflict. It appears in conflicts between wholes and their parts or between the parts in some whole. The **categorical imperative** forbids both logical and practical self-contradiction. Hence anti-reflexivity is forbidden. It is the darkness which is the privation of the **light** of the **Good**.

arrow

An arrow (Sect. 2.2 in Chapter 2) transforms some old thing or things into some new thing or things. Arrows can be atomic, molecular, biological, cosmic, and collective. Here is an atomic arrow: carbon + carbon → neon + helium. Here is a biological arrow: male + female → offspring. Here is a collective arrow: set + set → { set, set }. Arrows are the basic components of the **hardware** of **computers**. Arrows have variable probabilities. The **software** of some computer is its assignment of probabilities to its arrows.

axiology

Axiology is the general study of value. Axiology includes logical values (true, false), ethical values (good, evil), aesthetic values (beautiful, ugly). An axiological progression is one in which some value is always increasing.

axiotropy

Axiotropy (Sect. 4.3 in Chapter 3) is a natural power that **skews** the probabilities assigned to the arrows in computers. Axiotropy rewrites **software** by redistributing arrow probabilities. It skews them towards convergent algorithms which increase complexity; thus axiotropy shapes algorithms into **entelechies**. Axiotropy in our universe is closely linked with the **maximum entropy production principle** or something like it; it is a **thermodynamic force** which drives self-organization.

axis mundi

The *axis mundi* (Sect. 4.3 in Chapter 4) is the vertical axis of existence. It begins in the **earth** (that is, **being-itself**) and rises up through the **sky** of abstract object to the **sun** (that is, the **Good**). The **axis mundi** is a vertical number line that runs into the infinite. It provides numbers for the floors of **books** in the various **libraries**.

being-itself

Being-itself (Sect. 1.4 in Chapter 6) emerges from the self-negation of non-being or **nothing**. It is the purely logical category of existence. Since non-being is pure self-inconsistency, being-itself is pure self-consistency. The ontological demand for pure self-consistency appears in the **categorical imperative**, which rules that maximally self-consistent maxims are obligatory. Being-itself is the ground of being, which produces beings. It is the purely logical basis for **existential quantification**. Since non-being is the **Zero**, being-itself is the **One**. It is symbolized by **earth**.

biocosmic analogy

The biocosmic analogy (Sect. 3.1 in Chapter 4) states that universes are analogous to organisms. Just as organisms self-reproduce, so universes self-reproduce. The **animats** are the replicators in cosmic reproduction.

books

A book is a possible structure. So an atomic book is a possible atom, a biological book is a possible organism, a cosmic book is a possible universe. Collections of books form **libraries** stratified into complexity ranks on the *axis mundi*. Some old books get transformed by **arrows** into new books.

categorical imperative

The categorical imperative (Sects. 3.1–3.3 in Chapter 8) is a procedure in **deontic logic** for assigning deontic properties to **maxims** or policies. Given some maxim as input, it decides whether it is forbidden or obligatory.

celestial computers

A celestial computer (Sect. 2.3 in Chapter 2) consists of a star pumping entropy out of its solar system into a black hole, which is an **abyss**. Many celestial computers also include planets. Celestial computers run algorithms composed of systems of arrows. The probabilities of these arrows are skewed far from randomness. This non-random **skew** drives flows of matter in solar systems to self-organize in accordance with **maximum entropy production principle**. These flows are **axiotropic**.

channeling

Channeling (Sect. 3.1 in Chapter 7) is mentally simulating some alternative self in its alternative universe. You channel your **counterparts**. Channeling occurs when you become absorbed in your simulation so that you lose your location in logical space.

complexity

Types of wholes have complexities (Sect. 1.2 in Chapter 2). The complexity of some type of whole is proportional to the probability that the type will be destroyed by rearranging its parts. The type *human* is highly complex because, if you scramble the parts of a human, it is highly probable you will destroy its humanness. Hence complex types are improbable. Complex things are produced by **algorithms** like **entelechies** which are highly **skewed** away from random-ness. Since complexity is **intrinsically valu-able**, it is aroused into **actuality** by the **Good**, which is the ultimate source of all skew. Complexity follows the **accumulation of design**. Complexity has a precise math-ematical definition (Sect. 1.2 in Chapter 2) which relates it to **entropy** (Sect. 1.1 in Chapter 3).

computer   A computer (Sect. 2.3 in Chapter 2) contains some **hardware** and **software**. Its hardware is some system of arrows. Its software is its algo-rithm, which is an assignment of probabilities to these arrows. More matter flows through higher probability arrows.

cosmological arguments   A cosmological argument (Sect. 1 in Chapter 7) runs backwards along some ordered series to the original **alpha** of the series. Cosmological arguments do not run backwards to **God**. All alphas are entirely natural objects.

counterparts   Things in one universe in the **modal library** have counterparts in other universes in that library. A counterpart of any thing is an alter-native possible version of that thing. Your counterpart in some other universe is the thing in that universe which is most similar to yourself or which has all your essential features there. You can **channel** counterparts by mentally simulating them.

crane   A crane (Sect. 2.3 in Chapter 2) is any **computer** that runs an **entelechy**. A crane therefore runs an **algorithm** which is highly **skewed** towards increasing **complexity**. Most cranes are mindless **agents** that **moil teleo-nomically** towards their **ecstasies**.

dark books	The dark books (Sect. 5.3 in Chapter 4) are those books in the **modal library** which are not actualized by the **Good**. They lie outside of the **treasury** in the **night**.
day	The day (Sect. 1.3 in Chapter 5) includes all and only those **books** in the **modal library** that are actualized by the **Good**. The day is the **treasury**. **Dark books** are not in the day; they are in the **night**.
deontic logic	Deontic logic governs deontic properties like the forbidden, the permissible, the obligatory, and so on. Deontic properties apply to the **maxims** of **agents**. Models of deontic logics often involve **axiological** progressions of possible universes, so that deontic logic is closely related to **modal logic**. The **categorical imperative** is one procedure for assigning deontic properties to maxims.
design arguments	The design arguments reason from **complexity** to the existence of some **crane** which produced it. The organic design argument (Sect. 3 in Chapter 3) reasons from biological complexity to biological evolution. The cosmic design argument (Sect. 4 in Chapter 5) reasons from cosmic complexity to cosmic evolution. Design arguments do not conclude with **God**. All cranes are entirely natural objects.
earth	Earth (Sect. 1.4 in Chapter 6) symbolizes **being-itself**. Earth is one of the five elemental powers, along with **water**, **air**, **light**, and **fire**. These elements can be used symbolically in **meturgical** rituals or practices.
ecstasy	An ecstasy (Sect. 2.3 in Chapter 2) is the **finality** of some **crane**. Thus it is the limit of some computer that runs an **entelechy**. An

ecstasy is often a limit or ideal which stands outside of or transcends the progression of which it is a limit. The **Good** is the ultimate ecstasy of nature. A goal is an ecstasy of an intelligent agent. However, most ecstasies are not goals; they are **teleonomic** rather than **teleological**.

entelechy

An entelechy (Sect. 2.3 in Chapter 2) is a convergent algorithm which **teleonomically** increases **complexity**.

entropy

Entropy (Sect. 1.1 in Chapter 3) is not disorder. Entropy measures the flatness of a landscape. The landscape can be an energy distribution. A flat energy distribution has no more potential or free energy to spend doing work. Thus entropy is energy dispersion or exhaustion. The landscape can be a probability distribution. A flat probability distribution is random, and makes no distinctions among its states.

existential quantifier

The existential quantifier (Sect. 1.4 in Chapter 6) is the backwards "E", which refers to **being-itself**. To say that Socrates exists means that there exists some x which is identical with Socrates. This is written logically as (there exists x) (x is identical with Socrates). This is symbolized as $(\exists x)$ (x is identical with Socrates). Socrates is a being among beings, but the $\exists$ refers to being-itself, which makes beings be. The existential quantifier plays a crucial role in **mystical experience**.

finality

A finality (Sect. 2.3 in Chapter 2) is the limit of a finite or infinite series of computational steps. Any convergent **algorithm** runs to some finality.

fire

Fire (Sect. 5.1 in Chapter 6) symbolizes the actualizing power of the **Good**. Fire flows through the **arrows** in the **world tree**. It emerges when the **One**, which is the **earth**, is struck by the light of the **Good**. Fire symbolizes **axiotropy**. Fire is one of the five elemental powers, along with **water**, **earth**, **air**, and **light**. These elements can be used symbolically in **meturgical** rituals or practices.

God

The proper name "God" (Sect. 2.1 in Chapter 4) is used here only in its theistic sense. This sense is defined by the scriptures and social practices of the Abrahamic religions, namely, Judaism, Christianity, and Islam. Thus God is the person worshipped in those religions. The name "God" is not used here in any non-theistic sense. The nontheistic uses of the term "God" are examples of religious **hijacking**.

Good, the

The Good (Sects. 3–4 in Chapter 6) is the best proposition. The truth of the Good is the conclusion of the **ontological argument**. The Good actualizes all the universes in the **treasury**. The logical power of the Good is **light** and so the Good is symbolized by the **sun**. These universes in the treasury make the **day**. The Good is the ultimate source of the **skew** which permits **cranes** to run **entelechies**.

hardware

Hardware (Sect. 2.3 in Chapter 2) is a component of a **computer**. The hardware of some computer is its system of **arrows**. Thus the hardware of any atomic computer is the total system of nuclear reaction arrows that transform old atoms onto new atoms. Some arrows are more probable than others. The assignment of probabilities to some system of arrows is the **software** or **algorithm** running on that hardware.

hijacking   Hijacking (Sect. 6 in Chapter 1) refers to the theistic practice of binding concepts to **God**. Theists have hijacked metaphysical, spiritual, religious, and other concepts. For example, **mystical experience** is hijacked by saying it is awareness of God; **being-itself** and the **Good** are hijacked by identifying them with God. Spiritual naturalists seek to **reclaim** hijacked concepts by liberating them from their theistic bondage.

ideal   An ideal (Sect. 2.5 in Chapter 6) is the limit of an unsurpassable progression of increasingly valuable but surpassable things. An ideal universe is the limit of an infinite progression of increasingly valuable universes. Many **deontic logics** posit ideal universes. Ideals are symbolized by **stars**. The **Good** is an ideal proposition.

intrinsic value   Intrinsic value (Sect. 2.4 in Chapter 3) is the value that a thing has in itself, independent of any observer, user, or evaluator. It is the objective value of the being of the thing. The intrinsic value of a thing is correlated with its **complexity**.

library   A library (Sect. 2.1 in Chapter 2) is a collection of **books** organized into floors. The books are possible objects; that is, they are the abstract forms or structures of objects like atoms, molecules, organisms, universes. The floors rise along the *axis mundi*, which is the vertical number line. Books on higher floors are more **complex** and more **intrinsically valuable**. Libraries rise into the **sky** of abstract possibilities.

light

Light (Sect. 5.1 in Chapter 6) is the logical power of the **Good** to actualize the universes in the **treasury**. The light of the Good corresponds to the **sun**. Light is one of the five elemental powers, along with **water**, **earth**, **air**, and **fire**. These elements can be used symbolically in **meturgical** rituals or practices.

liturgy

A liturgy (Sect. 1.2 in Chapter 2) is a course of scientific topics which you work through for the sake of spiritual development. Liturgies reveal meaning and value in nature. A liturgy provides a spiritual interpretation of some science. There are complexity, atomic, molecular, biological, thermodynamic, cosmological, and ontological liturgies.

maxim

A maxim (Sect. 2.4 in Chapter 3) is a convergent **algorithm** that directs an **agent** to optimize some value. The maxims of agents are encoded in their **software**. The maxims of organisms are encoded in their genomes or brains. Ethical maxims vary from simple imperatives like "Keep promises" to complex policies. The **categorical imperative** assigns **deontic** properties like obligatory and forbidden to maxims.

metaphysical question

The metaphysical question (Sect. 1.1 in Chapter 6) asks: Why is there something rather than nothing? It cannot be answered by any existing thing like **God** or like some fundamental physical entity.

meturgy

Meturgy (Sect. 4 in Chapter 9) comes from the Greek words meaning *change-working*. Meturgical practices are transformational practices designed to enable embodied brains to **channel** better possible selves in better possible universes. Meturgical practices facilitate modal **shifting** to other universes; but they are not **religious**, since they do not exercise authoritarian control over shifting, but allow people to shift as they will.

modal library   The modal library (Sect. 4 in Chapter 4) is a **library** whose **books** are possible universes. These cosmic books are abstract mathematical structures. This library is a model for **modal logic**. The **treasury** includes exactly those cosmic books which are actualized by the **Good**. According to the **biocosmic analogy**, the **animats** replicate through the modal library much like organisms replicate through the biological library.

modal logic   Modal logic is the logic of possibility and necessity. Models of modal logic involve possible universes in the **modal library**. The modal logic used here is very similar to the **counterpart** theory of David Lewis.

moil   Moiling (Sect. 2.3 in Chapter 2) is the complexity-building work done by a **crane**, that is, a computer that runs an **entelechy**. The stars moil to fuse simpler atoms into complex atoms. The earthly biosphere moils to evolve complex organisms. Humans moil to create complex technologies. **Animats** moil to generate complex universes.

Mount Improbable   Mount Improbable is an abstract mountain which rises into the **sky** of increasing **complexity**. The points on Mount Improbable are **books** (that is, possible objects). Higher points are more complex. Specific Mount Improbables rise into specific **libraries**. The atomic Mount Improbable rises into the atomic library.

mystical experience	Mystical experience (Sect. 4 in Chapter 7) occurs when the experience of **being-itself** overwhelms the experience of beings. Ordinary experience can be expressed as $(\exists x)$ $(\exists y)$(the self x is aware of its object y). During mystical experience, the brain simulates only the **existential quantifiers**, which are the $\exists$s. Mystical experience was **hijacked** by theism but is here **reclaimed** for spiritual naturalism.
night	The night (Sect. 5.3 in Chapter 4) includes all and only those books in the **modal library** that are *not* actualized by the **Good**. These are the **dark books**. The night is the **modal library** minus the **treasury**. So the modal library divides into night and **day**.
nothing	Nothing (Sect. 1.3 in Chapter 6) is the absolute absence of any existing thing. It is the purely logical category of negation. As pure negativity, it negates itself. Nothing is pure self-inconsistency. Since self-inconsistency negates itself, it is ontologically forbidden. The **categorical imperative** entails that **maxims** which are practically self-inconsistent are forbidden. Nothing is symbolized by the **abyss**, which is the element of **water**.
omega point	An omega point (Sect. 5 in Chapter 2) is an infinitely intelligent **computer** which some futurists say will come into existence at the end of time. An omega point is an **ideal finality** for intelligent self-surpassing. Omega points are possible in our universe, which means that some other versions of our universe contain them.
One, the	The One (Sect. 1.4 in Chapter 6) is **being-itself**. Although it is inspired by Plotinus, it

should not be confused with the Plotinian One. The **existential quantifier** refers to the One. The One maximizes self-consistency, which is **reflexivity**. Its reflexivity grounds the **categorical imperative**. The One is symbolized by **earth**.

ontological argument   The ontological argument (Sect. 4 in Chapter 6) justifies the truth of the best proposition, which is the **Good**. The ontological argument does not conclude with God. The best proposition is a purely logical object.

reclaiming   Reclaiming (Sect. 6 in Chapter 1) is the act of liberating a **hijacked** concept from its theistic bondage to **God**. The **cosmological arguments** are reclaimed by showing that they do not conclude with God. **Mystical experience** is reclaimed by showing that it involves the simulation of pure **existential quantification** and not God.

reflexivity   Reflexivity (Sect. 2.5 in Chapter 3, Sect. 1.5 in Chapter 6) is positive self-relation. Logical reflexivity is self-consistency. Reflexivity also appears as the consistency of parts with wholes, such that the structure of some part mirrors, represents, or reflects the structure of the whole. Reflexivity is proportional to **complexity** and is **intrinsically valuable**. The **categorical imperative** selects maximally reflexive policies as obligatory and anti-reflexive policies as forbidden. Nature serves the **Good** by maximizing reflexivity.

religion   Religion primarily refers to the Abrahamic religions. The modal theory of religion (Sect. 3 in Chapter 7) states that religions are technologies for socially **shifting** to fictional universes. Religions employ authoritarian mechanisms to ensure social shifting to the same universe, which may be a **dark book**. **Meturgical** practices, by contrast, encourage shifting to fictional universes anywhere in the **treasury**.

shifting

Shifting (Sect. 3 in Chapter 7) occurs when embodied brains simulate other possible selves in other possible **universes**, and when these simulations are so engaging that the brain temporarily identifies its actual self with its simulated other self. The shifted brain **channels** an alternative self. Shifting can help to overcome the contingent negativities of actuality. Both **religious** and **meturgical** practices aim at shifting the self.

skew

The skew (Sect. 2.2 in Chapter 2) of any distribution of probabilities is its deviation from randomness. Skew measures the non-randomness of the **software** of some **computer**. Since **complexity** is improbability, any complexity-generating computation is highly skewed. Celestial computers are highly skewed. Skew requires an explanation; the ultimate explanation for skew is **the Good**.

sky

The sky (Sect. 2.4 in Chapter 6) is the totality of mathematical objects, such as numbers, sets, and structures. These are sorted into ranks or altitudes along the *axis mundi*, which rises from the **earth**, through the sky, to the **sun**. The sky is filled with **air**.

software

Software (Sect. 2.3 in Chapter 2) is a component of a **computer**. The software of some computer is the distribution of probabilities to the **arrows** in its **hardware**. Any two physical stars share the same atomic hardware, which is the total system of nuclear reaction arrows. But two stars may differ in the probabilities assigned to those arrows. Consider the arrow helium + helium $\rightarrow$ beryllium. Its probability in our sun is very low while its probability in some other star may be very high. The extent to which probabilities are non-randomly distributed over arrows is the **skew** of the computer.

spotlight
The spotlight (Sect. 1.4 in Chapter 8) is the **light** of presence, which seems to move through time along with our self-awareness. The spotlight is generated by the **fire** which animates all **actual** universes. This fire flows through the is-present-to relations, which **animats** create as they actualize the universes in the **treasury**.

stars
The stars (Sect. 2.5 in Chapter 6) symbolize **ideal** objects. A star is the limit of an absolutely infinite progression. It is the limit of an unsurpassable series of surpassable things.

string rewriting
String rewriting (Sect. 2.3 in Chapter 2) is a mathematical model of computing devised by Emil Post. It is equivalent to other models of computing devised by Alan Turing and Alonzo Church. It involves **arrows** which transform old strings into new strings: old strings $\rightarrow$ new strings.

sun
The sun (Sect. 4.2 in Chapter 6) is the traditional Platonic symbol for the **Good**. The sun is an **ideal** object, that is, it is the limit of an unsurpassable series of propositions. The sun is a **star**. The sun dwells at the top of the **sky**, and shines on the **world tree**. When the **light** of the **sun** strikes the earth, it generates the **fire** of **actuality**.

teleology	A teleological agent (Sect. 2.3 in Chapter 2) is an intelligent teleonomic agent which can mentally represent its finalities. Its finalities are its goals or purposes. Every teleological agent is teleonomic; but most teleonomic agents are not teleological.
teleonomy	A teleonomic agent (Sect. 2.3 in Chapter 2) runs a convergent **algorithm** which aims mechanically at some **finality**. An algorithm that repeatedly divides its input by two runs teleonomically to zero. Many teleonomic agents are entirely mindless; but some evolve intelligence and thereby become **teleological** agents.
treasury	The treasury (Sect. 5 in Chapter 4) is the class of cosmic **books** which are actualized by the **Good**. Hence the treasury is the **actual world** and the best world. The treasury is the **day**. Books in the treasury are optimal in the sense that they lie on paths in the **modal library** on which value only increases. The treasury contains some but not all of the books in the modal library. Books not in the treasury are **dark books**.
universe	A universe (Sects. 4–5 in Chapter 4) is a maximal physical structure, that is, a maximal whole organized by spatio-temporal-causal relations. A universe is a cosmic **book** in the **modal library**. All universes are possible objects; but an **actual** universe is energized by **fire**. Only the universes in the **treasury** are actual. They are actualized by the logical power of the **Good**. So the treasury is the **actual world**. No universe is the best. Every universe is surpassed by infinitely many better universes.

water

Water (Sect. 1.3 in Chapter 6) symbolizes the **nothing** of the **abyss**. Thus the abyss is an ocean, from which the earth rises like an island supporting the world tree. Water is one of the five elemental powers, along with **earth**, **air**, **light**, and **fire**. These elements can be used symbolically in **meturgical** rituals or practices.

world

A world (Sect. 3.3 in Chapter 6) is a class of **universes**. The empty world contains no universes. The **modal library** is the maximal world, containing all universes. The **treasury** is the class of actual universes and so is the **actual world**. Since the treasury is the best world, the actual world is the best of all possible worlds. However, no universe is best; every universe in the best world is surpassed by better universes.

world tree

The world tree (Sect. 5 in Chapter 6) is an infinitely branching tree of actual **universes**. Its root is the cosmic **alpha**, the original universe, which lies in the **earth**. Every universe and progression in the world tree is surpassed by better universes. The branches in the world tree are **arrows** linking cosmic books. **Animats** reproduce along these arrows. The world tree is identical with the **treasury**. Every universe in the world tree is animated by **fire**. The world tree is illuminated by the **light** of the **Good**, which shines on this tree like the **sun**. Hence the world tree is a burning shrine.

Zero, the

The Zero (Sect. 1.3 in Chapter 6) is the **nothing**; it is the **abyss**.

References

Aisemberg, G. (2008). Dewey's atheistic mysticism. *The Pluralist, 3*(3), 23–62.

Albert, D. (2012, March 25). On the origin of everything. Review of L. Krauss (2012) *A Universe from Nothing. New York Times, Sunday Book Review*, BR20.

Anderson, J. (2002). *The Airplane: A History of Its Technology*. Reston, VA: American Institute of Aeronautics and Astronautics.

Anderson, T. (2009). Understanding the alteration and decline of a music scene: Observations from rave culture. *Sociological Forum, 24*(2), 307–336.

Anscombe, G. E. M. (1958). Modern moral philosophy. *Philosophy, 33*(124), 1–19.

Atkins, P. (1992). *Creation Revisited: The Origin of Space, Time, and the Universe*. New York: W. H. Freeman & Company.

Atran, S. (2002). *In Gods We Trust: The Evolutionary Landscape of Religion*. New York: Oxford University Press.

Atran, S., & Norenzayan, A. (2004). Religion's evolutionary landscape. *Behavioral and Brain Sciences, 27,* 713–770.

Aunger, R. (2006). What's the matter with memes? In A. Grafen & M. Ridley (Eds.), *Richard Dawkins: How a Scientist Changed the Way We Think* (pp. 176–188). New York: Oxford University Press.

© The Editor(s) (if applicable) and The Author(s), under exclusive license to Springer Nature Switzerland AG 2020
E. Steinhart, *Believing in Dawkins,*
https://doi.org/10.1007/978-3-030-43052-8

Ayer, A. J., & Copleston, F. (1949). Logical positivism—A debate. In M. Diamond & T. Litzenburg (1975), *The Logic of God: Theology and Verification* (pp. 98–118). Indianapolis: Bobbs-Merrill.

Baer, R., et al. (2019). Doing no harm in mindfulness-based programs. *Clinical Psychology Review, 71,* 101–114.

Balaguer, M. (1998). *Platonism and Anti-Platonism in Mathematics*. New York: Oxford University Press.

Bamford, R. (2016). Nietzsche on experience, naturalism, and experimentalism. *Journal of Nietzsche Studies, 47*(1), 9–29.

Barash, D. (2006). What the whale wondered: Evolution, existentialism and the search for 'meaning'. In A. Grafen & M. Ridley (Eds.), *Richard Dawkins: How a Scientist Changed the Way We Think* (pp. 255–262). New York: Oxford.

Barrett, J. (2007). Cognitive science of religion: What is it and why is it? *Religion Compass, 1,* 1–19.

Basalla, G. (1988). *The Evolution of Technology*. New York: Cambridge University Press.

Batchelor, S. (1997). *Buddhism Without Beliefs: A Contemporary Guide to Awakening*. New York: Penguin.

Bekoff, M., & Pierce, J. (2009). *Wild Justice: The Moral Lives of Animals*. Chicago: University of Chicago Press.

Bell, E. et al. (2015). Potentially biogenic carbon preserved in a 4.1 billion-year-old zircon. *Proceedings of the National Academy of Sciences, 112*(47), 14518–14521.

Benitez, E. (1995). The good or the demiurge: Causation and the unity of good in Plato. *Apeiron, 28,* 113–139.

Bennett, C. (1988). Logical depth and physical complexity. In R. Herken, *The Universal Turing Machine: A Half-Century Survey* (pp. 227–257). New York: Oxford University Press.

Bennett, C. (1990). How to define complexity in physics, and why. In W. Zurek (Ed.), *Complexity, Entropy, and the Physics of Information* (pp. 137–148). Reading, MA: Addison-Wesley.

Bertz, S. (1981). The first general index of molecular complexity. *Journal of the American Chemical Society, 103,* 3599–3601.

Bishop, J. (2010). Secular spirituality and the logic of giving thanks. *Sophia, 49,* 523–534.

Black, M. (1964). The gap between "is" and "should". *Philosophical Review, 73,* 165–181.

Blackmore, S. (1999). *The Meme Machine*. New York: Oxford University Press.

Blackmore, S. (2011). *Zen and the Art of Consciousness*. Oxford: Oneworld.

Bloch, M. (2008). Why religion is nothing special but is central. *Philosophical Transactions of the Royal Society B, 353,* 2055–2061.

Bostrom, N. (2003). Are you living in a computer simulation? *Philosophical Quarterly, 53*(211), 243–255.

Bottcher, T. (2016). An additive definition of molecular complexity. *Journal of Chemical Information and Modeling, 56,* 462–470.

Bournez, O., Ibanescu, L., & Kirchner, H. (2006). From chemical rules to term rewriting. *Electronic Notes in Computer Science, 147,* 113–134.

Bower, J. (1988). *The Evolution of Complexity by Means of Natural Selection.* Princeton, NJ: Princeton University Press.

Brey, X. (2008). Technological design as an evolutionary process. In P. Vermaas, P. Kroes, A. Light, & S. Moore (Eds.), *Philosophy and Design* (pp. 61–76). New York: Springer.

Brown, G. W. (2013). *Missa Charles Darwin.* Performed by New York Polyphony. North Hampton, NH: Navona Records.

Brown, J. (1980). Counting proper classes. *Analysis, 40*(3), 123–126.

Bruton, E. (1979). *The History of Clocks and Watches.* New York: Rizzoli.

Bueno, O. (2003). Is it possible to nominalize quantum mechanics? *Philosophy of Science, 70,* 1424–1436.

Bulbulia, J. (2009). Religiosity as mental time-travel: Cognitive adaptations for religious behavior. In J. Schloss & M. Murray (Eds.), *The Believing Primate* (pp. 44–75). New York: Oxford.

Burd-Sharps, S., Lewis, K., & Borges Martins, E. (2008). *The Measure of America: American Human Development Report 2008–2009.* New York: Columbia University Press.

Catling, D. (2013). *Astrobiology: A Very Short Introduction.* New York: Oxford.

Chaisson, E. (2006). *The Epic of Evolution: The Seven Ages of our Cosmos.* New York: Columbia University Press.

Chapman, E., Childers, D., & Vallino, J. (2016). How the second law of thermodynamics has informed ecosystem ecology through its history. *BioSciences, 66,* 27–39.

Cimino, R., & Smith, C. (2014). *Atheist Awakening: Secular Activism and Community in America.* New York: Oxford.

Cirkovic, M. (2003). Physical eschatology. *American Journal of Physics, 71*(2), 122–133.

Cleland-Host, H., & Cleland-Host, J. (2014). *Elemental Birthdays: How to Bring Science into Every Party.* Solstice & Equinox Publishing.

Colledge, R. (2013). Secular spirituality and the hermeneutics of ontological gratitude. *Sophia, 52*, 27–43.

Colyvan, M. (2001). *The Indispensability of Mathematics*. New York: Oxford University Press.

Comte-Sponville, A. (2006). *The Little Book of Atheist Spirituality* (N. Huston, Trans.). New York: Viking.

Conselice, C., et al. (2016). The evolution of galaxy number density at $z<8$ and its implications. *The Astrophysics Journal, 830*(2), 83.

Copeland, B. J. (1998). Even Turing machines can compute uncomputable functions. In C. Calude, J. Casti, & M. Dinneen (Eds.), *Unconventional Models of Computation* (pp. 150–164). New York: Springer-Verlag.

Crosby, D. (2014). *More Than Discourse: Symbolic Expressions of Naturalistic Faith*. Albany, NY: SUNY Press.

Cummins, R. (1996). *Representations, Targets, and Attitudes*. Cambridge, MA: The MIT Press.

Cusack, C. (2010). *Invented Religions: Imagination, Fiction and Faith*. Burlington, VT: Ashgate.

Davidsen, M. (2014). *The Spiritual Tolkein Milieu*. Dissertation, University of Leiden.

Dawkins, R. (1976). *The Selfish Gene* (30th anniversary ed.). New York: Oxford University Press. ISBN 978-0-19-929115-1.

Dawkins, R. (1982). *The Extended Phenotype* (Oxford Landmark Science ed.). New York: Oxford University Press. ISBN: 978-0-19-878891-1.

Dawkins, R. (1986). *The Blind Watchmaker: Why the Evidence of Evolution Reveals a Universe Without Design*. New York: W. W. Norton. ISBN 0-393-31570-3.

Dawkins, R. (1988). The evolution of evolvability. In C. Langton (Ed.), *Artificial Life* (pp. 201–220). Boston, MA: Addison-Wesley.

Dawkins, R. (1992, April 20). A scientist's case against God. *The Independent*, 17.

Dawkins, R. (1993a). Worlds in microcosm. In N. Spurway (Ed.) *Humanity, Environment, God* (pp. 105–125). Cambridge, MA: Blackwell.

Dawkins, R. (1993b, September 11). The evolutionary future of man: A biological view of progress. *The Economist, 328*(7828), 87–90.

Dawkins, R. (1994, August 5). An atheist's vision of life. *The Spectator*, 17.

Dawkins, R. (1995a). *River out of Eden: A Darwinian View of Life*. New York: Basic Books. ISBN 978-0-465-06990-3.

Dawkins, R. (1995b). A reply to Poole. *Science and Christian Belief, 7*(1), 45–50.

Dawkins, R. (1995c, December 31). The real romance in the stars. *The Independent*. On line at www.independent.co.uk/voices/the-real-romance-in-the-stars-1527970.html. Accessed 16 February 2019.

Dawkins, R. (1996a). *Climbing Mount Improbable*. New York: W. W. Norton. ISBN 0-393-31682-3.

Dawkins, R. (1996b, February 4). Review of Richard Swinburne's *Is There a God? The Sunday Times*. Reprinted (2003) *Think, 2*(4), 51–54.

Dawkins, R. (1998a). *Unweaving the Rainbow: Science, Delusion, and the Appetite for Wonder*. New York: Houghton Mifflin. ISBN 978-0-618-05673-6.

Dawkins, R. (1998b, August 19). Who's afraid of the Frankenstein wolf? *London Evening Standard*, 11.

Dawkins, R. (2003a). *A Devil's Chaplain*. New York: Houghton Mifflin. ISBN 0-618-48539-2.

Dawkins, R. (2003b). I. The Science of Religion. II. The Religion of Science. *The Tanner Lectures on Human Values*. In G. Peterson (Ed.) (2005) *The Tanner Lectures on Human Values* (Vol. 25). Salt Lake City: The University of Utah Press.

Dawkins, R. (2003c). Now here's a bright idea! *Free Inquiry, 23*(4), 12–13.

Dawkins, R. (2004a). The sacred and the scientist. In B. Rogers (Ed.), *Is Nothing Sacred?* (pp. 135–137). New York: Routledge.

Dawkins, R. (2004b). Who owns the argument from improbability? *Free Inquiry, 24*(6), 11–12.

Dawkins, R. (2004c). *BBC Belief interview*. Online at www.bbc.co.uk/religion/religions/atheism/people/dawkins.shtml. Accessed 19 April 2019.

Dawkins, R. (2005). The theology of the tsunami. *Free Inquiry, 25*(3), 12–13.

Dawkins, R. (2006a). Intelligent aliens. In J. Brockman (Ed.), *Intelligent Thought*. New York: Random House.

Dawkins, R. (2006b). Afterword. In J. Brockman (Ed.), *What Is Your Dangerous Idea?* (pp. 297–301). New York: Simon & Schuster.

Dawkins, R. (2007). Happy Newton day! *New Statesman, 136*(4875–4877), 42–43.

Dawkins, R. (2008). *The God Delusion*. New York: Houghton-Mifflin. ISBN 978-0-618-68000-9.

Dawkins, R. (2009a). *The Greatest Show on Earth: The Evidence for Evolution*. New York: Free Press. ISBN 978-1-4165-9479-6.

Dawkins, R. (2009b). *Intelligence squared debate "Atheism is the new fundamentalism"*. Part 11 of 12. www.youtube.com/watch?v=lheDgyaItOA&list=PL45065FAA19FE4EE1&index=11.

Dawkins, R. (2010). *Giving thanks in a vacuum*. Global Atheist Convention. MCEC Melbourne. www.youtube.com/watch?v=kGGmuUvA2Mg.

Dawkins, R. (2011). *Interview with The King James Bible Trust (1611–2011)*. Online at www.youtube.com/watch?v=Ej1auSuVM-M. Accessed 30 March 2018.

Dawkins, R. (2012a). Afterword to Krauss. In L. Krauss (Ed.), *A Universe from Nothing: Why there is Something rather than Nothing* (pp. 187–191). New York: Free Press.

Dawkins, R. (2012b). *The Magic of Reality: How We Know What's Really True*. New York: Free Press. ISBN 978-1-4516-7504-7.

Dawkins, R. (2013). Why the word "spiritual" has been hijacked by religions. *Sydney Opera House Talks & Ideas*. Online at www.youtube.com/watch?v=Are53Pg0hZ8. Accessed 30 March 2018.

Dawkins, R. (2015a). *Brief Candle in the Dark: My Life in Science*. New York: HarperCollins. ISBN 978-0-06-228845-5.

Dawkins, R. (2015b). Evolvability. In J. Brockman (Ed.), *Life*. New York: HarperCollins.

Dawkins, R. (2017). *Science in the Soul: Selected Writings of a Passionate Rationalist*. New York: Random House. ISBN 978-0-399-59224-9.

Dawkins, R., & Hutton, W. (2012, February 18). What is the proper place for religion in Britain's public life? *The Guardian*. Online at www.theguardian.com/world/2012/feb/19/religion-secularism-atheism-hutton-dawkins. Accessed 21 May 2019.

Dawkins, R., & Jollimore, T. (2008). *An interview with Richard Dawkins*. Online at believermag.com/an-interview-with-richard-dawkins/. Accessed 19 April 2019.

Dawkins, R., & Krauss, L. (2007, June 19). Should science speak to faith? (Extended version). *Scientific American*. Online at www.scientificamerican.com/article/should-science-speak-to-faith-extended/.

Dawkins, R., & Lightman, A. (2018). *Richard Dawkins & Alan Lightman on Science & Religion*. Online at www.youtube.com/watch?v=eSCDfjTDVCk. Accessed 22 February 2019.

Dawkins, R., & Wong, Y. (2016). *The Ancestor's Tale* (Revised and Expanded ed.). New York: Houghton Mifflin Harcourt. ISBN 978-0-544-85993-7.

Dawkins, R., Dennett, D., Harris, S., & Hitchens, C. (2007). *The Four Horsemen*. New York: Random House. ISBN 978-0-593-08039-9.

de Botton, A. (2012). *Religion for Atheists: A Non-Believer's Guide to the Uses of Religion*. New York: Random House.

De Loore, C., & Doom, C. (1992). *Structure and Evolution of Single and Binary Stars*. Boston: Kluwer.

Dehaene, S., Piazza, M., Pinel, P., & Cohen, L. (2003). Three parietal circuits for number processing. *Cognitive Neuropsychology, 20*(3/4/5/6), 487–506.

Demorest, P., et al. (2010, October 28). A two-solar-mass neutron star measured using Shapiro delay. *Nature, 467*, 1081–1083.

Deneulin, S., & Shahani, L. (2009). *An Introduction to the Human Development and Capability Approach*. Sterling, VA: Earthscan.

Dennett, D. (1993). *Consciousness Explained*. New York: Penguin.

Dennett, D. (1995). *Darwin's Dangerous Idea: Evolution and the Meanings of Life*. New York: Simon & Schuster.

Dennett, D. (2004). Could there be a Darwinian account of human creativity? In A. Moya & E. Font (Eds.), *Evolution: From Molecules to Ecosystems* (pp. 273–279). New York: Oxford University Press.

Dennett, D. (2006). *Breaking the Spell*. New York: Viking Penguin.

Derex, M., et al. (2019). Causal understanding is not necessary for the improvement of culturally evolving technology. *Nature Human Behavior*. Online at www.nature.com/articles/s41562-019-0567-9. Accessed 29 April 2019.

Deutsch, D. (1985). Quantum theory, the Church-Turing principle and the universal quantum computer. *Proceedings of the Royal Society, Series A, 400*, 97–117.

Dewar, R. (2006). Maximum entropy production and non-equilibrium statistical mechanics. In A. Kleidon & R. Lorenz (Eds.), *Non-Equilibrium Thermodynamics and the Production of Entropy* (pp. 41–55). New York: Springer.

Dewar, R., Juretic, D., & Zupanovic, P. (2006). The functional design of the rotary enzyme ATP synthase is consistent with maximum entropy production. *Chemical Physics Letters, 430*, 177–182.

Dickinson, G. (1931). *J. McT. E. McTaggart*. Cambridge: Cambridge University Press.

Dil, E., & Yumak, T. (2018). Emergent entropic nature of fundamental interactions. Online at arxiv.org/abs/1702.04635. Accessed 19 April 2019.

Dillon, J. (2007). Iamblichus' defense of theurgy: Some reflections. *The International Journal of the Platonic Tradition, 1*, 30–41.

Dillon, J. (2016). The divinizing of matter: Some reflections on Iamblichus' theurgic approach to matter. In J. Halfwasse, et al. (Eds.), *Soul and Matter in Neoplatonism* (pp. 177–188). Heidelberg: University of Heidelberg Press.

Dobovisek, A., et al. (2011). Enzyme kinematics and the maximum entropy production principle. *Biophysical Chemistry, 154,* 49–55.

Doherty, B. (2004). *This Is Burning Man.* New York: Little Brown.

Drake, F. (1974). *Set Theory: An Introduction to Large Cardinals.* New York: American Elsevier.

Dyga, K., & Stupak, R. (2015). Meditation and psychosis: Trigger or cure? *Archives of Psychiatry and Psychotherapy, 3,* 48–58.

Dyson, F. (1985). *Infinite in All Directions.* New York: HarperCollins.

Dyson, G. (1997). *Darwin Among the Machines: The Evolution of Global Intelligence.* Reading, MA: Perseus Books.

Dyson, G. (2012). *Turing's Cathedral: The Origins of the Digital Universe.* New York: Vintage Press.

Ebbing, D., & Gammon, S. (2017). *General Chemistry* (Eleventh ed.). Boston, MA: Cengage Learning.

England, J. (2013). Statistical physics of self-replication. *Journal of Chemical Physics, 139*(121923), 1–8.

England, J. (2014). A new physics theory of life (interview with N. Wolchover). *Quanta Magazine.* Online at www.quantamagazine.org/a-new-thermodyn amics-theory-of-the-origin-of-life-20140122/.

England, J. (2015). Dissipative adaptation in driven self-assembly. *Nature Nanotechnology, 10,* 919–923.

Enoch, J. (1998). The enigma of early lens use. *Technology and Culture, 39*(2), 273–291.

Essinger, J. (2004). *Jacquard's Web: How a Hand-Loom Led to the Birth of the Information Age.* New York: Oxford University Press.

Fenton, J. (1965). Being-itself and religious symbolism. *The Journal of Religion, 45*(2), 73–86.

Field, H. (1980). *Science Without Numbers.* Princeton, NJ: Princeton University Press.

Field, H. (1985). Comments and criticisms on conservativeness and incompleteness. *Journal of Philosophy, 82*(5), 239–260.

Foot, P. (2001). *Natural Goodness.* New York: Oxford University Press.

Foucault, M. (1988). Technologies of the self. In L. Martin, H. Gutman, & P. Hutton (Eds.), *Technologies of the Self: A Seminar with Michel Foucault* (pp. 16–49). Amherst, MA: University of Massachusetts Press.

Fredkin, E. (2003). An introduction to digital philosophy. *International Journal of Theoretical Physics, 42*(2), 189–247.

French, S., & Ladyman, J. (2010). In defence of ontic structural realism. In A. Bokulich & P. Bokulich (Eds.), *Scientific Structuralism* (pp. 25–42). New York: Springer.

Froding, B. (2013). *Virtue Ethics and Human Enhancement*. New York: Springer.

Gardner, A. (2014). Life, the universe and everything. *Biology and Philosophy, 29*, 207–215.

Gardner, A., & Conlon, J. (2013). Cosmological natural selection and the purpose of the universe. *Complexity, 18*(5), 48–56.

Garrod, R., Widicus Weaver, S., & Herbst, E. (2008). Complex chemistry in star-forming regions: An expanded gas-grain warm-up chemical model. *The Astrophysical Journal, 682*, 283–302.

Gattringer, C., & Lang, C. (2009). *Quantum Chromodynamics on the Lattice*. New York: Springer.

Gauthier, F. (2004). Rave and religion? A contemporary youth phenomenon as seen through the lens of religious studies. *Studies in Religion, 33*(3–4), 397–413.

Gerson, L. (1994). *Plotinus*. New York: Routledge.

Gilmore, L. (2010). *Theatre in a Crowded Fire: Ritual and Spirituality at Burning Man*. Berkeley, CA: University of California Press.

Goodenough, U. (1998). *The Sacred Depths of Nature*. New York: Oxford University Press.

Goodman, N. D. (1990). Mathematics as natural science. *The Journal of Symbolic Logic, 55*(1), 182–193.

Greene, B. (2005). *The Fabric of the Cosmos*. New York: Vintage.

Gutmann, J. (1954). The 'Tremendous moment' of Nietzsche's vision. *The Journal of Philosophy, 51*(25), 837–842.

Hadot, P. (1995). *Philosophy as a Way of Life: Spiritual Exercises from Socrates to Foucault* (M. Chase, Trans. and Ed. A. Davidson). Malden, MA: Wiley-Blackwell.

Hadot, P. (2011). *The Present Alone Is Our Happiness: Conversations with Jeannie Carlier and Arnold I. Davidson* (M. Djaballah & M. Chase, Trans.). Stanford, CA: Stanford University Press.

Hahm, D. (1977). *The Origins of Stoic Cosmology*. Columbus: Ohio State University Press.

Harris, J. (2007). *Enhancing Evolution: The Ethical Case for Making Better People*. Princeton, NJ: Princeton University Press.

Harris, S. (2014). *Waking Up*. New York: Simon & Schuster.

Hartshorne, C. (1967). *A Natural Theology for Our Time*. La Salle, IL: Open Court.

Hartshorne, C. (1984). *Omnipotence and Other Theological Mistakes*. Albany, NY: State University of New York Press.

Harvey, S. (2017). *Playa Fire: Spirit and Soul at Burning Man*. San Francisco: HarperElixir.

Haught, J. (2008). *God and the New Atheism: A Critical Response to Dawkins, Harris, and Hitchens*. Louisville, KY: Westminster John Knox Press.

Hawking, S. (1988). *A Brief History of Time*. Toronto: Bantam Books.

Heidegger, M. (1998). What is metaphysics? In W. McNeill (Ed.), *Pathmarks* (pp. 82–96). New York: Cambridge University Press.

Henderson, B. (2006). *The Gospel of the Flying Spaghetti Monster*. New York: Villard.

Hitchens, C., & Blair, T. (2011). *Hitchens vs. Blair: Be It Resolved Religion Is a Force for Good in the World*. Berkeley, CA: Publishers Group West.

Horgan, J., & Ellis, G. (2014). Physicist George Ellis knocks physicists for knocking philosophy, falsification, free will. *Scientific American Blog*. Online at blogs.scientificamerican.com/cross-check/physicist-george-ellis-knocks-physicists-for-knocking-philosophy-falsification-free-will/. Accessed 14 March 2018.

Horowitz, J., & England, J. (2017). Spontaneous fine-tuning to environment in many-species chemical reaction networks. *Proceedings of the National Academy of Sciences, 114*(29), 7565–7570.

Hughes, J. (2013). Using neurotechnologies to develop virtues: A Buddhist approach to cognitive enhancement. *Accountability in Research, 20*, 27–41.

Hume, D. (1779/1990). *Dialogues Concerning Natural Religion*. New York: Penguin.

Hutson, S. (2000). The rave: Spiritual healing in modern Western subculture. *Anthropological Quarterly, 73*(1), 35–49.

Inge, W. (1918). *The Philosophy of Plotinus* (Vol. 2). London: Longmans Green.

Irvine, W. (2009). *A Guide to the Good Life: The Ancient Art of Stoic Joy*. New York: Oxford University Press.

Jackson, B. D. (1967). Plotinus and the *Parmenides*. *Journal of the History of Philosophy, 5*(4), 315–327.

Jaffe, R., & Taylor, W. (2018). *The Physics of Energy*. New York: Cambridge University Press.

Johnson-Laird, P. N. (1983). *Mental Models*. Cambridge, MA: Harvard University Press.

Johnson-Laird, P. (2005). Flying bicycles: How the Wright brothers invented the airplane. *Mind and Society, 4,* 27–48.

Johnston, I., Ahnert, S., Doye, J., & Louis, A. (2011). Evolutionary dynamics in a simple model of self-assembly. *Physical Review, 83*(6), 066105.

Johnston, S. (2008). Animating statues: A case study in ritual. *Arethusa, 41*(3), 445–477.

Juretic, D., & Zupanovic, P. (2003). Photosynthetic models with maximum entropy production in irreversible charge transfer steps. *Computational Biology and Chemistry, 27,* 541–553.

Kachman, T., Owen, J., & England, J. (2017). Self-organized resonance during search of a diverse chemical space. *Physical Review Letters, 119*(3), 038001.

Kanamori, A. (2005). *The Higher Infinite: Large Cardinals in Set Theory from Their Beginnings.* New York: Springer.

Kant, I. (1790/1951). *Critique of Judgment* (J. Bernard, Trans.). New York: Macmillan.

Kelly, K. (2010). *What Technology Wants.* New York: Viking.

Khovanova, T. (2010). *The sexual side of life.* By John H. Conway as told to Tanya Khovanova. Online at blog.tanyakhovanova.com/?p=260. Accessed 10 May 2019.

Kiteley, M. (1958). Existence and the ontological argument. *Philosophy and Phenomenological Research, 18*(4), 533–535.

Koestler, A. (1969). *The Invisible Writing.* New York: The Macmillan Company.

Kosslyn, S. (1994). *Image and Brain: The Resolution of the Imagery Debate.* Cambridge, MA: MIT Press.

Kotz, J., Treichel, P., & Townsend, J. (2009). *Chemistry & Chemical Reactivity* (Vol. 2). Belmont, CA: Thompson.

Kouvaris, K., et al. (2017). How evolution learns to generalise. *PLoS Computational Biology, 13*(4), e1005358.

Kraay, K. (2011). Theism and modal collapse. *American Philosophical Quarterly, 48*(4), 361–372.

Krauss, L. (2012). *A Universe from Nothing: Why There Is Something Rather Than Nothing.* New York: Free Press.

Kurzweil, R. (2005). *The Singularity Is Near: When Humans Transcend Biology.* New York: Viking.

Lacewing, M. (2016). Can non-theists appropriately feel existential gratitude? *Religious Studies, 52,* 145–165.

Langer, E. (1975). The illusion of control. *Journal of Personality and Social Psychology, 32*(2), 311–328.

Leibniz, G. W. (1697). On the ultimate origination of the universe. In P. Schrecker & A. Schrecker (Eds.) (1988) *Leibniz: Monadology and Other Essays* (pp. 84–94). New York: Macmillan Publishing.

Leslie, J. (1989). *Universes*. New York: Routledge.

Lewis, D. (1978). Truth in fiction. *American Philosophical Quarterly, 15*(1), 37–46.

Lewis, D. (1983). *Philosophical Papers* (Vol. 1). New York: Oxford University Press.

Lewis, D. (1986). *On the Plurality of Worlds*. Cambridge, MA: Blackwell.

Lightman, A. (2018). *Searching for Stars on an Island in Maine*. New York: Random House.

Linde, A. D. (1986, August 14). Eternally existing self-reproducing chaotic inflationary universe. *Physics Letters B, 175*(4), 387–502.

Linde, A. D. (1994). The self-reproducing inflationary universe. *Scientific American, 271*(5), 48–55.

Liston, M. (1993). Taking mathematical fictions seriously. *Synthese, 95,* 433–458.

Lloyd, S. (2002, May). Computational capacity of the universe. *Physical Review Letters, 88*(23), 237901–237905.

Lokhorst, G.-J. (2006). Andersonian deontic logic, propositional quantification, and Mally. *Notre Dame Journal of Formal Logic, 47*(3), 385–395.

Lovejoy, A. (1936). *The Great Chain of Being*. Cambridge, MA: Harvard University Press.

Machta, J. (2011). Natural complexity, computational complexity, and depth. *Chaos, 21,* 0371111–0371118.

Mackendrick, K. (2012). We have an imaginary friend in Jesus: What can imaginary companions teach us about religion? *Implicit Religion, 15*(1), 61–79.

Maddy, P. (1983). Proper classes. *Journal of Symbolic Logic, 48*(1), 113–139.

Maddy, P. (1996). Ontological commitment: Between Quine and Duhem. *Nous 30*(Supplement: Philosophical Perspectives 10), 317–341.

Magill, J., & Galy, J. (2005). *Radioactivity Radionuclides Radiation*. New York: Springer.

Marenduzzo, D., Finn, K., & Cook, P. (2006). The depletion attraction: An underappreciated force driving cellular organization. *Journal of Cell Biology, 175*(5), 681–686.

Martin, O., & Horvath, J. (2013). Biological evolution of replicator systems: Towards a quantitative approach. *Origins of Life and Evolution of Biospheres, 43,* 151–160.

Martyushev, L. (2013). Entropy and entropy production: Old misconceptions and new breakthroughs. *Entropy, 15,* 1152–1170.

Martyushev, L., & Seleznev, V. (2006). Maximum entropy production principle in physics, chemistry, and biology. *Physics Reports, 426,* 1–45.

Mataxis, T. (1962, September). Change of life, computer style. *Army, 13*(2), 61–67.

Maudlin, T. (2014). *New Foundations for Physical Geometry.* New York: Oxford University Press.

McKinsey, J., Sugar, A., & Suppes, P. (1953). Axiomatic foundations of classical particle mechanics. *Journal of Rational Mechanics and Analysis, 2,* 253–272.

McLendon, H. (1960). Beyond being. *The Journal of Philosophy, 57*(22/23), 712–725.

Melia, J. (1998). Field's programme: Some interference. *Analysis, 58*(2), 63–71.

Millican, P. (2004). The one fatal flaw in Anselm's argument. *Mind, 113,* 451–467.

Moravec, H. (1988). *Mind Children: The Future of Robot and Human Intelligence.* Cambridge, MA: Harvard University Press.

Mortley, R. (1976). Recent work on Neoplatonism. *Prudentia, 7*(1), 47–62.

Mulgan, T. (2015). *Purpose in the Universe: The Moral and Metaphysical Case for Ananthropocentric Purposivism.* New York: Oxford University Press.

Murdoch, I. (1970). *The Sovereignty of the Good.* New York: Schocken.

Murdoch, I. (1992). *Metaphysics as a Guide to Morals.* London: Chatto & Windus.

Nagasawa, Y. (2018). The problem of evil for atheists. In N. Trakakis (Ed.), *The Problem of Evil: Eight Views in Dialogue* (pp. 151–175). New York: Oxford.

Nieder, A. (2005). Counting on neurons: The neurobiology of numerical competence. *Nature Reviews: Neuroscience, 6*(March), 177–190.

Nixey, C. (2018). *The Darkening Age: The Christian Destruction of the Classical World.* New York: Houghton Mifflin Harcourt.

Norenzayan, A. (2013). *Big Gods: How Religion Transformed Cooperation and Conflict.* Princeton, NJ: Princeton University Press.

Nozick, R. (1981). *Philosophical Explanations.* Cambridge, MA: Harvard University Press.

Nutman, A., et al. (2016). Rapid emergence of life shown by discovery of 3,700-million-year-old microbial structures. *Nature, 537,* 535–538.

O'Hagen, E. (2009). Animals, agency, and obligation in Kantian ethics. *Social Theory and Practice, 35*(4), 531–554.

Oppy, G. (2018). *Naturalism and Religion.* New York: Routledge.

Otto, R. (1958). *The Idea of the Holy* (J. W. Harvey, Trans.). New York: Oxford University Press.

Partner, M., Kashtan, N., & Alon, U. (2008). Facilitated variation: How evolution learns from past environments to generalize to new environments. *PLoS Computational Biology, 4*(11), e1000206.

Peirce, C. S. (1965). Collected papers of Charles Sanders Peirce. In C. Hartshorne & P. Weiss (Eds.), *Cambridge*. MA: Harvard University Press.

Penrose, R. (1979). Singularities and time-asymmetry. In S. Hawking & W. Israel (Eds.), *General Relativity: An Einstein Centenary Survey* (pp. 581–638). New York: Cambridge University Press.

Petigura, E., et al. (2013). Prevalence of Earth-size planets orbiting Sun-like stars. *PNAS, 110*(48), 19273–19278.

Pigliucci, M. (2013). New Atheism and the scientistic turn in the atheism movement. *Midwest Studies in Philosophy, XXXVII,* 142–153.

Pigliucci, M. (2017). *How to Be a Stoic: Using Ancient Philosophy to Live a Modern Life*. New York: Basic Books.

Pike, S. (2005). No Novenas for the dead: Ritual action and communal memory at the temple of tears. In L. Gilmore & M. Van Proyen (Eds.), *AfterBurn: Reflections on Burning Man* (pp. 195–214). Albuquerque, NM: University of New Mexico Press.

Post, E. (1943). Formal reductions of the general combinatorial decision problem. *American Journal of Mathematics, 65*(2), 197–215.

Potter, M. (2004). *Set Theory and Its Philosophy*. New York: Oxford University Press.

Poundstone, W. (1985). *The Recursive Universe: Cosmic Complexity and the Limits of Scientific Knowledge*. Chicago: Contemporary Books Inc.

Pritchard, L., & Dufton, M. (2000). Do proteins learn to evolve? The Hopfield network as a basis for the understanding of protein evolution. *Journal of Theoretical Biology, 202,* 77–86.

Quine, W. V. O. (1948). On what there is. In J. Kim & E. Sosa (Eds.) (1999) *Metaphysics: An Anthology* (pp. 4–12). Malden, MA: Blackwell.

Rahula, W. (1974). *What the Buddha Taught*. New York: Grove/Atlantic.

Ray, T. (1992). An approach to the synthesis of life. In C. Langton, C. Taylor, J. Farmer, & S. Rasmussen, *Artificial Life II* (Vol. 10, pp. 371–408). SFI Studies in the Sciences of Complexity. Reading, MA: Addison-Wesley.

Redfield, A. (2017). An analysis of the experiences and integration of transpersonal phenomena induced by electronic dance music events. *International Journal of Transpersonal Studies, 36*(1), 67–80.

Redfield, A., & Thouin-Savard, M. (2017). Electronic dance music events as modern-day ritual. *International Journal of Transpersonal Studies, 36* (1), 52–66.

Rees, M. (2001). *Just Six Numbers: The Deep Forces That Shape the Universe.* New York: Basic Books.

Rendell, P. (2002). Turing universality of the game of life. In A. Adamatzky (Ed.) (2001) *Collision-Based Computing* (pp. 513–539). London: Springer-Verlag.

Rescher, N. (1979). *Leibniz: An Introduction to His Philosophy.* Totowa, NJ: Rowman & Littlefield.

Rescher, N. (2010). *Axiogenesis: An Essay in Metaphysical Optimalism.* New York: Lexington Books.

Resnik, M. (1995). Scientific vs. mathematical realism: The indispensability argument. *Philosophia Mathematica, 3* (3), 166–174.

Robertson, D. (2015). *Stoicism and the Art of Happiness.* New York: McGraw Hill.

Robeyns, I. (2005). The capability approach: A theoretical survey. *Journal of Human Development, 6* (1), 93–114.

Rosenberg, A. (2011). *The Atheist's Guide to Reality.* New York: W. W. Norton.

Rothman, T., & Ellis, G. (1993). Smolin's natural selection hypothesis. *Quarterly Journal of the Royal Astronomical Society, 34,* 201–212.

Rowlands, M. (2012). *Can Animals Be Moral?.* New York: Oxford University Press.

Rubin, M. (1986). Spectacles: Past, present, and future. *Survey of Opthamalogy, 30* (5), 321–327.

Russell, B. (1967). *The Autobiography of Bertrand Russell* (Vol. 1). London: George Allen & Unwin.

Rutherford, D. (1995). *Leibniz and the Rational Order of Nature.* New York: Cambridge University Press.

Sagan, C. (1977). *The Dragons of Eden: Speculations on the Evolution of Human Intelligence.* New York: Ballantine Books.

Sagan, C. (1995). *Pale Blue Dot.* London: Headline.

Sandberg, A. (1999). The physics of information processing superobjects: Daily life among the Jupiter brains. *Journal of Evolution and Technology, 5* (1), 1–34.

Schmidhuber, J. (1997). A computer scientist's view of life, the universe, and everything. In C. Freksa (Ed.), *Foundations of Computer Science: Potential—Theory—Cognition* (pp. 201–208). New York: Springer.

Schneider, E., & Kay, J. (1994). Life as a manifestation of the second law of thermodynamics. *Mathematical and Computer Modelling, 19*(6–8), 25–48.

Science Tarot. (2019). Online at sciencetarot.com. Accessed 3 May 2019.

Seligman, A., Weller, R., Puett, M., & Simon, B. (2008). *Ritual and Its Consequences: An Essay on the Limits of Sincerity*. New York: Oxford University Press.

Sen, A. (1993). Capability and well-being. In M. Nussbaum & A. Sen (Eds.), *The Quality of Life* (pp. 30–53). New York: Oxford University Press.

Shapiro, S. (1983). Conservativeness and incompleteness. *Journal of Philosophy, 80*(9), 521–531.

Shapiro, S. (1997). *Philosophy of Mathematics: Structure and Ontology*. New York: Oxford University Press.

Sharp, L. (2019). *Animal Ethos: The Morality of Human-Animal Encounters in Experimental Lab Science*. Oakland: University of California Press.

Shaw, G. (2014). *Theurgy and the Soul: The Neoplatonism of Iamblichus* (2nd ed.). Kettering, OH: Angelico Press.

Sideris, L. (2017). *Consecrating Science: Wonder, Knowledge, and the Natural World*. Oakland: University of California Press.

Silk, J. (1997). Holistic cosmology. *Science, 277*(5326), 644.

Silk, J. (2001). *The Big Bang* (3rd ed.). New York: Henry Holt & Co.

Simonton, D. (2010). Creative thought as blind-variation and selective retention. *Physics of Life Reviews, 7*, 156–179.

Simonton, D. (2015). Thomas Edison's creative career. *Psychology of Aesthetics, Creativity, and the Arts, 9*(1), 2–14.

Skene, K. (2015). Life's a gas: A thermodynamic theory of biological evolution. *Entropy, 17*, 5522–5548.

Smith, Q. (1988). An analysis of holiness. *Religious Studies, 24*(4), 511–527.

Smolin, L. (1997). *The Life of the Cosmos*. New York: Oxford University Press.

Smolin, L. (2004). Cosmological natural selection as the explanation for the complexity of the universe. *Physica A, 340*, 705–713.

Soltis, D., & Soltis, P. (2019). *The Great Tree of Life*. Cambridge, MA: Academic Press.

Springel, V., et al. (2005, June 2). Simulations of the formation, evolution and clustering of galaxies and quasars. *Nature, 435*, 629–636.

St John, G. (Ed.). (2004). *Rave Culture and Religion*. New York: Routledge.

Steiner, M. (1998). *The Applicability of Mathematics as a Philosophical Problem*. Cambridge, MA: Harvard University Press.

Steinhardt, P., & Turok, N. (2007). *Endless Universe: Beyond the Big Bang*. New York: Doubleday.

Steinhart, E. (2003). Supermachines and superminds. *Minds and Machines, 13,* 155–186.

Steinhart, E. (2012). Royce's model of the Absolute. *Transactions of the Charles S. Peirce Society, 48*(3), 356–384.

Steinhart, E. (2014). *Your Digital Afterlives: Computational Theories of Life After Death.* New York: Palgrave Macmillan.

Steinhart, E. (2018). Spirit. *Sophia, 56*(4), 557–571.

Steinhart, E. (2019). Spiritual naturalism. In R. Nicholls & H. Salazar (Eds.), *The Philosophy of Spirituality* (pp. 312–338). New York: Brill.

Swenson, R. (2006). Spontaneous order, autocatakinetic closure, and the development of space-time. *Annals of the New York Academy of Sciences, 901,* 311–319.

Swenson, R. (2009). The fourth law of thermodynamics or the law of maximum entropy production (LMEP). *Chemistry, 18*(1), 333–339.

Swinburne, R. (2012). The argument from design. In L. Pojman & M. Rea (Eds.), *Philosophy of Religion* (6th ed., pp. 191–201). Boston: Wadsworth.

Sylvan, R. (2005). *Trance Formation.* New York: Routledge.

Taft, R., Pheasant, M., & Mattick, J. (2007). The relationship between non-protein-coding DNA and eukaryotic complexity. *BioEssays, 29*(3), 288–299.

Takahashi, M., & Olaveson, T. (2003). Music, dance, and raving bodies: Raving as spirituality in the Canadian rave scene. *Journal of Ritual Studies, 17*(2), 72–96.

Taylor, B. (2010). *Dark Green Religion: Nature Spirituality and the Planetary Future.* Berkeley, CA: University of California Press.

Taylor, S. M. (2007). *Green Sisters: A Spiritual Ecology.* Cambridge, MA: Harvard University Press.

Tegmark, M. (2014). *Our Mathematical Universe: My Quest for the Ultimate Nature of Reality.* New York: Random House.

Temkin, I., & Eldredge, N. (2007). Phylogenetics and material cultural evolution. *Current Anthropology, 48*(1), 146–153.

Tillich, P. (1951). *Systematic Theology* (Vol. 1). Chicago: University of Chicago Press.

Tipler, F. (1995). *The Physics of Immortality: Modern Cosmology, God and the Resurrection of the Dead.* New York: Anchor Books.

Tremlin, T. (2006). *Minds and Gods: The Cognitive Foundations of Religion.* New York: Oxford University Press.

Tzafestas, S. (2018). *Energy, Information, Feedback, Adaptation, and Self-Organization.* New York: Springer.

Unrean, P., & Srienc, F. (2012). Predicting the adaptive evolution of *Thermoanaerobacterium saccharolyticum*. *Journal of Biotechnology, 158,* 259–266.

Verlinde, E. (2016). Emergent gravity and the dark universe. *SciPost Physics, 2*(3.016), 1–41. https://doi.org/10.21468/scipostphys.2.3.016.

Vogelsberger, M., et al. (2014). Introducing the Illustris Project: Simulating the coevolution of dark and visible matter in the universe. *Monthly Notices of the Royal Astronomical Society, 444*(2), 1518–1547.

Wagner, A. (2014). *Arrival of the Fittest: How Nature Innovates.* New York: Penguin.

Wald, R. (2006). The arrow of time and the initial conditions of the universe. *Studies in History and Philosophy of Modern Physics, 37,* 394–398.

Watson, R., et al. (2016). Evolutionary connectionism. *Evolutionary Biology, 43,* 553–581.

Watson, R., & Szathmary, E. (2016). How can evolution learn? *Trends in Ecology & Evolution, 31*(2), 147–157.

Welch, P., & Horsten, L. (2016). Reflecting on absolute infinity. *Journal of Philosophy, 113*(2), 89–111.

Wigner, E. (1960). The unreasonable effectiveness of mathematics in the natural sciences. *Communications on Pure and Applied Mathematics, 13,* 1–14.

Willerslev, R., Vitebsky, P., & Alekseyev, A. (2014). Sacrifice as the ideal hunt: A cosmological explanation for the origin of reindeer domestication. *Journal of the Royal Anthropological Institute, 21,* 1–23.

Winslade, J. (2009). Alchemical rhythms: Fire circle culture and the pagan festival. In M. Pizza & J. Lewis, *Handbook of Contemporary Paganism* (pp. 241–282). Boston: Brill.

Wolfram, S. (2002). *A New Kind of Science.* Champaign, IL: Wolfram Media.

Women of Science Tarot. (2019). Online at shop.massivesci.com/products/women-of-science-tarot-deck. Accessed 3 May 2019.

Wood, C., & Shaver, J. (2018). Religion, evolution, and the basis of institutions: The institutional cognition model of religion. *Evolutionary Studies in Imaginative Culture, 2*(2), 1–20.

Wright, P. (1970). Entropy and disorder. *Contemporary Physics, 11*(6), 581–588.

Yen, J., et al. (2014). Thermodynamic extremization principles and their relevance to ecology. *Austral Ecology, 39,* 619–632.

Zamulinski, B. (2003). Religion and the pursuit of truth. *Religious Studies, 39,* 43–60.

Zeilinger, A. (1999). A foundational principle for quantum mechanics. *Foundations of Physics, 29*(4), 631–643.

Zenil, H., & Delahaye, J.-P. (2010). On the algorithmic nature of the world. In G. Dodig-Crnkovic & M. Burgin (Eds.), *Information and Computation.* Singapore: World Scientific.

Index

Printed in the United States
By Bookmasters